IoT and AI Technologies for Sustainable Living

This book brings together all the latest methodologies, tools and techniques related to the Internet of Things and Artificial Intelligence in a single volume to build insight into their use in sustainable living. The areas of application include agriculture, smart farming, healthcare, bioinformatics, self-diagnosis systems, body sensor networks, multimedia mining, and multimedia in forensics and security.

This book provides a comprehensive discussion of modeling and implementation in water resource optimization, recognizing pest patterns, traffic scheduling, web mining, cyber security and cyber forensics. It will help develop an understanding of the need for AI and IoT to have a sustainable era of human living. The tools covered include genetic algorithms, cloud computing, water resource management, web mining, machine learning, block chaining, learning algorithms, sentimental analysis and Natural Language Processing (NLP).

IoT and AI Technologies for Sustainable Living: A Practical Handbook will be a valuable source of knowledge for researchers, engineers, practitioners, and graduate and doctoral students working in the field of cloud computing. It will also be useful for faculty members of graduate schools and universities.

IoT and AI Technologies for Sustainable Living
A Practical Handbook

Edited by
Abid Hussain
Garima Tyagi
Sheng-Lung Peng

CRC Press is an imprint of the
Taylor & Francis Group, an **informa** business

First edition published 2023
by CRC Press
6000 Broken Sound Parkway NW, Suite 300, Boca Raton, FL 33487-2742

and by CRC Press
4 Park Square, Milton Park, Abingdon, Oxon, OX14 4RN

CRC Press is an imprint of Taylor & Francis Group, LLC

Library of Congress Cataloging-in-Publication Data
Names: Hussain, Abid (Economist), editor. | Tyagi, Garima, editor. |
Peng, Sheng-Lung, editor.
Title: IoT and AI technologies for sustainable living : a practical
handbook / edited by Abid Hussain, Garima Tyagi, Sheng-Lung Peng.
Description: First edition. | Boca Raton : CRC Press, 2023. |
Includes bibliographical references and index.
Identifiers: LCCN 2022020791 (print) | LCCN 2022020792 (ebook) |
ISBN 9780367507114 (hardback) | ISBN 9780367507268 (paperback) |
ISBN 9781003051022 (ebook)
Subjects: LCSH: Artificial intelligence—Industrial applications. |
Machine learning—Industrial applications. | Internet of things—Industrial
applications. | Sustainable living—Technological innovations. | Green technology.
Classification: LCC TA347.A78 I79 2023 (print) | LCC TA347.A78 (ebook) |
DDC 628—dc23/eng/20220803
LC record available at https://lccn.loc.gov/2022020791
LC ebook record available at https://lccn.loc.gov/2022020792

ISBN: 9780367507114 (hbk)
ISBN: 9780367507268 (pbk)
ISBN: 9781003051022 (ebk)

DOI: 10.1201/9781003051022

Typeset in Times
by codeMantra

Preface

Intelligence computing is an emerging research area that covers various fields such as machine learning, neural network, evolutionary algorithm, fuzzy system and computer science. Intelligence computing is used to investigate, simulate, analyze and solve real-world problems. Its ability to deal with uncertainty and imprecision has attracted researchers and practitioners to develop this field.

In recent years, computational intelligence has been used to deal with various problems related to daily life. The need for smart devices has increased continuously, ranging from personal business assistants to appliances for entertainment. Digital cameras, wireless communication, sensors and multimedia are increasingly common technologies in these devices. Sensors help a device to collect data and wireless communication enables the device to communicate with other devices in the network to exchange information as well as receive instruction. These functions have encouraged the development of the Internet of Things, which in many ways supports our daily lives.

Internet of Things refers to the interconnection of devices where the devices are able to exchange and consume data with minimal human intervention. It integrates new technologies, especially computing and communication technologies. The Internet of Things supports collaboration between industries, researchers and policymakers. It has created a new market that significantly changes the economy and society globally.

This book‘s 17 chapters discuss various issues related to further research, development and the application of Computational Intelligence and the Internet of Things. Each chapter addresses specific issues such as robotics, industrial applications, government enhancement, security and weather.

Preface

Contents

Preface v
Editors ix
Contributors xi

1 **Rapid Application Development in Cloud Computing with IoT** **1**
Nirupma Singh and Abid Hussain

2 **Integration of IoT with Artificial Intelligence in Health Care** **29**
N.K. Shiju, N. Ramachandran, Salini Suresh and Anubha Jain

3 **Significant Role of IoT in Agriculture for Smart Farming** **43**
Rohit Maheshwari, Mohnish Vidyarthi and Parth Vidyarthi

4 **Next Era of Computing with Machine Learning in Different Disciplines** **57**
R. Sabitha, J. Shanthini and S. Karthik

5 **Self-Diagnosis in Healthcare Systems Using AI Chatbots** **79**
Bhawna Nigam, Naman Mehra and M. Niranjanamurthy

6 **Digital Water: New Approach to Build Efficient Water Management Systems** **93**
Priyanka Verma and Anubha Jain

7 **Online Recommendation Using Machine Learning (ML) and NLP** **107**
Mohnish Vidyarthi, Parth Vidyarthi and Rohit Maheshwari

8 **Natural Language Processing and Translation Using Machine Learning** **119**
Garima Tyagi

9 **Text and Multimedia Mining through Machine Learning** **135**
T. Kohilakanagalakshmi, K.R. Radhakrishnan, Salini Suresh, and Ruchi Nanda

10 Application of IoT and Block Chaining for Business Analysis 155
K. Sheela, M. Mubina Begum and C. Priya

11 Applications of Body Sensor Network in Healthcare 171
K.R. Radhakrishnan, T. Kohilakanagalakshmi, Salini Suresh, T.M. Thiyagu and Amita Sharma

12 Sentimental Analysis with Web Engineering and Web Mining 197
Rahul Malik and Sagar Pande

13 Big Data in Cloud Computing - A Defense Mechanism 219
N. Ramachandran, Salini Suresh, Sunitha, Suneetha V, and Neha Tiwari

14 Sound and Precise Analysis of Web Applications for Injection Vulnerabilities 237
Chitsutha Soomlek, Krit Kamtuo and Ekkarat Boonchieng

15 Multimedia Applications in Forensics, Security and Intelligence 269
M. Mubina Begum and C. Priya

16 Advancements and Innovation in Digital Marketing and SEO 279
Anubha Jain, Chhavi Jain, Rahul G. Kargal and Salini Suresh

17 Advanced Wireless Solutions (Case Studies on Application Scenarios) 317
Jyoti Prabha

Editors

Abid Hussain is an Associate Professor in the School of Computer Application at Career Point University. He obtained a post-graduate degree in Computer Application from the University of Kota. He also received his PhD in Computer Application from Singhania University, Jhunjhunu, Rajasthan. His areas of interest are Cloud Computing, Network Security, Open Source Technologies, Web Engineering and Cyber Security. He is also a Research Supervisor in Computer Science & Application at Career Point University. He has published research papers in national and international journals of computer science. He also worked as a reviewer for various national and international conferences.

Garima Tyagi is an Assistant Professor in the School of Computer Application at Career Point University. She received post-graduation degrees in Chemistry from Rohilkhand University and Computer Applications from JNRV University respectively. She completed her Executive MBA in HR from EIILM University. Her research areas are VOIP, NLP, Algorithms and Soft Computing. Besides her research interest in Computer Science, she has also done a significant amount of research in the fields of TQM, BPR and HRM.

Sheng-Lung Peng is a Professor in the Department of Creative Technologies and Product Design, National Taipei University of Business, Taoyuan, Taiwan. He received a BS degree in Mathematics from National Tsing Hua University, and MS and PhD degrees in Computer Science from the National Chung Cheng University and National Tsing Hua University, Taiwan, respectively. He is an honorary Professor at Beijing Information Science and Technology University of China, and a visiting Professor at Ningxia Institute of Science and Technology of China. He is now serving as the regional director of the ICPC Contest Council for Taiwan and director of the Institute of Information and Computing Machinery, Information Service Association of Chinese Colleges and Taiwan Association of Cloud Computing. He is also a supervisor of the Chinese Information Literacy Association and the Association of Algorithms and Computation Theory. His research interests are in designing and analyzing algorithms for Bioinformatics, Combinatorics, Data Mining, and Networks. Dr. Peng has edited several special issues in journals such as *Soft Computing*, *Journal of Internet Technology*, *Journal of Computers* and *MDPI Algorithms*. He is also a reviewer of more than ten journals such as *IEEE Transactions on Emerging Topics in Computing*, *Theoretical Computer Science*, *Journal of Computer and System Sciences*, *Journal of Combinatorial Optimization*, *Journal of Modeling in Management*, *Soft Computing*, *Information Processing Letters*, *Discrete Mathematics*, *Discrete Applied Mathematics*, and *Discussiones Mathematicae Graph Theory*. He has published more than 100 international conferences and journal papers.

Contributors

M. Mubina Begum
Department of Computer Science
Vels Institute of Science,
Technology and Advanced
Studies (VISTAS)
Chennai, India

Ekkarat Boonchieng
Department of Computer Science
Chiang Mai University
Chiang Mai, Thailand

Abid Hussain
School of Computer Applications
Career Point University
Kota, India

Anubha Jain
Department of Computer
Science & IT
IIS (deemed to be University)
Jaipur, India

Chhavi Jain
Department of Management Studies
IIS (deemed to be University)
Jaipur, India

Krit Kamtuo
Buzzebees, Co., Ltd
Bangkok, Thailand

Rahul G. Kargal
Department of Management Studies
Dayananda Sagar College of
Engineering
Bangalore, India

S. Karthik
Department of Computer Science and
Engineering
SNS College of Technology
Coimbatore, India

T. Kohilakanagalakshmi
Dayananda Sagar Institutions
Bangalore, India

Rohit Maheshwari
Computer Science Engineering
Career Point University
Kota, India

Rahul Malik
Lovely Professional University
Jalandhar, India

Naman Mehra
Department of Electronics &
Communication, Institute of
Engineering & Technology (IET)
Devi Ahilya Vishwavidyalaya
Indore, India

Ruchi Nanda
Department of Computer Science & IT
IIS (deemed to be University)
Jaipur, India

Bhawna Nigam
Department of Information Technology,
Institute of Engineering & Technology
(IET)
Devi Ahilya Vishwavidyalaya
Indore, India

M. Niranjanamurthy
Department of Computer Applications
M.S. Ramaiah Institute of Technology
Bangalore, India

Sagar Pande
Lovely Professional University
Jalandhar, India

Jyoti Prabha
Computer Science and Engineering
University College of Engineering & Technology
Rajasthan, India

C. Priya
Department of Information Technology
Vels Institute of Science, Technology and Advanced Studies (VISTAS)
Chennai, India

K.R. Radhakrishnan
St. Joseph's College of Engineering
Chennai, India

N. Ramachandran
Computer Science &Technology
Indian Institute of Management
Kozhikode, India

R. Sabitha
Department of Computer Science and Engineering,
Avinashilingam Institute for Home Science and Higher Education for Women
Coimbatore, India

J. Shanthini
Department of Computer Science and Engineering,
SNS College of Technology
Coimbatore, India

Amita Sharma
IIS (deemed to be University)
Jaipur, India

K. Sheela
Department of Computer Science
Vels Institute of Science, Technology and Advanced Studies (VISTAS)
Chennai, India

N.K. Shiju
Web Applications
Indian Institute of Management
Kozhikode, India

Nirupma Singh
School of Engineering and Technology
Career Point University
Kota, India

Chitsutha Soomlek
Department of Computer Science
Khon Kaen University
Khon Kaen, Thailand

Suneetha V
Department of Computer Applications
Dayananda Sagar College
Bangalore, India

Sunitha
Department of Computer Applications
Dayananda Sagar College
Bangalore, India

Salini Suresh
Department of Computer Applications
Dayananda Sagar College of Arts Science and Commerce
Bangalore, India

T.M. Thiyagu
College of Engineering
Anna University
Chennai, India

Neha Tiwari
IIS (deemed to be University)
Jaipur, India

Garima Tyagi
School of Computer Application
Career Point University
Kota, India

Priyanka Verma
IIS (deemed to be University)
Jaipur, India

Mohnish Vidyarthi
Computer Science Engineering
Career Point University
Kota, India

Parth Vidyarthi
Computer Science Engineering
Career Point University
Kota, India

1

Rapid Application Development in Cloud Computing with IoT

Nirupma Singh and Abid Hussain
Career Point University

Contents

1.1 Introduction to Rapid Application Development 3
1.2 Features of Rapid Application Development 3
1.3 The Rapid Application Development Model 4
1.4 Rapid Application Development Model 4
1.5 Steps in the High-Speed Application Development Process 5
 1.5.1 Phase 1: Planning for Exigency Fulfilment 5
 1.5.2 Phase 2: User Design 6
 1.5.3 Phase 3: Rapid Structure 6
 1.5.4 Phase 4: Cutover 7
1.6 RAD Model Pros and Benefits 7
 1.6.1 Does the RAD Model Suit Your Organization? 7
1.7 Benefits of RAD Model 8
1.8 RAD vs. Other Software Development Models 10
 1.8.1 RAD Model vs. Traditional System Development Lifecycle 10
 1.8.2 RAD vs. Agile 10

DOI: 10.1201/9781003051022-1

1.9 When to Use RAD Methodology? 11
1.10 A Radical Approach to Traditional Application Development 12
1.11 Cloud Platform for RAD 13
1.11.1 Mendix, a Cloud Platform That Supports Rapid Developers 13
1.11.2 Cloud Platform Function Enables Rapid Application Development 14
1.12 IoT with Cloud Computing for Rapid Application Development 14
1.13 IoT Cloud Application – Architecture 15
1.14 Best Practices for Developing a Robust IoT Cloud Application 17
1.14.1 Database Design Issues 17
1.14.2 Server Extensions and Application Cloning 17
1.14.3 IoT Security Applications in the Cloud 17
1.14.4 Thinking about Cloud Database Design 18
1.15 Three Ways of Achieving Rapid Application Development in IoT 18
1.15.1 Access to the Rapid Development of IoT Applications 18
1.15.1.1 Hardware Development vs Toy Development 19
1.16 The Ability to Simplify IoT Development 19
1.16.1 Three Ways to Quickly Develop IoT Applications 20
1.16.1.1 Option 1: Use Existing Hardware Platforms to Meet Application Requirements 20
1.16.1.2 Option 2: Use the Hardware Platform to Activate the Application 22
1.16.1.3 Option 3 – Use Development Tools to Create Pre-Designed IoT Applications on COTS Hardware 24
1.16.2 What Do You Think? 25
1.17 Global Rapid Application Development Market Is Expected to Reach USD 95.2 Billion by 2025: FIOR Markets 26
Bibliography 27

Abstract

Project management teams have been following the traditional rigorous methods for design, process and documentation for decades. Flexible management changes this direct approach. No wonder this method has become more and more common over time. Project managers have also attracted people's attention, especially in areas such as software development, where technology, goals and tasks are among constantly changing teams. The rapid application development (RAD) model can improve project efficiency, create products on time and improve customer satisfaction. RAD has become a major trend in the field of software development. This is a relatively new method that ensures the seamless and rapid development of new applications to meet the growing development needs of companies from different industries. In the field of project management, "RAD" is a Dujour method. The RAD model is based on native and iterative development procedures and does not contain specific designs. The software development process itself involves the programming required to develop the product. The RAD model reduces the total development time by reusing and developing components at the same time, thereby speeding up delivery. WORK can only work if highly qualified engineers use it, and the customer promises to achieve the initial goal within a certain period of time.

1.1 INTRODUCTION TO RAPID APPLICATION DEVELOPMENT

Rapid application development (RAD) is an application development model in which functions are created in parallel. The way to do this is to make each part look like a subroutine. The subroutines are then assembled into a model, also called a "prototype". Between these processes, application developers can more easily create, customize, and even modify model components faster.

In addition, RAD has a higher priority in terms of rapid release and revision of prototypes. In addition, the rapid development of applications places more emphasis on the use of software and user feedback. In continuous development, software development is a very dynamic field. In addition, for some projects and market needs, it is very important to implement product changes quickly or to develop the product itself as soon as possible. In this case, RAD will outperform all other software development methods and models.

Although the RAD framework can be considered one of the types of flexible software development methods, it is different. It focuses more on current software programs and implements user feedback or customer requirements in the same process, rather than following strict plans. Obviously, this method is very beneficial for many companies, so let's continue to learn more about the rapid development of this software and why it is important for your product. *k* requires a recording mechanism and a strict planning.

For decades, the project management team has followed traditional methods of rigorous planning, process and documentation. Agile management has overturned this method, and it is no surprise that it has become increasingly popular over time. According to a 2017 PricewaterhouseCoopers (PwC) study, the success rate of flexible design is 28% higher than that of traditional design.

Project managers have noticed this, especially in areas such as software development, where technology, goals and objectives are constantly changing.

1.2 FEATURES OF RAPID APPLICATION DEVELOPMENT

One of the most common problems for software development teams is spending a lot of time planning step-by-step and not delivering the product on time. As a result, according to the Wellington survey, more than 32% of organizations never complete the project on time.

RAD methodology can help to solve this problem.

The main benefit of RAD development is the vision to manage our software like clay instead of steel. The idea is that it should always be malleable. As you learn more about the definition of RAD, you should remember this aspect. The RAD methodology first appeared in the 1980s and stems from the agile approach. It may not be new, but it is still popular with teams looking for a flexible application development approach that allows them to meet the growing needs of customers, users and businesses.

What you need to know is that RAD takes precedence over original releases and iterations, and emphasizes software usage and user feedback on stringent design and registration requirements.

1.3 THE RAPID APPLICATION DEVELOPMENT MODEL

The RAD method can help to solve this problem. The main advantage of RAD development is that it can handle software projects like clay instead of steel. They should always adapt to the idea. Please keep this in mind when learning about the definition of RAD.

The RAD method first appeared in the 1980s and was derived from agile methods. Although this is nothing new, it is still popular in teams to find agile methods to meet the growing needs of customers, users and enterprises for application development.

It is important to understand that when developing applications rapidly, prototype versions and iterations take precedence. They emphasize the use of software and user feedback, rather than strict planning and recording requirements.

1.4 RAPID APPLICATION DEVELOPMENT MODEL

The RAD method is a powerful alternative to the traditional waterfall expansion model, and the traditional waterfall expansion model is not always effective. From the beginning, RAD was obtained from the spiral model.

However, over time, RAD has changed. It adapts to the developer's time constraints and maintains some basic development guidelines. When we realize that speed is always important in this model, the meaning of RAD becomes more clear. RAD model is one of the ways to write software (you can read more about agile development). There are few long-term plans for the process, and more emphasis is placed on the adaptability of the development workflow. The RAD model uses information gathered from seminars and other focus groups to determine customer expectations for their products. The first product is also being tested. In this way, you can create a final product and continue to use product parts that have been proven effective (Figure 1.1).

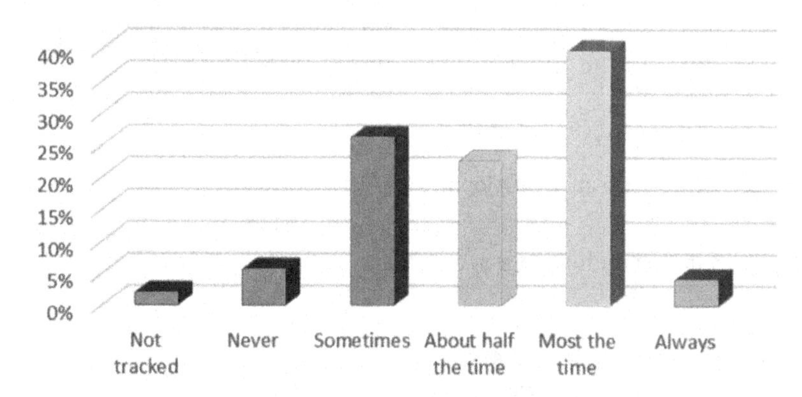

FIGURE 1.1 Project completion graph.

1.5 STEPS IN THE HIGH-SPEED APPLICATION DEVELOPMENT PROCESS

RAD is a popular agile project management strategy in software development. The main advantage of the RAD approach is the rapid turnaround of designs, which makes it an attractive choice for developers working in fast-paced environments such as software development. RAD is focused on minimizing the planning stages and developing prototypes to the greatest extent, thus making this rapid progress possible. By reducing planning time and emphasizing prototype iterations, RAD enables project managers and stakeholders to accurately measure progress and communicate in real time on new issues and changes. This can increase efficiency, speed up development and achieve effective communication.

The process can be separated in several ways, but in general, RAD follows four main steps (Figure 1.2).

1.5.1 Phase 1: Planning for Exigency Fulfilment

This stage corresponds to the project site meeting. Compared with other project management methods, the design phase is short, but this is an important step for the complete success of the project.

In this phase, developers, customers (software users) and team members communicate with each other to set project goals and expectations, as well as to solve current and potential problems during the creation process.

Rapid Application Development (RAD)

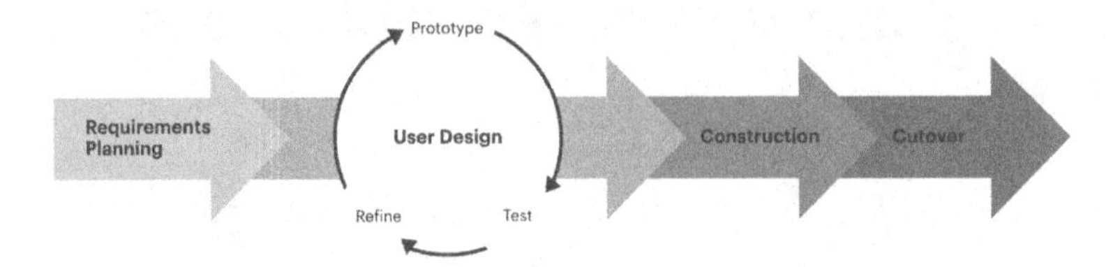

FIGURE 1.2 Rapid application development (RAD) methodology.

The basic analysis at this stage is as follows:

- Find current issues
- Define project requirements
- Fulfil the requirements with the consent of each party

It is important that everyone has the opportunity to evaluate the goals and expectations of the project and their weight. With the approval of all key stakeholders and developers, the team can avoid communication problems and expensive change requests. With the experience to make sure there is no potential for something to slip through the cracks.

1.5.2 Phase 2: User Design

Once the design is defined, you can proceed with development and build your design through different iterations of the prototype. This is the essence of the RAD approach, which distinguishes it from other project management strategies. At this stage, the client works with the developers to ensure that their requirements are met at every step of the design process. Just like customizable software development, it allows users to test prototypes of any product at any step to ensure that they meet their expectations.

All errors and nodes are fixed in the iterative process. The developer designs a prototype, the customer (user) tests it, and then they meet to discuss about the effectiveness and ineffectiveness of the methods. This method allows developers to modify the model until they are satisfied with the design. Software developers and customers can learn from experience and ensure that certain things will not slip.

1.5.3 Phase 3: Rapid Structure

This third phase is important because the client still gives input throughout the process. They can suggest alterations, changes or even new ideas that can solve problems as they arise. In the third stage, the prototype and Beta system is obtained from the design stage and transformed into a working model. Most problems and changes are resolved in the complete iterative design phase, allowing developers to create the final working model faster than traditional project management methods. This stage is divided into several small steps:

- Prepare for rapid construction
- Developing programs and applications
- Coding
- Unit, integration and system testing

Software development teams including programmers, testers and developers work together at this stage to ensure that everything runs smoothly and that the final result meets customer expectations and goals. The third step is very important. In fact, customers can make their own contributions throughout the process. They can propose changes and new ideas that can solve the problems that arise.

1.5.4 Phase 4: Cutover

This is the assembly phase when the final product is released. These include data transfer, testing, transition to new systems and user training. All final changes are made as the developer and customer continue to check for system errors.

1.6 RAD MODEL PROS AND BENEFITS

RAD is one of the most successful software development programs available today which provides numerous benefits to both the software development team and its customers.

Here are some benefits:

- It allows you to break down your project into smaller, easier-to-use tasks.
- A work-based structure allows project managers to optimize team effectiveness by delegating tasks according to the capacity and experience of their members.
- Customers deliver suitable products in less time.
- Regular communication and continuous feedback between team members and stakeholders improve the efficiency of the design and construction process.

With a short planning phase and a focus on repetitive design and construction, RAD teams can achieve more in less time without compromising customer satisfaction.

1.6.1 Does the RAD Model Suit Your Organization?

It depends! RAD methodology is an effective strategy for many projects and teams. However, the following key factors need to be considered before implementation:

Do you have a team of talented and experienced programmers and engineers who can work on this continuous development process?

The RAD process is very intensive and requires practical consideration by the team.

Are your customers open to adopt this approach?

Without the client's approval from the beginning, the project is more likely to fail due to communication interruptions.

Is your client willing to work on a specific procedure to complete the model?

The difference between the RAD method and the traditional method is that RAD follows strict deadlines. All parties involved must agree on a timetable for the success of the project.

Do you have the right tools and software to effectively apply this method?

The success of RAD depends on the project manager's ability to fully describe each development stage and communicate effectively with team members and stakeholders in real time (Figure 1.3).

The flexible and easy-to-use tools are required for agile projects so that you and your team can focus on time from start to finish. Whether you are a flexible fanatic or you win, the Agile RAD model increases project efficiency if you have the right team and dedicated stakeholders. And there is no doubt that it is possible to produce it, get results on time and increase customer satisfaction.

1.7 BENEFITS OF RAD MODEL

Reduce risk – The ability to quickly create and share prototypes enables companies to view features in the early lifecycle of the application and avoid repetitive work that could disrupt the entire project. The Timebox method of RAD method can reduce costs and risks.

Rapid growth – RAD ensures rapid growth by applying repeated versions and reusing code on low-model-based platforms. This allows the company to focus on overloaded processes and documents and provide solutions on time.

Improve quality – RAD includes prototypes and functional controls throughout the project life cycle to verify and develop requirements based on user feedback, thereby improving software quality.

Increase productivity – At every step of the customer's busy schedule, developers can often promote their work. This ensures that the final product is delivered to the developer and significantly improves the productivity of the developer.

Improve customer satisfaction – Customers participate in the development cycle, thereby increasing customer satisfaction and reducing the risk of non-compliance with business requirements.

Budget-friendly – The framework for RAD requires short and flexible sprints, which are repeated as many times as required by the project. This iterative approach can detect errors and logic issues before they affect delivery. It also helps by greatly reducing application development costs, shortening cycle time, increasing productivity and reducing resources.

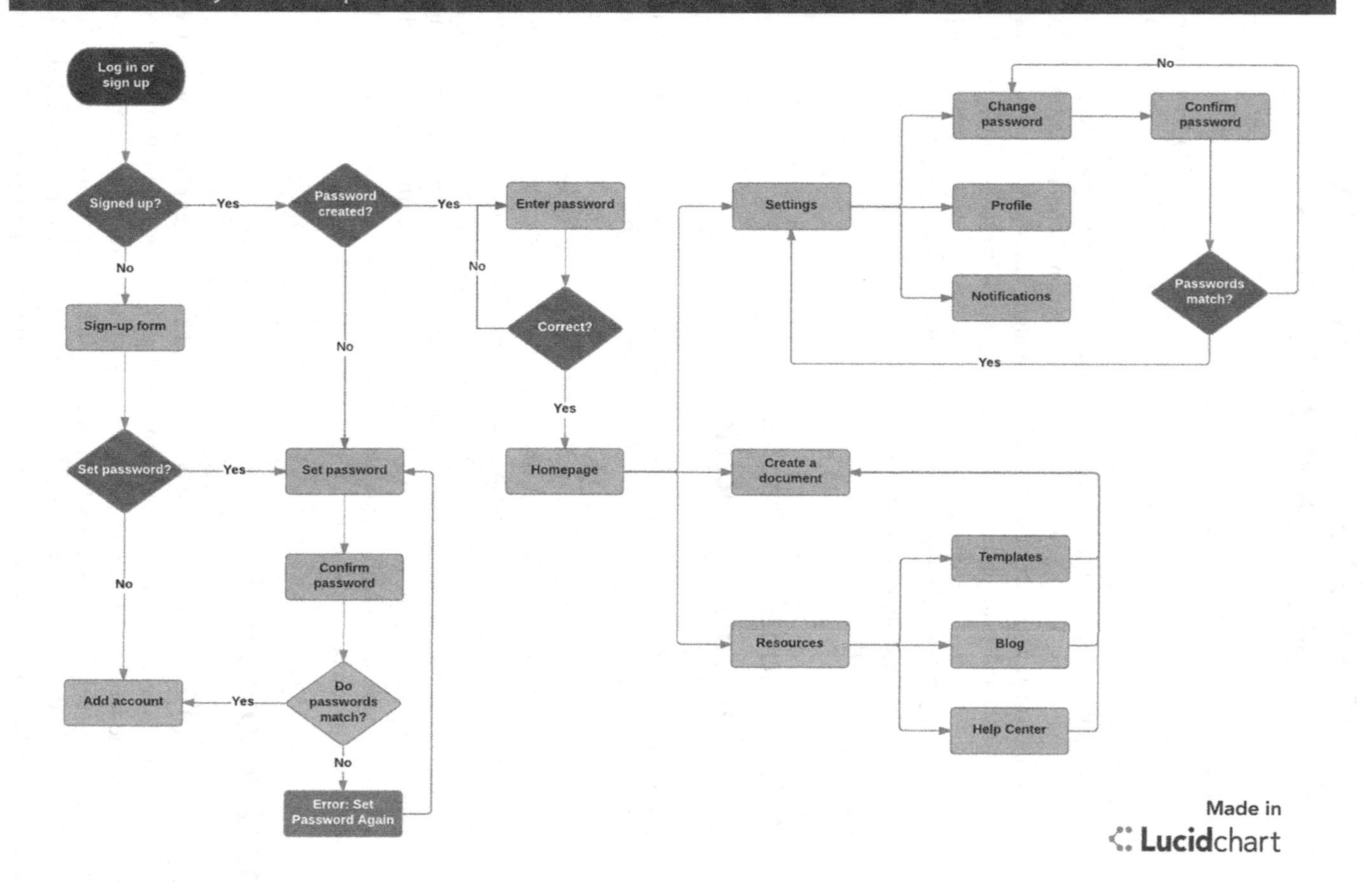

FIGURE 1.3 User journey flow.

1.8 RAD VS. OTHER SOFTWARE DEVELOPMENT MODELS

Growth rate – The main difference between RAD and other models is how application development focuses on speed. Other models tend to focus more on delivering a product that is beneficial to customers when they continue to use them.

Member of the team – You can quickly develop applications without a team of too many members. This allows for fast communication, frequent meetings and fast information transfer. For example, although other development models differ, a flexible team prefers to divide large teams into different major ones.

End user – For RAD, it is important to involve the end user throughout the development process. Other models typically limit user research to the beginning of the development cycle, while user testing is limited to the end of the development cycle.

1.8.1 RAD Model vs. Traditional System Development Lifecycle

Completing the traditional system development lifecycle (SDLC) requires a lot of planning and analysis before the actual coding process begins. Such a waterfall model can be difficult for clients because they invest their time and resources in projects that have not had a significant MVP for some time. Software changes after development are time-consuming and may not be completed once the product reaches a certain point of development. RAD models are much more efficient because they provide customers with models that work much earlier. Customers can quickly check out the prototype, talk to investors in the meantime, show the look of the product and make the changes much easier. Of course, speed is not always the best choice, especially when the product is in the final stages of development and its functionality becomes more complex. In the early stages, this can be a good start.

1.8.2 RAD vs. Agile

Agile has proved to be one of the most common models in the genre of software development methods developed so far. The main differences between these two models are: this work is the forerunner of agile, but agile does not only cover development models. The working model does not specify a specific timeline for repetition. On the contrary, the focus is clearly on the growth rate in RAD. Unlike agility, end-user feedback is important to the work process in RAD. Moreover, work principles focus on functionality and customer satisfaction. Agile emphasizes technical aspects and sound design. The RAD is not limited to organizational structures like agile.

1.9 WHEN TO USE RAD METHODOLOGY?

RAD models can be very successful when quick delivery of a product is needed for a customer. It is also the best model to choose when there are changes need to be made to the prototype throughout the process before the final product is completed. It is important to know that the RAD model is only valid when there are plenty of knowledgeable developers and engineers on hand prepared to work on the progress of the product. The customers must also remain committed to the process and the schedule in place for the completion of the model. When either of these two components is not available, the RAD formula can fail.

For sure, the RAD model is not applicable to every project and team. But in some cases, it might be quite beneficial. So for the last point, let's discover in what cases the rapid application can be helpful:

When you can reliably test your prototypes.

Do you have the means to test your design prototypes with users who can give full feedback on the prototypes you make? If so, then RAD is a great model to follow.

When you have the budget.

Compared to other development models, RAD is relatively inexpensive. There are always outlying instances, though.

When you need a project done rapidly.

If you have a tight deadline, RAD is your best bet. If you're under pressure to deliver something that works, then opting for a RAD platform is always suitable.

How to apply the RAD model.

As long as the product being worked on can be easily divided into separate units, the RAD model can be implemented. If the division is not possible, the RAD model may not work.

These are few situations where an RAD model may be successful:

- When the system can be modularized and then distributed in a divided form.
- When there are many designers available for the modelling.
- When there is money in the budget for using automated code-generating tools.
- When there is an expert available to decide which model should be used (RAD or SDLC).
- When a prototype is expected by customers within 2–3 months.
- When changes to the product are planned throughout the development process.

When customers need to deliver products immediately, the RAD model can be very effective. This is also an ideal model for choosing whether to change the original in the process before completing the final product. It is important to note that the RAD model is only applicable if you have a large number of experienced programmers and technicians who want to engage in further product development. In addition, the customer must continue to participate in the process and plan to complete the model. If either of these two components is not available, RAD printing may fail. Of course, the RAD model is not suitable for all projects and teams. But in some cases, it can be very useful. Finally, let us know when it is a good idea to use RAD.

When you can reliably test the prototype.

Can users try to design a prototype to provide complete feedback on the prototype they created? In this case, RAD is an important criterion.

If you have a budget.

Compared with other development models, this series is relatively cheap. However, in some cases it is always inaccessible.

When you need to complete a project quickly.

If time is limited, this is a quick way to develop applications. When you are under pressure to deliver useful content, choosing a RAD platform is always a good choice.

How to use the RAD model.

If you can easily divide the product being processed into individual units, you can use the RAD model. If you cannot divide, the RAD model may not work properly.

In some cases, the RAD model can be successful:

- When the system can be constructed modularly and distributed separately.
- When there are many designers available in the model.
- If you have enough money to use AutoCAD.
- If you have an expert, you will determine which model (RAD or SDLC) to use.
- If the customer wants to get the original within 2–3 months.
- When there are plans to make product changes during the development process.

1.10 A RADICAL APPROACH TO TRADITIONAL APPLICATION DEVELOPMENT

Designing a custom application for your main device will create a native application for your mobile device. You can apply changes at any time without having to browse an unlimited number of originals.

The following are the three steps to speed up application delivery:

- Development
- Cast in weeks
- Process

Develop your application quickly.

Start with the OTC application, and customize it according to your needs. You can also use the drag-and-drop interface to import spreadsheets and build applications from scratch.

Do more with less: Use built-in integrations between applications and between Zoho services and standalone tools (such as QuickBooks, Zapier and Xero).

Cast in a few weeks.

Don't wait for months of distribution time. Cloud applications can be used to say goodbye to downtime, infrastructure costs and maintenance work.

Get the original app on any device including Android, iPhone and iPad. Customize or just use actions and scroll the screen. You don't need a professional programmer.

Update your changes quickly.

Improve your application, and try new features in a sandy environment. Users can only see the changes if they choose to publish. There are no downtime, slow installation and waiting for commissioning. Are you satisfied with the update? Backup allows you to undo changes, so you can go back to any part of the app development process.

1.11 CLOUD PLATFORM FOR RAD

More and more IT teams are migrating to cloud platforms to meet the growing demand for new business applications. The cloud platform eliminates interference from the infrastructure and improves the efficiency of the development kit.

However, not all cloud platforms provide the same level of abstraction services. For teams that need to deliver applications quickly, look for a cloud platform that simplifies the application life cycle and minimizes reliance on traditional coding languages.

As your business grows, these cloud technologies can be easily added and save money in the long run because you don't have to pay for unnecessary resources. You can configure and deploy programs without building or maintaining expensive infrastructure. The price is extremely competitive, and now is the right time to start exploring your options. Call us now to get more information about cloud computing support.

1.11.1 Mendix, a Cloud Platform That Supports Rapid Developers

Mendix's cloud platform abstracts technical components throughout the application life cycle, thereby accelerating the delivery of customized solutions by IT teams. By reducing the cost and complexity of managing application development infrastructure, Mendix users can focus on implementing their best-in-class solutions faster than ever.

Mendix provides the highest level of abstract services within the platform. The service uses visual models instead of traditional coding languages, so users can arrange components in a readable format to cope with complex and highly customized applications. Mendix's application platform allows people to be more diverse and expand the full funding of talented developers by modelling applications instead of coding them.

1.11.2 Cloud Platform Function Enables Rapid Application Development

Not all cloud platforms provide the same efficiency. Although some platform service providers focus on developers by providing a development array suitable for traditional development methods, most business plans can save time during the entire life cycle of an application. You need an agile approach.

The RAD model provides a faster way to transform business ideas into effective applications. The new approach is based on cloud software, which eliminates infrastructure outages, allowing users to focus on delivering the right solutions faster and get a faster return on investment. There are three main components to look for when evaluating cloud platform providers:

- Comprehensive application life cycle support – Will the vendor provide abstract services at all stages of the application life cycle to enable the organization to adopt more iterative and flexible processes to adapt to changing needs?
- Democratize application development – Does the vendor use visual modelling tools to enable a large number of users with more field experience to build their applications?
- IT business collaboration – Does the supplier improve IT business collaboration through project management, social collaboration and end-user feedback mechanisms?

The Mendix platform is a leader in the field of cloud computing. Learn more about our rapid application delivery platform.

1.12 IoT WITH CLOUD COMPUTING FOR RAPID APPLICATION DEVELOPMENT

In very complex Internet of Things (IoT) applications, the ecosystem is usually based on the cloud. Cloud computing includes the following functions that are different from distributed computing technology:

- Ability to rise and fall
- Utilities-based payment model
- Safety

- Provide self-service
- API (application programming interface)
- Grade

Reducing investment costs and business flexibility are the most important business factors for the rapid adoption of IoT cloud applications in industries and enterprises.

1.13 IoT CLOUD APPLICATION – ARCHITECTURE

The IoT cloud includes the following key elements (Figure 1.4):

Cloud IoT application development service – Cloud IoT application with complete API and other interfaces. Push and extract data/commands between IoT sensor nodes/devices and downstream applications.

Integrated IoT software – An example of it is the MQTT cluster. It receives information from sensors and sends it to cloud services for processing. MQTT does not guarantee data loss and creates asynchronous communication between IoT devices and cloud services. Provide the necessary infrastructure database, security patches and other software/algorithms.

Analyser – Relational database, machine learning (ML) and artificial intelligence software algorithm.

Integrate into IoT-based mobile/web/desktop applications or a series of business intelligence applications.

Development of IoT sensor nodes and cloud interfaces:

  - Use MQTT, CoAP, AMQP, Websocket, etc. to develop a secure and powerful communication interface between IoT and cloud devices.
- IoT device connections are constantly monitored by IoT applications in the cloud. It also alerts end users about connection problems and proposes solutions.
  - Apply security processing mechanisms for communication/data transmission between IoT sensor nodes/devices and cloud services.

Development of IoT applications for end users:

- End-user role management – Only after successfully logging in, the user can access the assigned IoT sensor node/device. End users are defined based on the roles and responsibilities of IoT cloud application management and management.
  - IoT device management – IoT device ID is created, stored in the cloud and assigned to each user to prevent unauthorized access.
  - Mobile application/IoT application – End users can view, monitor and control the settings, status and process of devices stored in the IoT cloud through mobile/web/desktop applications.

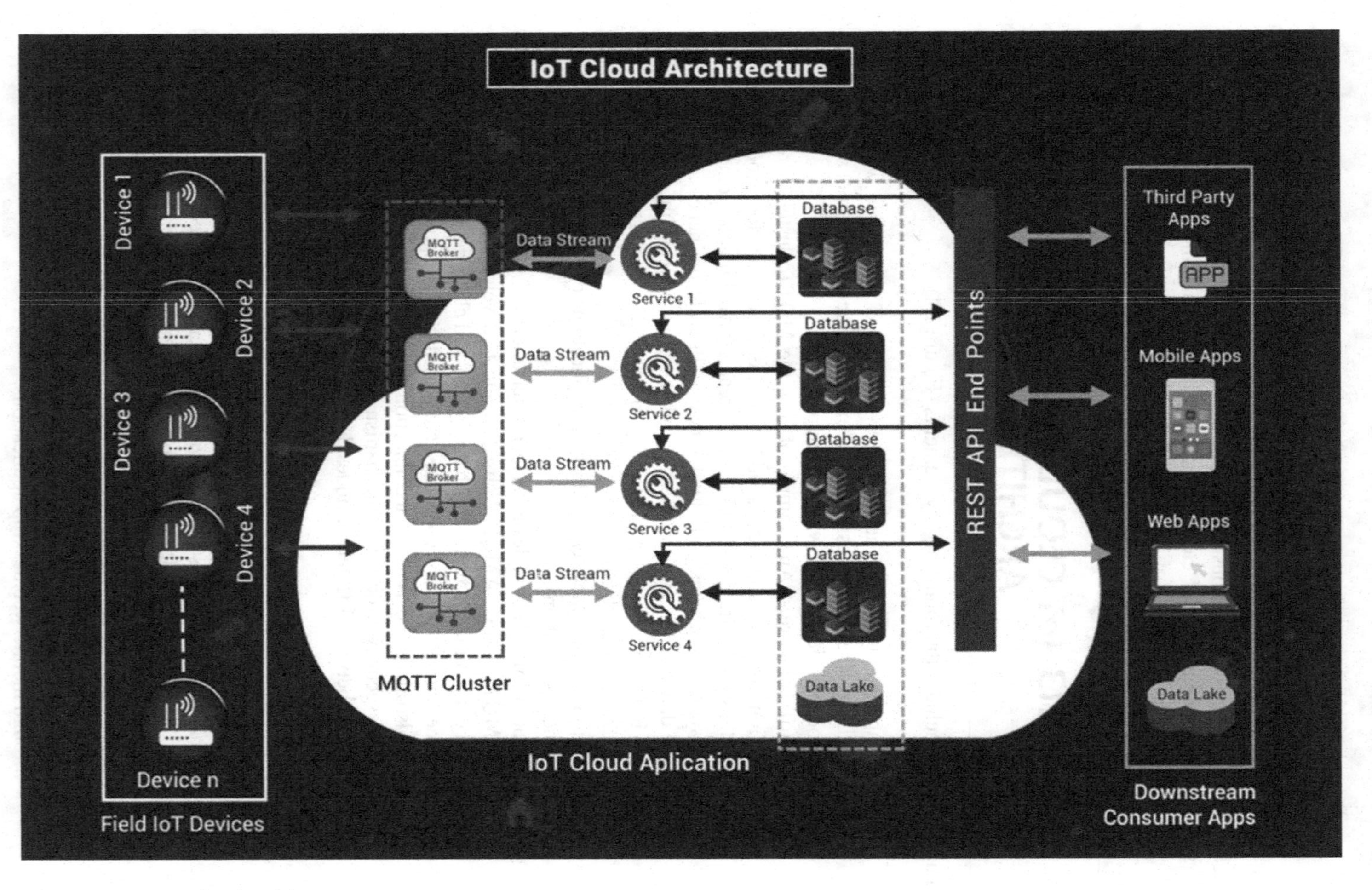

FIGURE 1.4 IoT cloud architecture.

Database design and data management:

- Powerful database architecture and design (relational/non-relational), but there is no guarantee for data loss.
- Database optimization, used to expand and manage large amounts of data.
- Store and manage all IoT device data to track distributed devices in real time.

Analysis and report

- Intuitively develop mobile, web and desktop applications for monitoring/managing connected devices.
- Develop ML algorithms based on business needs and integrate them into the IoT cloud.
- Enable cloud-based IoT applications to process historical data to predict device behaviour and determine predictable maintenance, resource usage or other business application models.
- Support custom graphics and text reports for analysis and monitoring.

1.14 BEST PRACTICES FOR DEVELOPING A ROBUST IoT CLOUD APPLICATION

1.14.1 Database Design Issues

Contact your IoT platform development partner to understand the scale of the project: the number of IoT nodes or sensor devices you need to set up, the amount of data, the importance of data, requirements, business intelligence, etc.

IoT development partners' IoT applications should design databases to optimize data management.

1.14.2 Server Extensions and Application Cloning

Analyse and predict the current and future number of end users of IoT applications.

Configure cloud server bandwidth based on traffic. In other words, it is automatic scaling using AWS-EC2.

Apply the cloning of IoT applications in the cloud to avoid system overload due to increased traffic.

1.14.3 IoT Security Applications in the Cloud

Enable TLS/SSL certificate – even if an intruder intercepts IoT network packets, it can prevent intruders from remotely accessing IoT devices/nodes.

Verify that your cloud IoT application developer has implemented packet data encryption and decryption to protect you from network data surveillance.

1.14.4 Thinking about Cloud Database Design

- Apply security protocols between your IoT device and the cloud platform to prevent data loss.
- Based on use cases, ML algorithms can be developed for time series analysis and data reporting. These details must be provided during the database design phase.
- Specify the time required to store the collected data. This is an important aspect to consider when designing a database.
- Include NoSQL databases to achieve better scalability and better performance. When you need to collect data about equipment phase changes, SQL databases are more reliable.

1.15 THREE WAYS OF ACHIEVING RAPID APPLICATION DEVELOPMENT IN IoT

Internet-related applications are growing rapidly, and new and innovative uses are being developed every year. However, one of the challenges faced by IoT companies is to reduce the cost and workload of IoT hardware prototypes. The development of IoT products is very complicated due to its interdisciplinary technical aspects. It is a combination of software, conventional circuits and hardware machines. All of these will lead to growth cycles and costs, and if this option is not enabled, the opportunity to relocate early may fail. Let us look at some of the most obvious and most reluctant options for dealing with this challenge.

1.15.1 Access to the Rapid Development of IoT Applications

For more than 20 years, the software industry has rapidly entered the field of application development. One of the earliest development tools that adopted this idea was Visual Basic. Visual Basic is a development environment for quickly creating GUI-based desktop applications. In recent years, web and mobile applications have adopted many of the same methods in the form of frameworks and libraries for rapid development. Why to abandon the IoT? The rapid development of the IoT may be a critical moment in the market. In addition, model-based frameworks or methods can reduce development risks and simplify code design and maintenance. For experienced embedded developers, quick

FIGURE 1.5 Hardware development vs. toy development.

access does not seem to be much. You obviously want more design flexibility. However, in this rapidly changing market environment, new applications will appear every 2 days. In order to keep up with this situation, it is particularly important to be flexible and quick to try new features and ideas. When they decided to develop applications quickly, they replaced design flexibility with flexibility to quickly test the new sales features of the product.

1.15.1.1 Hardware Development vs Toy Development

Imagine the development of the IoT, just like the development of Penguin Games. You can draw precise shapes and create perfect-looking penguin toys, or combine existing LEGO blocks to quickly create penguin-like toys (Figure 1.5).

1.16 THE ABILITY TO SIMPLIFY IoT DEVELOPMENT

At a very high level, IoT device components can be divided into a simple three-tier array consisting of hardware, firmware and applications (Figure 1.6).

Each stack level has its own development challenges. But before we start, let us look at the key steps in the entire IoT product development process (Figure 1.7).

The development of product hardware elements is similar to the natural method of creating objects. Hardware components require more time and effort than software components. Restarting takes longer and increases the cost. This is inevitable because product development is an iterative process, and products are not accepted based on customer needs and business models. Most companies follow some accepted methods to accelerate hardware development. However, these methods have room for further optimization. Let us look at them.

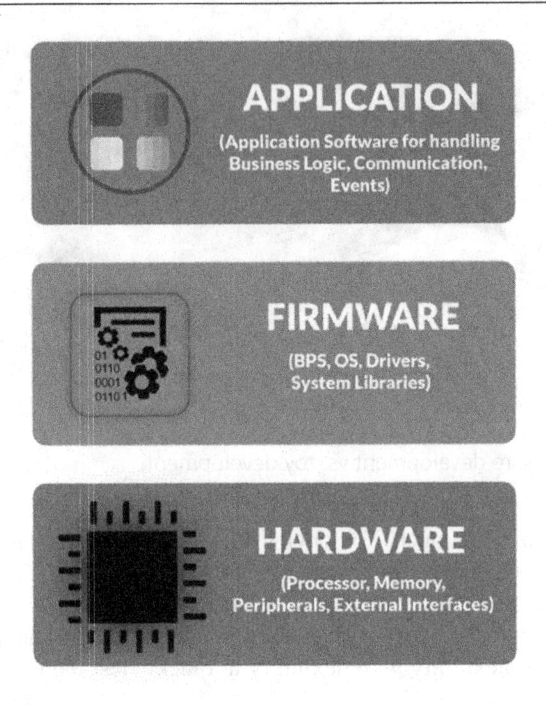

FIGURE 1.6 Three-layered stack of IIT components.

1.16.1 Three Ways to Quickly Develop IoT Applications

There are many ways to deal with the challenges of IoT development. The common idea is to always create a multi-step model to save time and effort. The following is a detailed analysis of the three options that may be suitable for you.

1.16.1.1 Option 1: Use Existing Hardware Platforms to Meet Application Requirements

This is the most common method used by companies to accelerate their growth cycles. Don't start with the chip, but start with the original circuit board or EVK (evaluation kit). There are popular hardware platforms, such as TI Launchpad, which is suitable for many IoT-enabled products.

Choosing the original board eliminates the complexity of the board's hardware design, assembly and shipping procedures, so you can skip the development cycle and focus on creating application software. In addition, the bill of materials selection stage is significantly reduced because you only need to worry about the peripherals connected to the card, not the details of the card itself. Then, the mapping becomes a template for creating software applications.

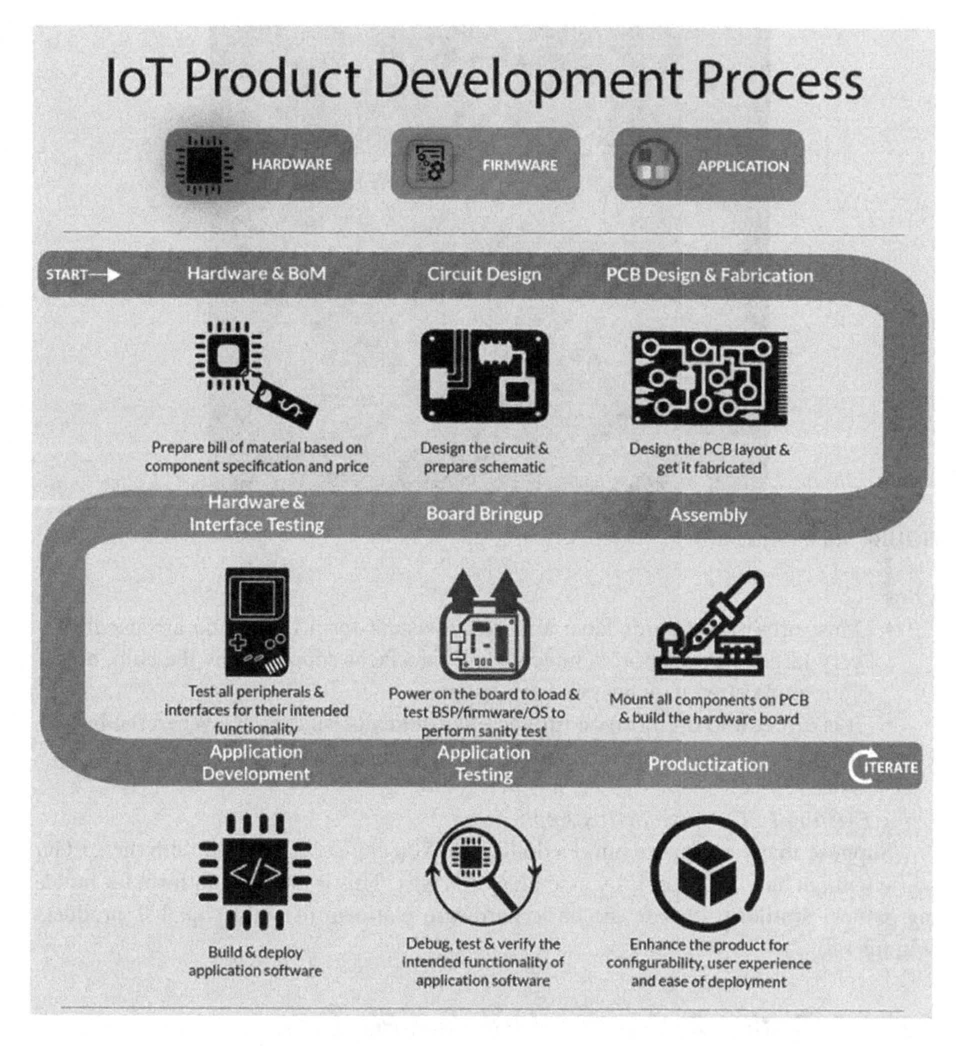

FIGURE 1.7 The major steps involved in developing an IoT hardware product.

In this way, the company can reduce the time and risks associated with hardware development. In addition, the choice of fixed cards supported by famous ODMs ensures long-term support. This strategy helps offset the cost of designing and developing your own hardware, in the entire life cycle of the product.

Pros

- Hardware redesign is a very expensive activity that can significantly reduce the risk of hardware design errors.
- Most hardware kits are handled by the card manufacturer, so you can focus on the software.
- The functions are easy to program and test, so you can use the software to try new programmable ideas.

FIGURE 1.8 Lego house.

Cons

- Most prototype boards have a more consistent form factor and are usually very large. This support connects to all interfaces supported by the chip, but your application may not need this support.
- It is difficult to use in some limited environments (such as B). When the battery is low.

Using Existing LEGO Sets for Toy Application

Suppose that you want to build a dollhouse. You can put it together with the LEGO house without having to carefully assemble each part. This is another platform for building games. Similarly, choose the basic hardware platform to build your IoT products (Figure 1.8).

1.16.1.2 Option 2: Use the Hardware Platform to Activate the Application

This method is very similar to Option 1, except that the hardware vendor also provides Standard Development Kit (SDK) and sample applications. These types of platforms (also called application activation platforms) include a complete set of software applications that are embedded on the basic hardware board and tailored for specific purposes. Depending on the configuration options provided by the manufacturer, you can use ready-to-use software with little change.

STMicroelectronics SensorTile is one of the latest trends in this category. This is a small and compact IoT device with a large number of sensors that can make any object smart. You can hear actions, environmental parameters, sounds, surrounding activities and other information. The best thing about this kit is that it comes with various software packages that allow you to read data directly from the sensor and send data through the available communication interface. If you want to make the most common consumer products such as shoes and bottles smarter, you can connect this device to it. You can use various applications to create applications that connect to the SensorTile Bluetooth interface. In this way, almost all stages of hardware product development can be eliminated, except for testing and marketing.

Pros

- Except for the embedded business logic of embedded applications, all content can be used, including SDK and software demonstrations.
- By supporting the software package provided by the manufacturer, you can easily test new features.

Cons

- Finding the ideal hardware platform that meets the needs of your application requires a lot of time and research.
- This is a way to cut cookies when creating products. Subtle changes in application requirements may lead to the search for new platforms. Therefore, there is a high risk of material insurance in case the demand changes in the future.

Using LEGO Minifigs

Lego minifigures represent certain creatures. You can move your limbs a little to change your clothes slightly. But that's it. Figure 1.9 represents the specific purpose of the game in terms of appearance and cannot be flexibly changed.

FIGURE 1.9 Lego minifig.

Similarly, the application activation platform can meet the needs of certain applications that cannot be flexibly changed.

1.16.1.3 Option 3 – Use Development Tools to Create Pre-Designed IoT Applications on COTS Hardware

The idea of Commercial off-the-shelf (COTS) has been popular in the personal computer industry for a long time. The COTS generation is represented as Common. Therefore, computer hardware technology developed from the 1960s to the 1980s. Companies such as IBM and Intel played an important role in standardizing computer hardware architecture (x86) and innovating individuals.

A similar revolution began in 2012, when the Raspberry Pi entered the market as a credit card-sized computer. Many companies have used it as their IoT hardware platform, and some companies have also used it for manufacturing applications. You can use Raspberry Pi as Option 1 of the popular Raspbian operating system and build your own programs on it. However, there is another way.

Development tools such as ForwardLoop CLIfloop show developers how hardware and COTS operating system virtualization can accelerate firmware development. In addition to Floop CLI, developers also use built-in Docker images to standardize how they interact with different devices. Developers spend all their time and energy on creating embedded software, because they only need to create a basic image of the operating system once. This combination of Docker images and standardized COTS-based hardware abstracts most of the differences between devices.

Pros

- Using standardized images, developers can focus on writing software instead of responding to hardware and operating system issues.
- The proliferation of hardware and COTS operating systems provides more support and shorter debugging cycles.

Cons

- Only works on MPU-based hardware that supports a complete operating system.
- This method requires an earlier understanding of Docker.

Using LEGO with Sugru Adhesive Mould

Sugru is a self-curing silicone rubber that can be moulded into various shapes. You can create different applications by combining them with a kind of Lego component. Just as you can use different Sugru erasers to create multiple applications for the same LEGO block, you can also use Docker images to create multiple IoT applications using the same standard COTS hardware (Figure 1.10).

The idea of COTS IoT hardware has not yet been fully developed because IoT is still an early technology compared to computers. However, it tends to move in this direction. For example, we are talking more and more about the IT axis, and more

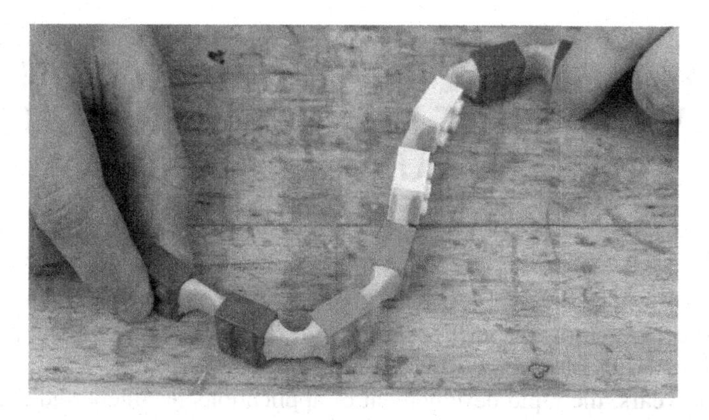

FIGURE 1.10 Sugru adhesive mould.

and more applications are realizing the advantages of edge computing in the cloud. Therefore, the COTS method can be used to design different variants of advanced computing equipment.

The Emergence of COTS in Network Technology

If you have been following the development of network technology, you may have heard of SDN (Network Defined Network) and NFV (Network Function Virtualization).

Ten years ago, routers and network devices used in communications infrastructure were built using proprietary hardware. Due to fundamental changes in the design, management and distribution of network equipment, the idea of SDN has blurred this space. SDN uses COTS hardware with a standard software framework called NFV, which allows operators to perform various network operations (VNF network operations) on a common basic platform as virtual instances of applications.

Similar types of changes may occur in the IoT. Imagine a company building a series of IoT devices for related applications. They can take advantage of this method through standard COTS hardware and Docker-based software. In this way, all products have a common core platform, allowing developers to focus on the application software built into each product.

1.16.2 What Do You Think?

Undoubtedly, none of these options can match the flexibility provided by traditional hardware design flows. However, for small and medium batch production, these are the best solutions to reduce overall development time and cost. Consider the cost of product iterations, because it may take up to five iterations to complete a product to be accepted by the customer. IoT standardization allows many proposals to circumvent traditional methods and is based on the COTS method outlined in Option 3. It is also interesting to see if the company will use this method in its internal IoT standardization efforts.

1.17 GLOBAL RAPID APPLICATION DEVELOPMENT MARKET IS EXPECTED TO REACH USD 95.2 BILLION BY 2025: FIOR MARKETS

According to a report released by FIOR Markets, the fast-growing global application market is expected to grow from US$6.4 billion in 2017 to US$95.2 billion in 2025, with an annual growth rate of 40.1% during the forecast period.

In recent years, the rapid development of applications has increased dramatically. The development of innovation and the integration of advanced technology into the enterprise must be an important factor in the rapid development of application development. A RAD method allows developers and designers to use the discoveries and knowledge gained during the development process to make plans and completely change the direction of the software.

Due to the growing demand for rapid adaptability and scalability, the global RAD market is expected to grow rapidly during the forecast period. In addition, the improvement of enterprise mobility will lead to an increase in the use of smartphones by enterprises, thereby affecting demand. However, retailers' excessive reliance on personalized settings may hinder growth of the market during the forecast period. At the same time, digital transformation and the development of information technology can consolidate the market in the next few years.

The main players in the fast-growing global application market include Salesforce, Appian, Mendix, Zoho Corporation, IBM, AWS, Google, Microsoft, Oracle, OutSystems, LANSA, Ninox, Oro, Matsoft, Kony and Pegasystems, ServiceNow, WaveMaker, KiSSFL, Radzen, AmpleLogic, FileMaker, QuickBase, AppSheet and K2. In order to participate in the fast-growing global application market, major players are currently focusing on strategies such as recent developments, mergers and acquisitions, product innovations, consortia, partnerships and partnerships.

In 2017, the largest market share of the codeless part was US$3.66 billion.

The type segment is divided into uncoded code and low code. Codeless has the largest market share, with a 2017 valuation of US$3.66 billion. No code is dominant because these platforms follow a visual development approach. This makes it scalable and available to those who understand the technology.

The service industry occupied the market with the highest share of 60.10% in 2017.

The component part contains tools and services. The service industry occupies a dominant position in the market, with the highest market share in 2017 at 60.10%. The need for technical support and system integration for suppliers and customers who need fewer variants to meet their needs, as well as the benefits of improving the performance of the distributed bicycle platform, apply to all service departments.

The cloud computing sector accounted for 61.10% of the market share in 2017.

Part of the development model includes cloud and local. In 2017, the cloud segment accounted for 61.10% of the market share. With the increase in cloud applications, large and small enterprises are transitioning from the traditional indoor installation model to the cloud-based development model.

Small and medium-sized enterprises have the largest market share, valued at approximately US$4.14 billion in 2017.

Organizational departments include large and small enterprises. The market share of SMEs is the largest, with approximately US$4.14 billion in 2017. Small businesses use the RAD platform to quickly increase productivity and modify their business applications for profit.

The media and entertainment industry dominated the industry with the largest share (25.60%) in 2017.

Industry verticals include banking, financial services, insurance, manufacturing and automobiles, retail and consumer goods, telecommunications and information technology, public and state-owned sectors, healthcare, media and entertainment, and education. The media and entertainment field dominates with the largest share of 25.60% in 2017. The increasing popularity of media and entertainment has created a demand for new applications, which has led to market growth.

The regional segmentation of the RAD market is analysed.

Market analysis areas include North America, Europe, South America, Asia-Pacific, Middle East and Africa. In 2017, the Asia-Pacific region led the fast-growing global application market with a market value of US$2.66 billion. North America is the fastest-growing region due to the existence of large global companies, early adoption of cloud solutions and more investment in data management solutions to make better manufacturing decisions.

The global market value of fast-growing applications is analysed (in billions of dollars). All sectors on a global, regional and national scale are also analysed. The survey includes an analysis of more than 30 countries/regions in each department. The research includes five Porter performance models: attractiveness analysis, raw material analysis, supply and demand analysis, competitive position network analysis, and distribution and marketing channel analysis.

*****

BIBLIOGRAPHY

Agarwal, Ritu, Jayesh Prasad, Mohan Tanniru, and John Lynch. "Risks of rapid application development." *Communications of the ACM* 43, no. 11es (2000): 1–es.

Cai, Hongming, Boyi Xu, Lihong Jiang, and Athanasios V. Vasilakos. "IoT-based big data storage systems in cloud computing: perspectives and challenges." *IEEE Internet of Things Journal* 4, no. 1 (2016): 75–87.

Fior Market Research LLP. *Global Rapid Application Development Market Is Expected to Reach USD 95.2 Billion by 2025: Fior Markets*. GlobeNewswire News Room. Fior Market Research LLP, February 19, 2020. https://www.globenewswire.com/news-release/2020/02/19/1987450/0/en/Global-Rapid-Application-Development-Market-is-Expected-to-Reach-USD-95-2-Billion-by-2025-Fior-Markets.html.

Koehler, David W. "Adopting a rapid application development methodology." *Cause/Effect* 15, no. 3 (1992): 20.

"Three ways of achieving rapid application development in IoT." *RadioStudio*, July 22, 2019. https://radiostud.io/adopting-rapid-application-development-in-iot/.

Wang, Yuanbin, Yuan Lin, Ray Y. Zhong, and Xun Xu. "IoT-enabled cloud-based additive manufacturing platform to support rapid product development." *International Journal of Production Research* 57, no. 12 (2019): 3975–3991.

2 Integration of IoT with Artificial Intelligence in Health Care

N.K. Shiju and N. Ramachandran
Indian Institute of Management

Salini Suresh
Dayananda Sagar College

Anubha Jain
IIS (deemed to be University)

Contents

2.1 Introduction 30
2.2 The Terms AI and IoT 30
2.3 New Trends in the Healthcare Domain 32
  2.3.1 Early Contributors 32
  2.3.2 Current Trends 32
    2.3.2.1 Patient Care 33
    2.3.2.2 Diagnosis 34
    2.3.2.3 Virtual and Augmented Reality with AI and IoT in Healthcare 34
    2.3.2.4 Applying AI and IoT in Air Quality Assessment (AQA) 35
2.4 How COVID 19 Use AI and IoT in Treatment? 36
2.5 Disadvantages of AI and IoT in the Healthcare Domain 38
2.6 Regulations from the Health Insurance Portability and Accountability Act 38

DOI: 10.1201/9781003051022-2

2.6.1 Transport Encryption 38
2.6.2 Backup 39
2.6.3 Authorization 39
2.6.4 Integrity 39
2.6.5 Storage Encryption 40
2.6.6 Disposal 40
2.6.7 Business Associate Agreement 40
2.7 Conclusion 40
Bibliography 41

2.1 INTRODUCTION

At present in our lives, many interconnected electronic devices and computer-aided solutions collect vast amounts of data. This data is known as Big Data, which is the current trend and is growing enormously day by day. Analyzing these data, fabricating valuable reports and researching these data have significance in our life. In the Internet of Things (IoT), each physical device or equipment is connected to a shared network, and data can be collected and stored. There are various gears in the healthcare industry that collects patient data. Artificial Intelligence (AI) and its capability to handle vast amounts of data and analyze them in a fraction of seconds make them very much compatible with medicine and the healthcare system. Figure 2.1 shows a large picture of Big Data.

This study explores the position where AI and IoT stand in the healthcare industry. In Section 2.2, we describe the terms AI and IoT; Section 2.3 describes new trends in the healthcare domain; Section 2.4 describes how AI and IoT are used in the COVID 19 pathway; Section 2.5 describes the disadvantages of AI and IoT in the healthcare domain; Section 2.6 walks through major regulations from Health Insurance Portability and Accountability Act. Finally, Section 2.7 concludes the study we carried out.

2.2 THE TERMS AI AND IoT

IoT and AI may sound menacing to some, but we all use these technologies daily, from reading emails to driving directions. IoT is a computing process in which a system of interrelated, interconnected objects collects data over the network without human intervention. AI is intelligence performed by machines, unlike natural intelligence, which involves realization and emotions usually displayed by humans or animals. In Figure 2.2, many devices are interconnected with some networks; this explains what IoT is.

In recent times, the world witnessed an ever-growing path in all industries with AI. For example, we communicate to the overall supply chain management system from mobile phones and get goods at our doorstep. It entirely changed the way we communicate, interact and consume goods and services. The healthcare industry is also expanding with AI.

FIGURE 2.1 Big Data and various tools and technologies.

FIGURE 2.2 IoT – interconnected, interrelated objects collect data.

Data from various sources support that technologies such as AI and IoT simplify the lives of patients, physicians and the entire medical industry by taking care of multiple tasks that are typically done by many humans that too with minimal errors, less cost and quickly (Greco et al., 2020; Akmandor & Jha, 2018).

2.3 NEW TRENDS IN THE HEALTHCARE DOMAIN

The healthcare industry's current trend is a personalized patient-oriented treatment (PPOT) approach, improving the traditional healthcare system. AI and IoT play vital roles in this new trend. Integrating AI, IoT and machine learning (ML) in the healthcare domain outstretched the scope for serious research in these areas. ML supports various techniques and tools to solve diagnostic and prognostic problems in the medical industry (Magoulas & Prentza, 2001). Prediction of disease progression, therapy planning and overcall patient care will be easy if we integrate AI and ML (Qi et al., 2017).

There are many smart healthcare systems in the market capable of collecting large amounts of near real-time data. It includes mainly wearable sensor-enabled devices. When connecting these devices with a network and store data in a shared space, then processing with the unconditional capabilities of AI and ML may produce beneficial PPOT reports. In general, as the healthcare system becomes more reliant on computer technologies, AI, IoT and ML methods can provide overwhelming support to physicians and can even eliminate issues related to human fatigue and acclimation. It may give a result in a quick diagnosis (Greco et al., 2020; Acampora et al., 2013).

2.3.1 Early Contributors

The analysis started with PC-based solutions to monitor patients remotely, such as collecting data from ECG and accelerometers, then advising physicians about heart-rate variations and predicting critical situations. Then this process further improved with microcontrollers and single-board computers. In health monitoring, the patient is supposed to be allocated space such as hospital and home (Wang et al., 2014).

2.3.2 Current Trends

Now the conditions have changed. IoT technologies are getting matured. It put the foundation strong for people to study more on the applications and use technologies to develop different solutions that utilize network architectures and software platforms. This integrated software or hardware mainly assists healthcare services such as regular epidemic ailment monitoring, elderly and pediatric care and fitness management. IoT and AI are integrated into every step of healthcare. From Figure 2.3, we can easily understand the way IoT and AI are integrated with the healthcare system. It is highly

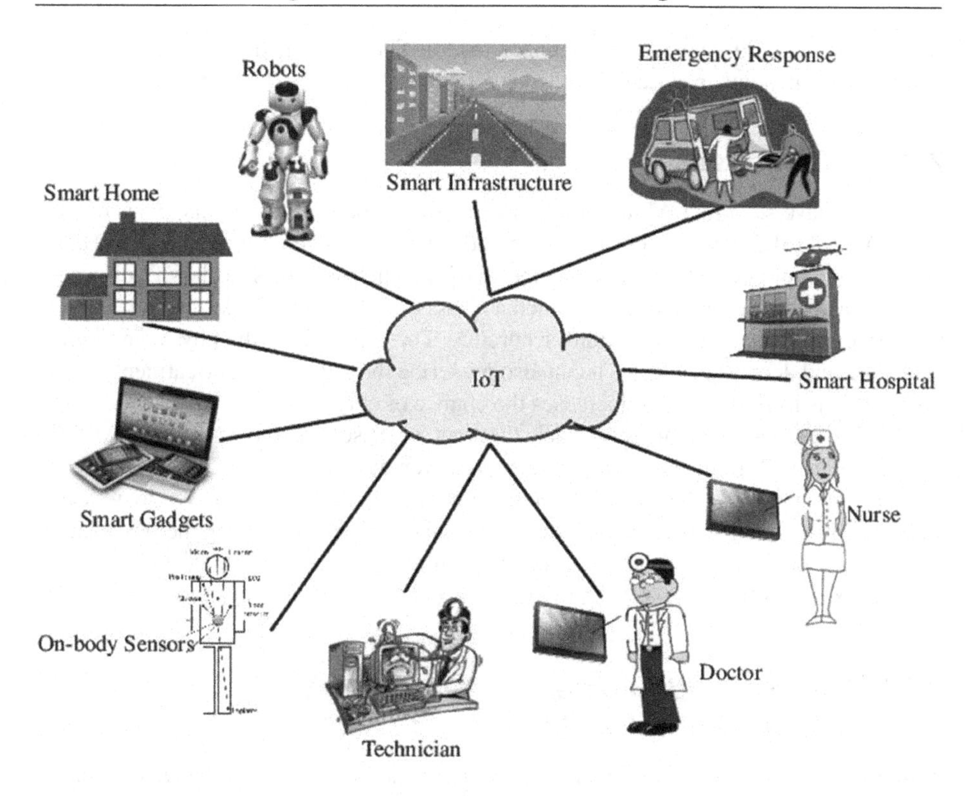

FIGURE 2.3 Pictorial representation of IoT usages in the healthcare domain.

coupled with everyone's life. From wearable devices to robotic help in treatment, IoT and AI are related (Akmandor & Jha, 2018; Fichman et al., 2011).

Numerous wearable fitness devices are limited to counting the number of steps humans walked a day. It offers more prospects for personal health care by analyzing users' rhythmic heartbeats, pre- and post-stages of some specific diseases, and exploring the implications of various conditions or medicines over the human body (Chen et al, 2017). It also detects atrial fibrillation in an early stage. It sends reports to doctors and analyzes blood pressure and oxygen level; very few scenarios in a wearable device can help a person in personal health care management.

2.3.2.1 Patient Care

Patient care is an inevitable service in any hospital. When improving the quality of patient care, automatically, the hospital's reputation increases as well. Future smart hospitals will rely on an ecosystem where patients get proper care even after discharge from the hospital. This ecosystem may expedite constant monitoring, storing and analyzing patients' health data (Sobhan Babu et al., 2016). Robotics and AI will be involved in personal patient care and be vital in the healthcare industry. Robots are also used in the medical sector for various operations such as surgical support and medical labs. The robots used in the medical industry will perform advanced tasks in the healthcare domain in the coming years (Puaschunder, 2020). They may help physicians examine patients, treat patients in rural areas through Telemedia, transport equipment, etc.

2.3.2.2 Diagnosis

Physicians have seen that AI and IoT deliver abundant benefits in the diagnosis of diseases. AI and IoT solutions can suggest whether a patient has some cynical medical condition even before evident symptoms appear. AI also helps doctors to classify diseases. In-depth analysis of medical images such as MRIs, X-Rays and CT scans with AI and ML can even spot certain conditions accurately. These tools help the physician to analyze more and decide accurately. It can also prescribe the most suitable treatment for the patient that is too early, which increases the chance of cure.

All these various scenarios spectacle that each of these technologies such as AI, ML and IoT are incredible and have substantial values even when they operate individually, but the potential upswings when we use them together across the human journey by wearing a device daily, which is used to track the daily events. It helps in alerting the physician if something wrong happens to the user, diagnosing the illness, treatment to the point of sickness and extend to personal patient care (Akmandor & Jha, 2018; Sobhan Babu et al., 2014).

2.3.2.3 Virtual and Augmented Reality with AI and IoT in Healthcare

Augmented reality (AR) is a technology that uses cameras and sensors to create an interactive experience with the digital world. It includes multiple sensory mishmashes, including pictorial, acoustic, kinesthetic communication, pressure, pain, or warm sensation on the body, and may even have a sense of smell. At the same time, virtual reality (VR) gives users a wholly immersed experience by creating an entire virtual world and shutting down the physical world. Users have to wear a specific device to get the full feel of the VR.

Now both these technologies are being used together as integration of AI and IoT technologies. These technologies are getting powered up when we use them together. The advanced AR and VR features with AI and IoT are now known as mixed reality (MR) (The Medical Futurist, 2017). It has a lot of scope in the healthcare industry.

AR is not a new concept, but now it is made practical by the camera industry's latest enhancements and research. We are still in the early stages of utilizing these in the healthcare industry. Now many physicians are interacting with the AI applications every day to improve patient education. High cost, limited functionality and acuity issues keep the VR at the back row in the industry (Madison, 2018).

The early invented AR product Google Glass has failed in the public market, mainly due to its pricing but got home in the healthcare industry. Now many companies are working hard to develop proper AR equipment for the medical field.

AR has a great scope in the medical education industry. Physicians or all the people related to the healthcare domain have to learn a lot about anatomy and body competencies. AR solutions give leverage to visualize and even intermingle with three-dimensional body structures. It makes AR an excellent tool for physicians to learn. No only physicians but also patients should be aware of some essential treatments and post-treatment

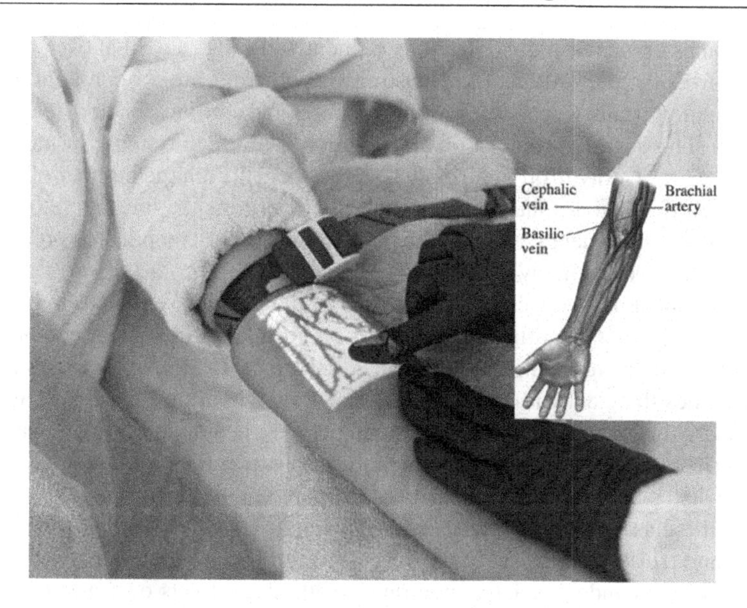

FIGURE 2.4 Vein projection system.

operations in many cases. Health workers can make use of AR capabilities in this scenario too. Physicians can explain and create awareness in patients about surgical procedures and how medicines will work on bodies (Madison, 2018).

In major surgeries today, physicians can use techniques to visualize the area they will operate in an AR way and can do the operations, which may more likely improve the accuracy, and patients will benefit from this.

An effortless but useful and practical application of AR now is the patient's vein projection system. It is often difficult to find the vein to inject medicine or get blood from a patient, and it creates a shoddier experience for the patient. Here comes the application of simple AR. There are vein projection applications that can display a map of patients' veins on their bodies, helping the physician find the right vein for the first time (Accuvein, 2017). Figure 2.4 shows a sample vein projection system.

AR can greatly assist physicians in their service. As the information provided in a 3D space in their vision, they can instantly access the cases, which may help their operations. From a student's perspective, they can quickly learn complex inner structures by seeing them in AR. In the future, an AR robotic surgery may also be possible. AR and VR with AI and IoT help stroke patients to overcome motor deficiencies by providing better-simulated environments and therapy techniques.

2.3.2.4 Applying AI and IoT in Air Quality Assessment (AQA)

Air pollution is one of the most slaughter bustle happening in a modern lifetime. Air pollution means the level of presence of contaminated gases in many forms upsurges in the earth's atmosphere. Measuring the impact of air pollution on the common man's health becomes an increasingly dangerous element. Now data says that over 70% of the population lives in cities. Most of the existing air pollutants are concomitant with serious assassinating health threats such as lung cancers and cardiovascular diseases (Costa et al., 2014). The degree of impact on air pollution is too high because of more people being exposed to it. Any drop in air pollution thus benefits a large number of lives. We have to design regional air quality plans (AQP) to determine the impact of air pollution with air quality assessment (AQA) on the public. It is quite a complicated process without IoT, and AI involves many people and massive data.

An assessment has to be conducted, which should be unbiased. It has to produce an estimate of the impact of mitigation measures on air quality in a given population and their health. It can be termed Health Impact Assessment (HIA). An HIA uses available hygienic studies that deal with the analysis of patterns of various diseases. This survey also considers the environmental changes and people's health data, which evaluate the probable effects of the proposed plan or policy on a population's health. It will also include the distribution of the change across the population (Costa et al., 2014).

When these assessments propose an AQP using IoT and AI technologies, it may be consistent and trustworthy. It analyses every aspect of the association between ambient air pollution levels and the relative contrary health after-effects over generations. The findings of studies on millions and trillions of patterns and health conditions provide a scientific basis for these decisions.

All these operations in the healthcare industry need the help of AI and IoT. When integrating AI and IoT capabilities for these analyses, the result may be more accurate and speedy. The targeted populations' health data can be collected with various devices connected to their living places or wearable devices. The gadgets need to be connected to the internet system to be collected and stored in a shared cloud space for analysis. Figure 2.5 describes a suggested sensor network.

2.4 HOW COVID 19 USE AI AND IoT IN TREATMENT?

AI and IoT are playing vital roles in the combat against COVID-19. It starts from elementary pandemic detection to the level of vaccine development and distribution. Every nation uses various AI-based applications to generate quick and efficient analysis and prediction reports from the data collected with IoT devices. In our day-to-day life, we can see many IoT-based thermal scanners and mask detection devices to help collect various public data.

COVID-19 quickens the use of telehealth resources in the healthcare domain. Till now, we all were familiar with the in-person visit for the treatments. But now situations changed, and people are looking for alternate ways to reduce contact with people, especially with patients. It reveals the importance of telehealth methods, facilitating direct communication with patients and physicians (Chamola et al., 2020). By wearing healthcare devices, patients help health workers and physicians to have real-time information

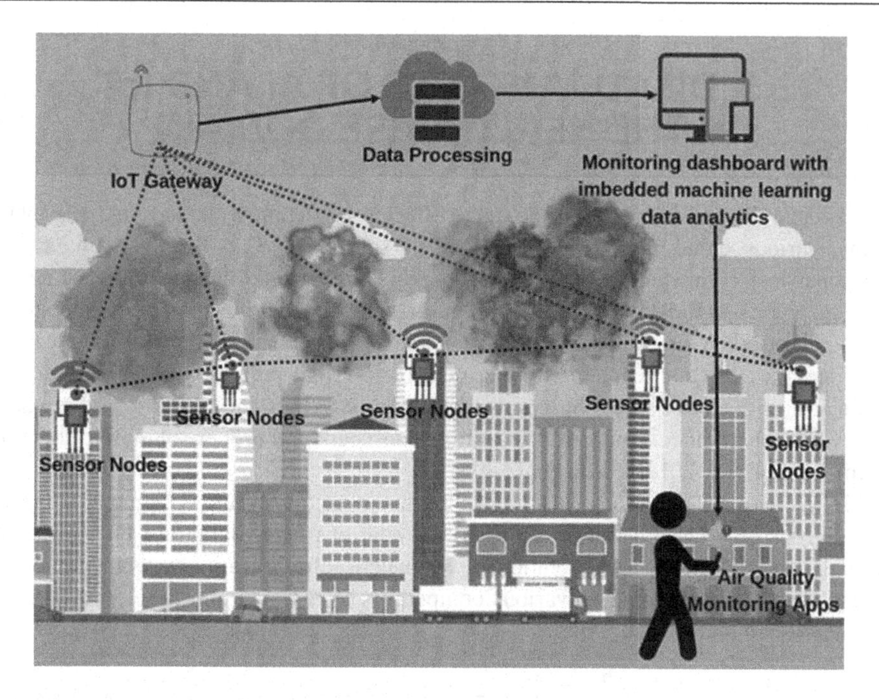

FIGURE 2.5 IoT and AI for air quality monitoring.

about them even when they are at home or in different locations. Telehealth's growth implies that the trend is likely to continue even after the pandemic. Telemedicine apps in mobile phones and electronic health records (EHR) are the critical features of telehealth. Cloud-based server solutions are a must for all the above facilities.

AI and ML play vital roles in vaccine development. When evolving a new vaccine, the aim is to include a resistant viral component that causes an immune system response. AI and ML together can help to identify these kinds of trashes to accomplish the goal. ML can efficiently analyze Big Data, which may improve the quality and speed of innovations that humans would not achieve alone. Every innovation needs human brains, but with machine backing, it process becomes quick and proficient. The data says, that the immunologists have recognized over one million occurrences of proteins on an cell's exterior that is discoverable by T-cells, which is splendid. Thanks to ML, C-19 vaccine development from different countries is happening and continuing quickly (Chamola et al., 2020).

When we scan a group of people with an ordinary thermometer, it may take a long time to get the proper result and complete the process. Instead, when we use infrared thermometers or thermal scanning and screening systems, which do not require any contact, we may change the sight. AI can quickly run through a group of people and identify one with high temperatures; it will be useful to isolate the symptomatic people.

ML in facial recognition technology has advanced enough to recognize people with a mask that too, with an accuracy of 95%. Also, AI can analyze CT scan images and identify pneumonia caused by COVID-19 in the chest through national training data for ML.

2.5 DISADVANTAGES OF AI AND IoT IN THE HEALTHCARE DOMAIN

For efficient AI and IoT data on the data collected by different devices and ML platforms, various cloud-based Big Data analysis solutions need an uninterrupted and robust network connection. When the patient is in a life-threatening condition and needs swift responses to the data collected, a high degree of network accessibility and stoutness is required. A single incidence of disconnection or delay in response may lead to fatal penalties in emergencies. Some critical factors that pull AI and the Internet of Medical Things (IoMT) are robust networks, cloud-based systems, security and privacy. It prevents the architecture from beginning a consistent and proper solution to the aim (Akmandor & Jha, 2018; Costa, et al., 2014).

In the medical industry, privacy is an extremely critical issue. For quick diagnosis and analysis of disease patterns, a massive amount of patient data must be transacted through different cloud-based solutions. Everywhere intruders are always trying to get access to it, it is beneficial for various competitors to research and develop their own business (Shah and Chircu, 2018).

2.6 REGULATIONS FROM THE HEALTH INSURANCE PORTABILITY AND ACCOUNTABILITY ACT

There are stringent regulations from the Health Insurance Portability and Accountability Act – 2020 (HIPAA – 2020) for every architecture or solution in the healthcare industry. Even though cloud-based solutions and AI can bring more productive systems capable of supporting the physicians in many ways, guidelines to secure electronically Protected Health Information (ePHI) are rigorous, and conforming to them is difficult. Some minimal requirements should be satisfied by AI and IoT solutions, when integrating with the healthcare industry. Some of them are Authorization, Transport Encryption, Storage Encryption, Backup, Disposal, Integrity and Business Associate Agreement (The summary of the HIPAA privacy rule). We will walk through what each point intends to do.

2.6.1 Transport Encryption

The first step to making a HIPAA compliant solution is to keep the sensitive data encrypted during broadcast across the solutions. Currently, there are protocols available for this. They are Secure Socket Layer (SSL), Transport Layer Security (TLS) and HyperText Transfer Protocol Secure (HTTPS) protocols. These protocols ensure that any data transferred between different users on the internet or between systems connected in a network makes it impossible to read. These protocols use encryption techniques and algorithms to obscure data in transport, which prevents hackers from hijacking it as it is sent over the network (The summary of the HIPAA privacy rule). When we say to send data over the web, whenever a user submits data to a system like personal data, health information, credit card numbers or other financial materials, the data transports over the network the user uses and reaches the application. Encrypting these data and securing access with complicated passwords can prevent sharing patient's data, and thus, the HIPAA security requirement may comply with communication.

2.6.2 Backup

Storing the collected massive amount of patient health data for a specified duration is as important as managing them. The various analysis algorithms required data over a long period to make a proper decision. These Big Data usually stores over Cloud storage, keeping the Big Data in multiple forms and places. It also ensures a robust mechanism to incorporate various historical data. Cloud platforms serve as the most potential framework for mobile communications, IoT, healthcare, multimedia and social networks to process enormous amounts of business data. Hosting providers have services for backup and recovery so that the data will be safe and won't be lost in case of any issues. If the application sends data to somewhere else such as email, this mail should be stored and backed upsecurely and accessible only to the approved staff.

2.6.3 Authorization

The available collected patient health data is very critical and needs to be handled with extreme attention. All the data should not be available to all physicians or admins in a healthcare institution. There are different levels of personnel and also have different levels of data access patterns. In many countries, patient health and disease laws are also in place, strictly restricting sharing disease-related or patient-oriented data with unauthorized people. Here comes the role of authorization in a solution. By creating different user levels and data, access arrangements authorization can be achieved in a solution.

2.6.4 Integrity

Keeping the collected patient data consistent across various significant data analysis processes must be achieved to satisfy the HIPAA compliance. The information collected, stored and floated over networks for analysis should be kept undamaged or altered in an undesired manner with or without intention. It should be ensured that the system has no unauthorized access to its data at the authorization level.

To make a software solution HIPAA compliant, it should prove proper authorization methods. All the data transfers should always be encrypted; all the stored data, backup and logs should be encrypted. All these physical access should also be appropriately monitored and authorized. It will be great if developers can adequately validate the system's input parameters to avoid further data processing issues. The data will be more accurate for further analysis by implementing this in user-level validations at user-level data entry.

2.6.5 Storage Encryption

As per the Transport Encryption compliance, the patient health data we collected should be floated across different platforms in an encrypted manner for analysis to avoid intruders to get data. Storage of the collected data is as essential as transferring data. When dealing with sensitive personal health data, the system should make sure that it is available to authorized persons only. When preventing unauthorized users' access, the system should also consider the stored data, databases, event logs and backups.

If the server for the solution is a shared one, sharing the server space with other hosts in the same hosting provider, the data must remain encrypted and inaccessible across the hosts. Server architecture has to incorporate industry-standard encryption techniques like AES and RSA algorithms with robust keys (preferably 256 bits for AES, and at least 4,096 bits for RSA).

2.6.6 Disposal

Patient health data collected is enormous as it has been stored for an extended period. Every piece of data will again have backs ups, and these backups may still have archives. These data may eat up the storage space in the long run if we keep these without an expiry policy. These data and all the related decryption keys should permanently get disposed of at a particular time. Whenever possible, solution should have a disposal policy to remove duplicated and irrelevant data.

2.6.7 Business Associate Agreement

Whenever a vendor develops or deals with health-related sensitive patient data, they must associate or comply with business associates. A hosting provider to whoever has access to patient health data must follow the minimum requirements to make the application HIPAA compliant.

2.7 CONCLUSION

It contributes to a greater understanding of two powerful technologies in this era and their contribution to the healthcare industry. IoT solutions in the healthcare domain are progressing from simple wearable devices that can collect, distribute and visualize data in different ways, to attract fitness-centric communities and physicians to analyze the cardio functions of the user. These simple devices turned to the complex and compact architectures that can analyze these Big Data in various ways and recognize different activities and even can make a crucial decision on the patient's health. We can see the role of ML and AI on the Big Data collected by IoT devices here.

The implementation of these techniques and solutions positively demands robust and uninterrupted internet connectivity. Cloud computational systems are getting better day by day to handle these patient data. In many cases, healthcare analysis needs real-time decision-making tools, requiring ongoing and robust internet. A solution with early predictive and minimal risk capabilities will be more preferred in systems like air pollution detection or patient's critical care section; it should have a meager response time to allow health workers to react to all kinds of health conditions proactively.

BIBLIOGRAPHY

Acampora, G., Cook, D.J., Rashidi, P. and Vasilakos, A.V. 2013. A survey on ambient intelligence in healthcare. *Proceedings of the IEEE* 101(12):2470–2494.

Accuvein. 2017. Accuvein vein visualization. https://www.accuvein.com (Accessed December 2017).

Ahmadi, H., Arji, G., Shahmoradi, L., Safdari, R., Nilashi, M. and Alizadeh, M. 2018. The application of internet of things in healthcare: A systematic literature review and classification. *Universal Access in the Information Society* 18:837–869.

Ahmadi, H., Arji, G., Shahmoradi, L., Safdari, R., Nilashi, M. and Alizadeh, M. 2019. The application of internet of things in healthcare: A systematic literature review and classification. *Springer* 18:837–869.

Akmandor, A.O. and Jha, N.K. 2018. Smart health care: An edge-side computing perspective. *IEEE Consumer Electronics Magazine* 7(1):29–37.

CableLabs. 2017. The near future: A better place. https://www.cablelabs.com/thenearfuture/ (Accessed December 2017).

Chamola, V., Hassija, V., Gupta, V. and Guizani, M. 2020. A comprehensive review of the COVID-19 pandemic and the role of IoT, drones, AI, blockchain, and 5G in managing its impact. *IEEE Access* 8:90225–90265.

Chandy, A. 2019. A review on IoT based medical imaging technology for healthcare applications. *Journal of Innovative Image Processing (JIIP)* 1:51–60.

Chen, M., Ma, Y., Li, Y., Wu, D., Zhang, Y. and Youn, C. 2017. Wearable 2.0: Enabling human-cloud integration in next generation healthcare systems. *IEEE Communications Magazine* 55(1):54–61.

Costa, S., Ferreira, J., Silveira, C. et al. 2014. Integrating health on air quality assessment—Review report on health risks of two major European outdoor air pollutants: PM and NO2. *Journal of Toxicology and Environmental Health, Part B* 17:307–340.

Datta, S.K., Bonnet, C., Gyrard, A., Ferreira da Costa, R.P. and Boudaoud, K. 2015. Applying internet of things for personalized healthcare in smart homes. *24th Wireless and Optical Communication Conference (WOCC)*:164–169.

Fichman, R.G., Kohli, R. and Krishnan, R. 2011. Editorial overview - the role of information systems in healthcare: Current research and future trends. *Information Systems Research* 22(3):419–428.

Greco, L., Percannella, G., Ritrovato, P., Tortorella, F. and Vento, M. 2020. Trends in IoT based solutions for health care: Moving AI to the edge. *Pattern Recognition Letters* 135:346–353.

Madison, D. 2018. The future of augmented reality in healthcare. *HealthManagement.org* 18(1):1377–7629.

Magoulas G.D. and Prentza A. 2001. Machine learning in medical applications. *Machine Learning and Its Applications* 2049:300–307.

Mahmud, R., Koch, F.L. and Buyya, R. 2018. Cloud-fog interoperability in IoT-enabled healthcare solutions. *19th International Conference on Distributed Computing and Networking* 32:1–10.

Miah, S.J. and Ahsan, K. 2013. An approach of purchasing decision support in healthcare supply chain management. *Operations and Supply Chain Management* 6(2):43–53.

Puaschunder, J.M. 2020. Die legale Situation von künstlicher Intelligenz, Robotern und Massendaten im Gesundheitswesen (The Legal Situation of Artificial Intelligence, Robotics and Big Data in Healthcare). Available at SSRN: https://ssrn.com/abstract=3525046 or http://dx.doi.org/10.2139/ssrn.3525046.

Qi, J., Yang, P., Min, G., Amft, O., Dong, F. and Xu, L. 2017. Advanced internet of things for personalised healthcare systems: A survey. *Pervasive and Mobile Computing* 41:132–149.

Shah, R. and Chircu, A. 2018. IoT and AI in healthcare: A systematic literature review. *Issues in Information Systems* 19(3):33–41.

Sobhan Babu, B., Srikanth, K., Ramanjaneyulu, T. and Lakshmi Narayana, I. 2014. IoT for healthcare. *International Journal of Science and Research* 5(2):322–326.

Sobhan Babu, B., Srikanth, K., Ramanjaneyulu, T. and Lakshmi Narayana, I. 2016. IoT for healthcare. *International Journal of Science and Research* 5(2):322–326.

The Medical Futurist. 2017. Mixed reality in healthcare – The HoloLens review business. http://medicalfuturist.com/mixed-reality-healthcare-hololens-review/ (Accessed December 2017).

The summary of the HIPAA privacy rule. http://optionsforrehab.com/downloads/Orientation/Summary%20of%20the%20HIPAA%20Privacy%20Rule.pdf.

Wang, X., Gui, Q., Liu, B., Jin, Z. and Chen, Y. 2014. Enabling smart personalized healthcare: A hybrid mobile-cloud approach for ECG telemonitoring. *IEEE Journal of Biomedical and Health Informatics* 18(3):739–745.

Significant Role of IoT in Agriculture for Smart Farming

3

Rohit Maheshwari, Mohnish Vidyarthi
and Parth Vidyarthi
Career Point University

Contents

3.1 Introduction 44
3.2 Why There Is a Need for Smart Farming? 45
3.3 Agriculture Sensors 45
3.3.1 Location Sensors 46
3.3.2 Optical Sensors 46
3.3.3 Electro Chemical Sensor 47
3.3.4 Mechanical Sensors 47
3.3.5 Dielectric Soil Moisture Sensors 47
3.3.6 Airflow Sensors 47
3.4 Sensor Output Applied 48
3.5 Smartphone Apps 50
3.6 Applications of IoT in Agriculture 50
3.7 Global Implications 54
3.8 Conclusion 54
Bibliography 55

DOI: 10.1201/9781003051022-3

ABSTRACT

If we talk about a technology that proved to be a boon for mankind, it is none other than IoT (Internet of Things) as we all are bound to the industries and agriculture where this technology has already spread its tentacles. According to the survey, it is an estimation that by 2025 the associated agriculture market will get hiked about 4.7 billion US dollars worldwide; it was 1.8 billion US dollars in 2019. When we ask ourselves what is the area where IoT is being used humongously, it will be in smart watches, Home Automation, Driverless Cars, etc. But in agriculture, it has impacted the most. As per statistics, the global population will reach almost 10 billion by 2050. Therefore, in order to consume the needs of such a gigantic population, it's become the need to head toward the latest technology in agriculture, but still we are facing some challenges such as weather condition, sudden changes in climate and environmental influence. To cope with such calamities, IoT has proved to be a game changer and helped in satiating the needs of eatables. As the industrial IoT in agriculture has been introduced globally, so do the advanced sensors been used in the market. There are various ways to connect the sensors to the cloud either by cellular or satellite network, through which we can have real time data just by using the sensors in projects. There are other applications also in IoT such as water monitoring system in which farmers can look into the tank for checking the water level in real life for the welfare of irrigation process, seed and plant monitoring system, in which farmers can monitor their farms from the very beginning phase one of seeds sprouting to the fully-grown plant can be nourished. Using IoT in farming, farmers have not only enhanced the yields in multifold but also decreased the cost of farming. With the proper usage of IoT technology, the farmers have gained a tremendous amount of benefit not only financially but also physically by doing smart farming. Now, the water management is much easier, water sprinkling system, where sprinkler automatic sprinkles the water once the sensors found a lack of humidity in the soil. In a nutshell, we can say that IoT enhances decision-making capabilities of the farmers with more precise data in real time and is like another wave of green revolution through smart farming.

3.1 INTRODUCTION

When a notion pops up in our mind, that the technology that can be a boon for human beings, it'll be Internet of Things (IoT) in sure.

The IoT in straightforward terms is "It's a network of devices, machines, objects connected with Internet that are able to gather and exchange data".

An IoT system contains sensors or devices that communicate to the cloud using Internet. When the data has been uploaded within the cloud, the software starts to process it, and then, it goes for the action stage such as sending a message of alert to the user automatically just by correcting the devices or sensors without manually anything done by the user (Figure 3.1).

FIGURE 3.1 Internet of Things.

3.2 WHY THERE IS A NEED FOR SMART FARMING?

When someone asks you a question, can you tell me the areas where the IoT has been used enormously? Then you must answer them as "Smart Watch, Home Automation, and Driverless Cars etc." But the area where the IoT impacted the most is in agriculture. According to the survey, the world's population will be estimated at around 10 billion by 2050. So, to consume the needs of such a mammoth people it will be tough to quench the hunger of every person; therefore, we need to look forward to the technology especially in agriculture, weather condition, Abrupt changes in climate and environment influence, so to overcome with such issues, IoT has set its benchmark as a game changer and helped in providing basic need of food.

3.3 AGRICULTURE SENSORS

Various types of sensing technology used in agriculture for precise execution of work such as providing data for monitoring the fields by the farmers and optimization of crops by keeping in mind the factors such as environmental change, by using some sensors as discussed below.

3.3.1 Location Sensors

Location sensors are generally used via global positioning system satellites to figure out the latitude, longitude and altitude in feet. At least three satellites are a minimum requirement for triangulation of a position. Exact positioning is the foundation of precision agriculture. The Global Positioning Systems ICs like NJR NJG1157PCD-TE1 are the best examples of location sensors (Figure 3.2).

3.3.2 Optical Sensors

Optical sensors use light for the measurement of soil. There are different frequencies of light reflectance which are as follows: near-infrared, mid-infrared and polarized light spectrums. We can place sensors on vehicles or on aerial devices such as drones or satellites. The motives to develop optical sensors are to determine clay, moisture or humus content, and organic content or matter in the soil. It's interesting to note that soil reflectance and plant color are the types of data which are used as a variable from optical sensors that can be stored and processed. For a real-life example, Vishay is a type of optical sensor used as a photo detector and photodiodes, which are basic infra of optical sensor (Figure 3.3).

FIGURE 3.2 Location sensor.

FIGURE 3.3 Optical sensor.

3.3.3 Electro Chemical Sensor

When it comes to precision agriculture, the important information that provides electrochemical sensors is pH and soil nutrient levels. It detects particular ions in the soil. In the current scenario, we used to gather, process and map soil chemical data using "sleds" which are mounted on sensors.

3.3.4 Mechanical Sensors

Mechanical sensors are used to measure the tightness of the soil. This sensor has some pins outwards direction, which are meant to dig into the soil and store resistive forces using load cells or strain gauges. And akin technology is predominantly been used on tractors for pulling out requirements for ground. Tensiometers such as Honeywell FSG15N1A are used to figure out the roots under water absorption and are very beneficial for irrigation purposes (Figure 3.4).

3.3.5 Dielectric Soil Moisture Sensors

Dielectric soil moisture sensors are used to measure the moisture or humus levels by using the dielectric constant present in the soil (Figure 3.5).

3.3.6 Airflow Sensors

Airflow sensors are used to count permeability. We can measure the motion in either singular location or dynamic. The pressure that is a sought output is asserted in a predetermined amount of air in the soil at a prescribed depth. There are numerous soil properties, compaction, structure, soil type and moisture level, which are unique in their ways.

FIGURE 3.4 Mechanical sensor.

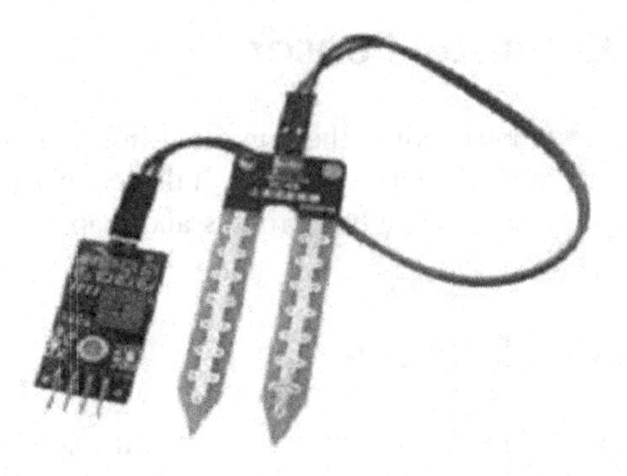

FIGURE 3.5 Soil moisture sensor.

3.4 SENSOR OUTPUT APPLIED

The technology that we use are action oriented and data driven so that it can be processed and implemented to optimize the crop yields by minimizing environmental effects.

1. Yield Monitoring System are used to kept on crops harvester vehicles such as corn harvesters. They are used to provide crop weight by time, distance or GPS location stores within 30cm.
2. Yield mapping is used in spatial coordinate data from global positioning system sensors mounted on harvesting equipment. These monitoring data are used to create yield maps by combining the coordinates.
3. Variable rate fertilizer is a tool that uses yields maps and optical surveys of plant health which are described by their color to control granular, liquid and gaseous fertilizer materials. These types of controllers are either being controlled manually or automatically using an on-board computer that are guided by global positioning system.
4. Weed mapping is used in operator interpretation and to generate maps by pointing locations on global positioning system receiver and data logger. These devices then mapped in yield maps, fertilizer maps and spray maps. When the visual recognition improves so as the manual entry will be replaced with automatic visual mounted device.
5. Variable spraying is used to control herbicide spray booms ON and OFF and customize the amount we need to spray. Once the weed area has been identified and properly mapped, then the volume of spray mix will be determined.
6. Topography and boundaries can be stored using high precision global positioning system, which are very accurate in topographic reference and presentation of any field. These types of maps are very useful for yield and weed maps interpretation. For farm planning, these types of data are very optimal especially in field boundaries, existing roads and wetlands.

7. Salinity mapping is done using salinity meter on sled towed fields affected by salinity. These types of mapping interpret issues with change in salinity over the time.
8. Guidance system is used to track a vehicle with precision of 30cm or less. These types of systems have replaced conventional spraying and seeding devise. And autonomous types of vehicles are currently under construction, which might change the future course of action.
9. IoT agriculture sensors are used to detect the leaks and levels of tanks for optimum usage of water and can avoid its scarcity. Precise and reliable alerts have saved the costs and energy of the peasants.
10. Wireless IoT agriculture sensors for humidity and temperature monitoring helps in water usage optimization and maximizes the crop fields, by alerting peasants in slight change of a weather conditions.
11. Waterproof and compact battery powered GPS tracking devices for livestock are used for location of herds, movement, speed and inactivity of herds, approximate temperature and many more (Figure 3.6).

FIGURE 3.6 (a) Guidance system, (b) livestock monitoring, (c) temperature monitoring and (d) water level monitoring.

3.5 SMARTPHONE APPS

Countless cell phone applications have commenced teaching IoT expectations, information accumulations and fast handling, giving data to low-area farmers with respect to seeding, weeding, watering and treating. These apps store data as a piece of information in handheld sensors, remote sensors, climate stations and creating important proposals with top to a bottom investigation. Various applications have been made for explicit thought processes for low-area farmers:

- Disease detection and diagnosis – In this application, photographs were taken first of the farfetched plant and afterward sent to the expert for observation.
- Manure calculator – In this leaf shading and soil sensors can figure out what supplements are required.
- Soil study – First we have to catch soil pictures with pH and substance information utilizing sensors, which permits farmers to follow along and alter soil conditions.
- Water study – It assists with distinguishing the leaf area index utilizing photographs and splendor, logging can assist peasants with identifying water prerequisites.
- Yield harvest readiness – It catches photographs with cameras with ultraviolet and white lights with accurately anticipated readiness.

3.6 APPLICATIONS OF IoT IN AGRICULTURE

- **Precision Farming**

  With regard to expanding domesticated animals and harvests by the peasants who used to rehearse controlled and exact cultivating that should be possible by utilizing precision farming. In this kind of cultivating the primary character is IT and various sensors, control frameworks, mechanical technology, self-sufficient vehicles and so forth.

  As we moved forward, rapid speed, cell phone gadgets and solid, handy satellites by the organizations are some special key features of the exact horticulture pattern.

  This sort of agribusiness is one of the most conspicuous applications of the IoT, particularly in horticulture streams, and different organizations are giving their best to upgrade this innovation around the world. Harvest Metrics is an exact agri-business organization that particularly centers around present-day arrangements while idealizing in the water system also.

The Crops Metrics provides services such as VRI optimization, soil moisture probe and virtual optimizer PRO. Variable rate irrigation optimization maximizes profit in irrigation fields with topographic or soil variability, improves yields and increases water use efficiency.

The technology that is named soil moisture probe is used in local agronomy support and is recommended for optimal use of water. The virtual optimizer PRO inculcates numerous tech for water management in one central, cloud-based and powerful location designed for consultations and beneficiaries to take advantage of precision irrigation using a simplified interface.

- **Agricultural Drones**

As time passes on so does the technology, and if we talk about drones in agriculture, it's a perfect example of change in agriculture. Nowadays just horticulture is where drones are broadly utilized and acknowledged in light of the fact that drones upgrade the assortment of farming practices. There are essentially two different ways of drones being utilized—one is ground-based and the other elevated-based drones, which are intended for crop wellbeing evaluation, water system, crop observing, splashing, planting, soil and field investigation.

There are different advantages additionally of utilizing drones that incorporate harvest wellbeing imaging, coordinate GIS mapping and usability, spare time and build yields. With legitimate arranging and procedure dependent on genuine information assortment and preparing, drone innovation will be exceptionally valuable in cutting-edge makeover, particularly in farming.

The genuine case of utilizing drones in farming is Precision Hawk, which is an association that utilizes and stores significant information for social affairs through a nexus of sensors that utilize images, maps and studies of agribusiness lands. Where this drone flies broadcasting in real time and takes information by checking and watching the farms, at that point, peasants enter the insights regarding the study directed by the automaton in the wake of choosing height and ground goals.

From the information gathered by the drone, we can bring understanding into plant records, plant checking, yield forecast, plant tallness the executives, shade spread mapping, field water ponding, and so forth.

The drones likewise gather different information such as multispectral, warm and visual symbolism during the flight and afterward arrives back to its unique position (Figure 3.7).

- **Livestock Monitoring**

Now, we might face a problem whether a large field owner can gather the information with respect to the area, prosperity and soundness of the steers. IoT can do this all the more deftly and without any problem. The task of the technic is to figure out the sick and debilitated animals present in the group or herd so that different animals won't become ill and malady won't spread in different animals because of which work cost gets lower and farmers can distinguish the specific area of their cows with the assistance of IoT.

FIGURE 3.7 Drones in farming.

How about we talk about a genuine cow monitoring system, namely JMB North America, an association that screens dairy animals. The fundamental errand of the system is to distinguish the pregnant bovines and give them legitimate treatment until calf birth. From the yearling, a sensor is controlled by a battery that screens whether the eater breaks or not; once it breaks, it sends the data to the peasants or proprietor with the goal that he will turn out to be progressively engaged.

- **Smart Greenhouses**

  The greenhouse technic of cultivation helps in upgrading the yields of vegetables, organic products, crops, and so forth in multifold as it controls the ecological boundaries utilizing manual intercession or by utilizing a relative control instrument. As manual mediation relates to production loss, energy loss and work cost, it's less compelling. A IoT-based smart greenhouse can be structured astutely through which one can control the atmosphere and dispense with the requirement for manual mediation.

  Presently on the off chance that one needs to control the environment in a smart greenhouse, various kinds of sensors are utilized by the plant's prerequisite. We can make a cloud server from remote getting to the framework when it's associated with the IoT. This causes the end of steady manual monitoring, and inside the greenhouse, the cloud server has been utilized to process the information and control the activity. This structure gives a practical and ideal answer for laborers with the least client mediation.

  How about we talk about a genuine case of a smart house; its name is Illuminum Greenhouse which is a trickle establishment and agri-tech association that instills the most recent innovation for administrations. This association utilizes sunlight-based force for IoT sensors, and with these sensors, the water utilization and nursery state can be observed through SMS alarms to the laborers with an online entryway. The programmed water system is done in suck sorts of nurseries.

The sensor utilized in the house gives information such as light level, weight, moistness and temperature. These sensors not just store information but even can control the actuators consequently opening a window, turn on lights, control a warmer, turn on a mister or turn on a fan that is totally controlled by means of Wi-Fi (Figure 3.8).

- **Aid Pest Management**

  In any case, when the workers do check for bug pervasions without anyone else which causes them a great deal of time, numerous issues may complete unnoticed as it's been physical which causes enormous misfortune.

  At the point when we need constant data IoT sensors to give it about yield wellbeing and irritations. Also, in the event we talk about low-goal picture sensors that will be utilized in an enormous region. These pictures are taken by the camera that we can't see with the unaided eye.

  In nutshell, there are high-resolution sensors that are utilized to catch the light energy discharged by a plant, likewise called a spectral signature.

  A laborer who, for the most part, gives more consideration to the explanation of invasions are almost certain when choosing to execute IoT gadgets for bug control.

We should take a model, IoT sensors are utilized to store information from those fields in which they have been utilized, which permits clients to recognize what will be the best counteraction strategies to getting the job done. For the most part, climate designs make the nuisances progressively infectious, and IoT sensors are utilized to stop the bugs ahead of time.

FIGURE 3.8 Smart greenhouse.

In smart farming, gadgets can be privatized to the peasants whether the treatment they have utilized on current irritations is fruitful or not and which zones are left unaffected by the precautionary measures, with the goal that other administration techniques can be applied. These sorts of cultivation showed best results in horticulture fields.

3.7 GLOBAL IMPLICATIONS

Talking about the developing countries, 500 million small area farmers produce 81% of the food expended, where precision agriculture is continuously getting all the more generally available around the globe. Presently, there is an on the off chance that the laborer needs to figure the wellbeing of plants; at that point, some handheld gadgets are unequivocally utilized for the best manures. According to the review, it's assessed that by 2026, the agribusiness market will climb to about 4.8 billion US dollars universally, it was 1.9 billion US dollars in 2019.

Taking care of the issues by looking at peasants utilizing smart cultivation, exactness horticulture not just expands food requests but even builds the existence costing of the laborers too.

The advantages offered by smart farming are as follows:

- It is utilized in bringing down fuel cost and utilization of vitality by decreasing CO_2 discharge.
- It decreases NO_2 (nitrous oxide) discharges from the dirt by improving nitrogen composts.
- It reduces synthetic utilization of utilizing exact compost and nuisance control.
- It disposes of supplement exhaustion through observing and keeping up soil wellbeing.
- It controls soil compaction by negligible gadget traffic.
- It maximizes water utilization productively.

3.8 CONCLUSION

Smart farming has expanded overall interest for food by utilizing a few advances that make it simple to utilize and less expensive to buy and apply information, in addition, to utilize assets all the more productively. Maybe enormous peasants have more advantages of utilizing such innovation, yet nowadays little homesteads are additionally getting advantages of utilizing cell phones, important applications and little estimated apparatus. What more we got by utilizing these advancements are sans contamination, a dangerous atmospheric deviation and preservation.

BIBLIOGRAPHY

Biz Intellia "5 Applications of IoT Applications in Agriculture", https://www.biz4intellia.com/blog/5-applications-of-iot-in-agriculture.

By Dataflair Team (September 15, 2018) "IoT Applications in Agriculture", https://dataflair.training/blogs/iot-applications-in-agriculture.

Digital Matter "IoT Agriculture Sensors", https://www.digitalmatter.com/Solutions/IoT-Agriculture-Sensors.

IoT for All (January 3, 2020) "IoT Applications in Agriculture", https://www.iotforall.com/iot-applications-in-agriculture.

Steven Schriber for Mouser Electronics "Smart Agriculture Sensors: Helping Small Farmers and Positively Impacting Global Issues, Too", https://www.mouser.in/applications/smart-agriculture-sensors.

4 Next Era of Computing with Machine Learning in Different Disciplines

R. Sabitha
Avinashilingam Institute for Home Science and Higher Education for Women

J. Shanthini and S. Karthik
SNS College of Technology

Contents

4.1 Introduction 59
4.2 Overview 60
 4.2.1 Anaemia Classification 61
 4.2.2 Introduction to CDSS 62
4.3 Problem Statement 64
4.4 Literature Review 64
4.5 Agent-Based CDSS for Anaemia Prediction 65
 4.5.1 Agent Systems 65
 4.5.2 Agents 66
 4.5.3 Multi-Agent Systems (MAS) 66

DOI: 10.1201/9781003051022-4

4.5.4 Intra-Agent Communication 67
4.5.5 JADE (Java Agent Development Framework) 67
4.5.5.1 Agent Class 68
4.5.5.2 JADE Agent 68
4.5.5.3 Agents Behaviour 68
4.5.5.4 Unlock an Agent 68
4.6 Agent-Based Architecture 69
4.7 Experimentation and Exploration 72
4.8 Conclusion and Future Work 77
Bibliography 77

ABSTRACT

Data mining refers to the extraction of useful information/knowledge from huge volumes of data that are in raw form. Various techniques such as clustering, classification, feature selection, association rule mining and outlier analysis are applied to the raw data for processing to obtain the desired useful knowledge. A variety of applications exists for data mining ranging from stock market to weather forecasting. One such application of data mining is its usability in the medical field known as healthcare mining (or) mining medical data. Almost all the hospitals use some hospital management system to manage healthcare in patients. Unfortunately, most of the systems rarely use huge clinical data where vital information is hidden. As these systems create a huge amount of data in varied forms, this data is seldom visited and remains untapped. Hence, lots of effort are required to make intelligent decisions in this scenario. Although computerised clinical guidelines may provide benefits for health outcomes and costs, their successful implementation is more challenging to investigate significant problems. Anaemia is a common haematological disease and it's a condition that develops when blood cells lack enough healthy red blood cells (RBCs) or haemoglobin. Anaemia is classified into different types based on complete blood cell (CBC) count test values. This work concentrates on the classification of the disease in top five disease categories that are commonly prevailing. In this work, artificial intelligence (AI) technologies and agent framework have made an attempt to assist in the diagnosis of the disease in question. In this work, a novel approach is proposed to develop a clinical decision support system (CDSS) for classifying the severity of anaemia disease diagnosis using data mining and AI techniques. The major goal is to build an expert system for diagnosing the presence of anaemic disease with an integrated automated classifier using AI techniques. Real-time dataset is employed in the work with major ranking attributes – mean corpuscular volume (MCV), mean corpuscular haemoglobin concentration (MCHC) and RBC. The dataset is pre-processed, and tests are run using Weka classifiers available in Weka 3.8. Experimental results demonstrate the effectiveness of the proposed CDSS in anaemia diagnosis. The purpose of this work is to develop a cost-effective analysis using JADE and J48 Decision Tree Data-Mining technology for facilitating database of the Decision Support System. Experimental analysis is done to identify the effectiveness of the proposed methodology.

4.1 INTRODUCTION

Data mining refers to the extraction of useful information/knowledge from huge volumes of data that are in raw form. Various techniques such as clustering, classification, feature selection, association rule mining and outlier analysis are applied to the raw data for processing to obtain the desired useful knowledge. A variety of applications exist for data mining ranging from the stock market to weather forecasting. One such application of data mining is its usability in the medical field known as healthcare mining (or) mining medical data. Data mining holds immense potential for the healthcare industry, thus enabling health systems to employ data and analytics to discover possibilities and practices to improve additional care and reduce cost. This creates a win/win strategy since various studies show that almost 30% of the overall cost is spent on such data analysis. Due to complex healthcare operations and a slower rate of technology adoption, there is a lag in implementing effective data-mining and analytic strategies.

Almost all hospitals employ a hospital management system to keep track of their patients' health. Unfortunately, most systems rarely utilise the massive clinical data that contains critical information. Because these systems generate a large volume of data in a variety of formats, this data is rarely accessed and hence remains untapped. Hence, lots of effort are required to make intelligent decisions in this scenario. Although computerised clinical guidelines may provide benefits for health outcomes and costs, their successful implementation is more challenging to investigate significant problems. The diagnosis of this disease using different features or symptoms is a complex activity. One of the effective solutions is to achieve an optimal trade-off between data ambiguity and good decision-making, which would further help in the integration of data-mining and artificial intelligence (AI) techniques.

Anaemia is a frequent haematological disorder that occurs when the body's red blood cells (RBCs) or haemoglobin levels are insufficient. According to the findings of the latest Global Nutrition Report 2017, India ranks last in the world in terms of the number of women affected by anaemia. The study shows that 51% of all women of reproductive age have anaemia. The reason for this is mainly due to the unawareness of the problem since the symptoms may not show up in the early stage. So the prediction of the disease is essential to identify the occurrence of the disease and thus aid in providing possible treatments to avoid malignancy. Anaemia is classified into different types based on Complete Blood Cell (CBC) count test values. This work concentrates on the classification of the disease in top five disease categories, which are commonly prevailing.

AI technologies and an agent framework were used in this study to aid in the diagnosis of the condition. Using data-mining and AI approaches, a unique way to develop a clinical decision support system (CDSS) for categorising the severity of anaemia disease diagnosis is suggested in this study. The major goal is to build an expert system for diagnosing the presence of anaemic disease with an integrated automated classifier using AI techniques. Real-time dataset is employed in the work with major ranking attributes mean corpuscular volume (MCV), mean corpuscular haemoglobin concentration (MCHC), and RBC. The dataset is pre-processed, and tests were run using Weka classifiers available in Weka 3.8. The proposed algorithm formalises the treatment of vagueness in decision support architecture, which also evaluates the performance measures among them. Experimental results demonstrate the effectiveness of the proposed CDSS in anaemia diagnosis. The purpose of this work is to develop a cost-effective analysis using JADE and J48 Decision Tree Data-Mining technology for facilitating database of the Decision Support System. Experimental analysis is done to identify the effectiveness of the proposed methodology.

4.2 OVERVIEW

Anaemia is a condition in which a person lacks sufficient healthy haemoglobin, the substance carrying oxygen in RBCs. This disease is very common and can result in the person feeling tired, weak, dizzy and short of breath. The disease burden varies greatly by nation, according to GlobalData's epidemiological analysis of anaemia, and it is extremely frequent even in developed countries.

Epidemiologists from GlobalData combed through the literature to establish the total prevalence of anaemia in the 16 major pharmaceutical markets (16MM: US, France, Germany, Italy, Spain, UK, Japan, Australia, Brazil, Canada, China, India, Mexico, Russia, South Africa and South Korea). The term "total prevalence" refers to both diagnosed and undiagnosed cases. Epidemiologists at GlobalData gathered information from studies that took blood samples from the general public and examined them for haemoglobin levels. Anaemia is characterised as haemoglobin levels that are below the World Health Organization's (WHO) standards for various age groups. The total prevalence of anaemia in the 16 MM is depicted in the graph below. The highest total prevalence of anaemia is 39.86% in India, while the lowest is 3% in Canada. The cumulative prevalence of the condition in the United States and the five EU countries (France, Germany, Italy, Spain and the United Kingdom) ranges from 5.6% to 10.74%, making it a common occurrence in these markets. Figure 4.1 depicts the global prevalence of anaemia in various countries.

Anaemia can be caused by a variety of illnesses, including sickle cell anaemia, iron deficiency anaemia, or anaemia caused by chronic kidney disease (CKD) or hypothyroidism. Iron deficiency anaemia is the most common type of anaemia, accounting for nearly half of all cases. The majority of anaemia's are treatable and preventable with a well-balanced, healthy diet. But, even in industrialised markets, why is there such a high prevalence of anaemia in the 16 MM? Among the 16 MM, India has the highest frequency of anaemia. Low haemoglobin levels are more common among Indian women, with nearly half of them having low haemoglobin levels. Poor eating habits (not eating enough fruits, vitamin C and legumes like beans and peas) and a lack of access to healthcare are the main causes of such a high frequency of anaemia among women, according to numerous research conducted in India. In India, iron supplementation efforts have not proved effective in reducing anaemia.

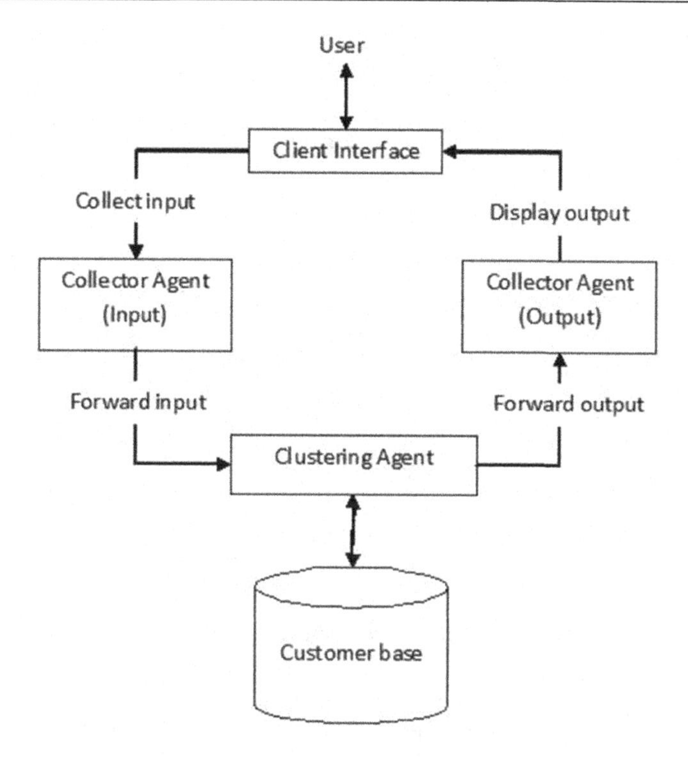

FIGURE 4.1 ACS model.

One reason for the high prevalence of anaemia in both industrialised and developing countries could be because the symptoms are non-specific in the early stages, making it difficult to identify. According to studies, more than half of persons with anaemia are completely unaware of their disease. Another reason could be the high prevalence of anaemia-causing illnesses, such as CKD. Anaemia develops in persons with CKD when renal disease reduces the generation of hormones that control haemoglobin formation. Anaemia in people with CKD is difficult to treat because it can progress to additional major consequences like heart disease. Because anaemia is frequently a symptom of more serious underlying conditions, it is critical to raise disease awareness and acquire a precise diagnosis of the cause.

4.2.1 Anaemia Classification

It is a medical disorder characterised by a decrease in the concentration of haemoglobin or RBCs in the human blood. A CBC count test is performed in a laboratory for patients. When attribute values such as age, gender, haemoglobin and haematocrit are below the normal range, it can be identified as anaemic categories. The classification of the types of anaemia according to CBC test values is illustrated in Figure 4.2 (Sanap et al., 2011).

Anaemia disease categorized into different types based on the CBC test values. In this model, anaemia types nomenclature illustrated (see Table 4.1) and classified

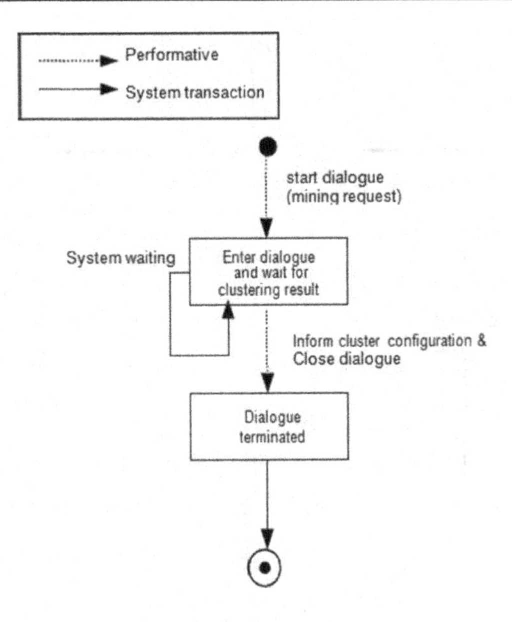

FIGURE 4.2 State transition diagram for collector agent.

TABLE 4.1 ANAEMIA Types Nomenclature

ACD	Anaemia of chronic disease
IDA	Iron deficiency anaemia
ARD	Anaemia of renal disease
THAL	Thalassemia
APA	Aplastic anaemia

according to MCV value into the three essential kinds of microcytic (MCV < 80) ft, normocytic (MCV = 80–100) ft and macrocytic (MCV > 100) ft anaemia and classified using MCHC (Mean corpuscular haemoglobin concentration) into normochromic (MCHC = 32–36) g/dL and hypochromic (MCHC < 32) g/dL anaemia. RDW (Red Cell Distribution Width) used to measure the anaemia and it is high if (RDW > 14.6) and normal if (RDW = 11.6–14.6) (Green, 2012; Sanap et al., 2011).

4.2.2 Introduction to CDSS

CDSS deals with applying data mining and AI to take clinical decisions where the system deals with taking health-related decisions and does predictions as humans. AI is a branch of computer science that includes the study and development of intelligent machines and software. Major AI researchers and textbooks define this field as "the study and design of intelligent agents", where an intelligent agent is a system that perceives its environment and takes actions that maximise its chances of success.

AI is especially useful in medicine when there is no conclusive evidence that one treatment option is better than another. New treatment strategies can be efficiently proposed based on the patient's profile, history, physical examination, diagnosis and previous therapy patterns.

AI is a set of algorithmic techniques for extracting useful patterns from large amounts of data. AI is data-driven, which is especially crucial in healthcare. AI is a set of tools and procedures that are used to process data in order to uncover hidden patterns, which will provide healthcare practitioners with a new source of information to help them make better decisions.

The most basic AI applications can be classified into two categories: classifiers ("if shiny, then diamond") and controllers ("if shiny, then diamond") ("if shiny then pick up"). However, before inferring actions, controllers classify situations; hence, classification is an important aspect of many AI systems. Classifiers are functions that find the closest match by comparing patterns. They are extremely appealing for usage in AI since they may be modified according to examples. Observations or patterns are the terms used to describe these examples. Each pattern in supervised learning belongs to a preset class. A class might be thought of as a decision to be made. A dataset is made up of all the observations and their class designations. When a new observation is received, it is categorised according to prior experience.

A classifier can be trained in a number of ways, including statistical and machine learning methods. The neural network, kernel methods such as the support vector machine, k-nearest neighbour algorithm, Gaussian mixture model, Naïve Bayes classifier and decision tree are the most extensively used classifiers. These classifiers' performance has been compared across a wide range of tasks. The performance of a classifier is highly dependent on the properties of the data to be classified. According to the "no free lunch" theorem, no single categorization method works optimally in all instances. Choosing the right classifier for the job is still more of an art than a science. Classification is the major AI idea used here. A set of unclassified cases is categorised by applying a class label to them.

1. Supervised Classification

   The list of possible classes is predetermined.

   - The training set, also known as the input data, is made up of many records, each with various properties or features.
   - A class label is assigned to each record.
   - The goal of classification is to examine the input data and create an appropriate description or model for each class based on the data's attributes.
   - This model is used to classify test data for which class descriptions are unknown.

2. Unsupervised Classification

   The set of possible classes is not known. After classification, we can try to assign a name to that class.

Unsupervised classification is called clustering. Various algorithms exist in this category.

4.3 PROBLEM STATEMENT

Health studies show that 51% of all women of reproductive age have anaemia. The reason for this is mainly due to the unawareness of the problem since the symptoms may not show up in the early stage. So prediction of the disease is essential to identify the occurrence of disease and thus aid in providing possible treatments to avoid malignancy. Anaemia is classified into different types based on CBC count test values. This work concentrates on the classification of the disease in top five disease categories that are commonly prevailing. In this work, AI technologies and agent framework has made an attempt to assist in the diagnosis of the disease in question. In this work, a novel approach is proposed to develop a clinical decision support system (CDSS) for classifying the severity of anaemia disease diagnosis using data-mining and AI techniques. The major goal is to build an expert system for diagnosing anaemia with an integrated automated classifier using AI techniques.

4.4 LITERATURE REVIEW

Many works used different data-mining algorithms to classify several types of diseases, such as anaemia disease for specific types based on data-mining algorithms (Elshami and Alhalees, 2012). In addition, many other researchers tried to find their own method. A person with anaemia probably is unaware of the problem may not appear. Millions of people may have anaemia and their health is exposed to risk. Therefore, the disease is significant and several studies carried out in this domain are mentioned in the literature (Yilmaz et al., 2013). Sanap et al. (2011) developed a system using the classification technique: C4.5 decision tree algorithm and SMO support vector machine WEKA. They implemented a number of experiments using these algorithms. The anaemia classification using a decision tree that gives clear results depends on CBC reports. Amin et al. (2015) have compared Naïve Bayes' J48 classifier and neural network classification algorithms using WEKA and worked on haematological data to specify the best and appropriate algorithm. The proposed model can predict haematological data, and the results showed that the best algorithm is J48 classifier with high accuracy, and Naïve Bayes is the lowest average in average errors. The study by Sanap et al. (2011) and Amin et al. (2015) proved that the C4.5 algorithm (as J48 in WEKA) results give higher accuracy than other classifiers. Dogan and Turkoglu (2008), based on the biochemistry blood parameters, designed a system to help physicians in the diagnosis of anaemia. The system was designed using the decision tree algorithm. The system used the characteristics of the haematology and classified the results into positive or negative anaemia. The results of this system accorded with physicians' decision. Siadaty and Knaus (2006) selected decision trees as a common and simple classifier that also has low computational complexity. The problem was the needed time to build a decision tree for a large dataset, which has become intractable. They solved the problem by developing a parallel model of ID3 algorithm. It is a thread-level parallelism decision tree that does the computations independently. The experiment was done on an anaemic patient's dataset. Kishore et al. (2015) presented a set of basic classification algorithms with a group of essential types of classification methods such as decision trees, Bayesian networks, k-nearest neighbour and support vector machine classifier. The study shows a comprehensive review of diverse classification algorithms in data mining. This research presents an investigation of five types of anaemia disease by using Naïve Bayes, multilayer perception, J48 decision tree and support vector machine data-mining algorithms depending on CBC data.

The best classification algorithms depend specifically on the problem domain (Kesavaraj and Sukumaran, 2013). Manal and Salma (2017) had specified the anaemia type for the anaemic patients through a predictive model conducting experimentation on some data-mining classification algorithms. The datasets were filtered and the undesirable variables were eliminated, then classification algorithms such as Naïve Bayes, multilayer perception, J48 and SMO were applied on these datasets using WEKA data-mining tool. The experiments proved that J48 decision tree algorithm gives the best potential classification of anaemia types. WEKA experimenter proves J48 decision tree algorithm has the best performance with accuracy, precision, recall, true positive rate, false positive rate and F-measure.

4.5 AGENT-BASED CDSS FOR ANAEMIA PREDICTION

4.5.1 Agent Systems

Over recent years, the computing trend has changed from a focus on individual standalone computer systems to a situation in which the real power of computers is realised through distributed, open and dynamic systems. In addition, the trend has been towards ever more intelligent systems which can automatically perform ever more complex tasks. However, in open, distributed and dynamic environments, the ability to cooperate and reach agreements with other systems is a necessary and overriding requirement. Agent technology provides an infrastructure that enhances our ability to solve problems in a collaborative manner by providing a "scaffolding" to support the required cooperation. At its simplest, a MAS is a software system that comprises a collection of software agents. These agents are capable of independent action, on behalf of their users or owners with some degree of autonomy. In other words, each agent can figure out for itself what it needs to do in order to satisfy its design objectives. The agents in MASs can interact with one another by exchanging messages using network protocols. The agents in a MAS typically have different goals and motivations, thus there is a chance of conflict with regard to the aims of individual agents. Hence, the operational requirement is that to successfully interact and resolve such conflicts, agents must have the ability to cooperate, coordinate and negotiate with each other (Wooldridge & Jennings 1995).

4.5.2 Agents

Agents are usually defined as computer entities that are suited to some environment and that are able to carry out their tasks in an autonomous manner. The environments in which agents operate may be dynamic, unpredictable and uncertain; thus, it is desirable for agents to exhibit some form of computer intelligence in order to interact with their environment. There are a number of capabilities that intelligent agents are expected to display (Wooldridge & Jennings 1995):

Reactivity – Intelligent agents should be able to anticipate tasks, according to their environment, so that they can respond in a timely fashion.
Proactiveness – Intelligent agents should be able to explore alternatives to achieve their design objectives.
Social ability – Intelligent agents should be capable of interacting with other agents (and possible humans) in order to satisfy their design objectives.

It is self-evident that the above capabilities can increase the degree of autonomy with which agents can perform tasks.

4.5.3 Multi-Agent Systems (MAS)

A MAS is a software system that employs a number of interactive agents to solve a problem in open and decentralised uncertain environments. A central feature of MAS is that there is no centralised control mechanism; agents are required to collaborate to achieve the design objective of a given MAS. A MAS has collective capabilities that an individual agent does not have. They are listed by Sycara (1998) as follows:

- A MAS distributes computing resources and capabilities among a network of interconnected agents to solve issues that are too big for a single agent to address. Resource constraints, performance bottlenecks and major breakdowns can all affect a centralised system.
- A MAS enables the integration and interaction of various existing legacy systems by wrapping them in an agent wrapper and incorporating them into an agent society.
- In a MAS, problems are modelled as autonomous interacting component agents, allowing these agents to work autonomously.
- Using a MAS can help in circumstances when expertise is dispersed both spatially and temporally. Expertise and resources can be exchanged based on an agent's social ability.

  Because of its distributed nature, MAS improves overall system performance, particularly in terms of computing efficiency, reliability, extensibility, robustness, maintainability, responsiveness, extensibility and reuse. However, for many data-mining activities, it is well established that there is no single "best" algorithm suited to all types of data. A data-mining

MAS where individual agents are equipped with data-mining algorithms produce results from which the best cluster configuration is obtained, such as that described in the following sections, is therefore clearly also of benefit.

4.5.4 Intra-Agent Communication

Agent communication languages (ACLs) were devised to allow agents to communicate by exchanging information and knowledge, especially in the form of complex objects, such as shared plans and goals or even shared experience and long-term strategies. Other means of exchanging information and knowledge among applications are remote procedure call (RPC), remote method invocation (RMI) and Computer Oriented BRidge Analysis (COBRA). Using these methods, software modules (objects) are able to communicate with one another regardless of where they are located in a network. Message passing ACL is the base of communication between agents. The class Agent's send method is used to convey messages. You must supply an object of type 'ACLMessage' to this method, which contains the recipient information, language, coding and message content. These messages are transmitted asynchronously, and they are stored in a message queue as they are received. There are two types of receiving ACL messages, blocking or non-blocking. For this provide methods blockingReceive() and *receive*()- respectively (Labrou et al., 1999).

ACLs provide additional functionality over the above for two reasons:

i. ACLs handle propositions, roles and actions (which have semantic associations), whereas RPC and RMI or COBRA have no such semantic association and
ii. ACL message describes a desired state in a declarative rather than a procedural form.

4.5.5 JADE (Java Agent Development Framework)

It is a software framework written entirely in the Java programming language. It facilitates the development of multi-agent systems by providing a middleware that adheres to the Foundation for Intelligent Physical Agents (FIPA) guidelines, as well as a set of graphical tools for debugging and deployment. The agent platform can be installed on several machines and configured remotely via a graphical user interface (GUI). JADE is a Java-based toolset for the building of agents. The JADE system promotes communication between agents and allows the system to discover services by supporting FIPA coordination between several agents and providing a standard implementation of the communication language FIPA-ACL. JADE is a middleware that facilitates the development of multi-agent systems under the standard FIPA for which purpose it creates multiple containers for agents, each of them can run on one or more systems (Bellifemine et al. 2005, 2007, 2008).

4.5.5.1 *Agent Class*

The agent class is a super class that allows the users to create JADE agents. To create an agent, one needs to inherit directly from *Agent*. Normally, each agent records several services that should be implemented by one or more behaviours.

This class provides methods to perform the basic tasks of the agents such as:

- Pass messages by objects ACLMessage, with pattern matching.
- Support the life cycle of an agent.
- Plan and execute multiple activities at the same time.

4.5.5.2 *JADE Agent*

These agents go through different states defined as follows:

1. Initiated – The agent has been created but has not registered yet the Agent Management System (AMS).
2. Active – The agent has been registered and has a name. In this state, it can communicate with other agents.
3. Suspended – The agent is stopped because its thread is suspended.
4. Waiting – The agent is blocked waiting for an event.
5. Deleted – The agent has finished, and his thread ended his execution, and there is nothing more in the *AMS*.
6. Transit – The agent is moving to a new location.

4.5.5.3 *Agents Behaviour*

The behaviour defines the actions under a given event. This behaviour of the agent is defined in the method *setup* using the method *addBehaviour*. The class Behaviour contains the abstract methods:

- **action()**: Is executed when the action takes place.
- **done()**: Is executed at the end of the performance.

A user can override the methods *onStart()* and *OnEnd()* property. Additionally, there are other methods such as block *()* and *restart()* used for modifying the agent's behaviour. A locked agent can be unlocked in different ways. Otherwise, the user can override the methods *onStart()* and *onEnd()* the agent possesses.

4.5.5.4 *Unlock an Agent*

1. Receiving a message.
2. When the timeout happens associated with block ().
3. Calling restart.

The reasons for choosing JADE were as follows:

Standard compliance – JADE complies with the FIPA specifications that enable end-to-end interoperability between agents on different platforms.

Well-known and widely used – JADE is the best-known and widely used toolkit for developing MAS applications. JADE is used by various communities of users both with respect to research activity and commercial applications. There is evidence for the use of JADE in many research papers, such as in Bellifemine et al. (2005, 2007, 2008).

Mature software framework – The JADE framework was released in 2003 and has been in use (although sometimes updated and maintained) ever since. There has been a continual improvement in JADE; while other development tools were created as part of short-time research activity, their further development and maintenance have thus typically ended after 2 or 3 years.

Extensibility – The JADE framework is distributed under an Open Source License; therefore, JADE code can be extended to enhance the platform with respect to some specific application or requirement. For example, the jade.mtp Application Program Interface (API) allows the users to add new message transport protocols.

User support and documentation – JADE code is distributed together with a programming guide and an administrator guide, examples and tutorials, and the javadocs for all the APIs.

Versatility – JADE provides APIs which are independent of underlying networks. With respect to JAVA technology, JADE also provides multi-platform support; the agent platform can be distributed across machines regardless of the operating system used.

Information hiding – Each JADE API is independent and provides some specific purpose – functionality; therefore, developers need to understand only the required functionalities used with respect to some specific implementation.

Transparent communication – The communication in the JADE framework is handled by the JADE run-time environment, and thus programmers need not be concerned with the mechanism used to actually deliver messages. A number of GUI-based tools are also provided in order to allow users to monitor and control run-time system activity.

A further reason why JADE was selected with respect to the work described in this thesis is the extensibility property of the FIPA ACL that allows the definition of user-defined performatives in order to support intra-agent communication in specific domains. User-defined performatives can make the syntax of the communication used by the agents more specific and descriptive of the particular task that the agents are communicating about (Bellifemine et al., 2005, 2007, 2008).

4.6 AGENT-BASED ARCHITECTURE

The model integrates multi-agent systems and the classification process. The agent-based framework proposed in this work is not aimed at improving the performance of individual data-mining techniques, but at facilitating the construction of hybrid intelligent systems for data-mining.

The agent-based model involves creating agents, namely, collector agent and clustering agent to perform anaemia prediction. The client interface acts as a mediator between the collector agents and user. The collector agent collects the input from the client interface and initiates the classification agent which clusters the data objects from the database based on the input received. Finally, the output is displayed through the collector agent. The agent-based system is implemented using JADE technology in Java. The model is shown in Figure 4.3 and the various modules in the ACS model are depicted below:

- **Client Interface Creation**
  Client Interface is being built using JAVA Swing that is compatible with the adaption of connectivity and agent communication. The input customer base is fed through the client interface to the clustering agent by the user.
- **Agent Creation**
  In this module, agents such as collector agent and clustering agent can be created. The input data from the client interface is sent to collector agent. The collector agent passes the input data to clustering agent. Clustering agent performs the clustering process that incorporates the various phases of SCS model described in Section 3.3. Clustering agent segments the set of data which is sent back to the collector agent and displayed to the user via the client interface. The model is depicted in Figure 4.3.
- **Agent Communication within the Framework**
  It may be possible to exchange the required information between the agents within the content of ACL messages. A set of performatives are used in order to enable the agents to negotiate between agents.

The performatives are classified into four categories (at a high conceptual level) as follows:

1. Holding a dialogue.
2. Performing a clustering task.
3. Informing about clustering results.

Table 4.2 illustrates the semantics of the defined performatives and gives the names of the performatives, the pre-conditions and the post-conditions. The pre-conditions indicate the necessary "state" for an agent to send a given performative. The post-conditions describe the state of the sender after the successful utterance of a performative, and the state of the receiver after the receiving and processing of a message.

The process can be illustrated using state transition diagrams. At the beginning of the process, a dialogue is opened by the client interface who wishes to *initiate a dialogue*. The collector agent joins the dialogue by issuing *join dialogue*. The collector agent initiates the classification agent and a request for data is sent to the data source *(request data)*. The data source then sends the data using the inform performative to a classification agent designated to perform the classification task *(inform data)*. The

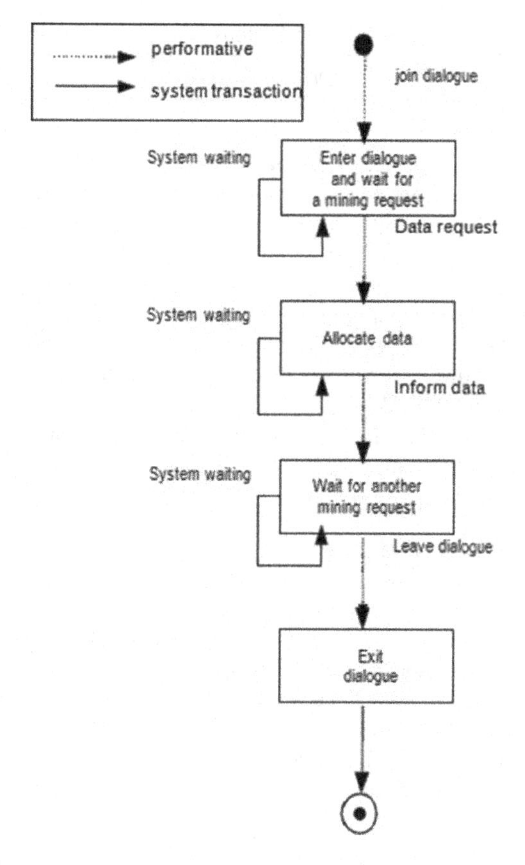

FIGURE 4.3 State transition diagram for data source.

classification task is thus performed, and the resulting configuration is generated. At the end of the communication, the classification agent informs the result to the user agent who made the initial mining request (inform configuration).

The agents use the above performatives as part of a protocol that governs the exchange of information between the agents. To illustrate the use of the protocol, Figures 4.4–4.6 provide some simple state transition diagrams that show dialogue moves (i.e. performatives that can be uttered) by the different agents and the subsequent choice of moves, which are then available in the new state.

A client interface establishes a dialogue by sending a mining (clustering) request as shown in Figure 4.2. This initiates the collector agent to receive the input. The request data performative is sent by the collector agent in order to commence a clustering process. The client interface waits until the final clustering result is obtained and then the dialogue will be closed.

The transition for data source is shown in Figure 4.3. After receiving the request data performative, the data is allocated to the clustering agent using the inform data performative, waits for other mining request and then leaves the dialogue.

TABLE 4.2 Performatives to Support ACS System

PERFORMATIVES	*PRE-CONDITIONS*	*POST-CONDITIONS*
Simple Dialogue		
Mining request (start dialogue)	An agent wishes to initiate a dialogue. No dialogue about this mining request is open.	Receivers (data source, collector agent (input and output) a clustering agent) obtain mining request.
Join dialogue	The agent sending the request is not yet a participant in the dialogue. The agent wants to join the dialogue to cooperate on a clustering task.	Sender maintains its intention to achieve its goal. The Sender agent is now a participant in the dialogue.
Close dialogue	The client interface has received the configuration.	The agents have all left the dialogue.
Task dialogue		
Request data	Receivers and Sender are in the same dialogue. Receivers have held a data source.	Receiver has information, such as receiver agent's name. Hence, the receiver knows who will obtain its data.
Inform data	Receivers and sender are in the same dialogue. Receiver does not currently have a data.	Receivers have obtained a set of items and started to perform the clustering task.
Inform cluster Configuration	The client interface does not know the result of the clustering process.	Receiver (a client interface) obtains an improved configuration.

When a classification agent enters a dialogue (see Figure 4.4), it waits for the mining request, once received it performs classification as per the J48 decision tree algorithm. Once completed, it waits for another mining request and finally leaves the dialogue.

4.7 EXPERIMENTATION AND EXPLORATION

This section presents the experimentation of the proposed agent model on anaemia dataset. Real-time dataset is employed in the work with major ranking attributes MCV, MCHC and RBC. The dataset is pre-processed, and tests were run using Weka classifiers employed using agent-based model supported by Weka 3.8.

The data is obtained from complete blood count (CBC) test results, which are conducted by collecting blood samples from 102 anaemia patients (102 instances) and constructing ANEMIA dataset. The dataset consists of six attributes and is defined in

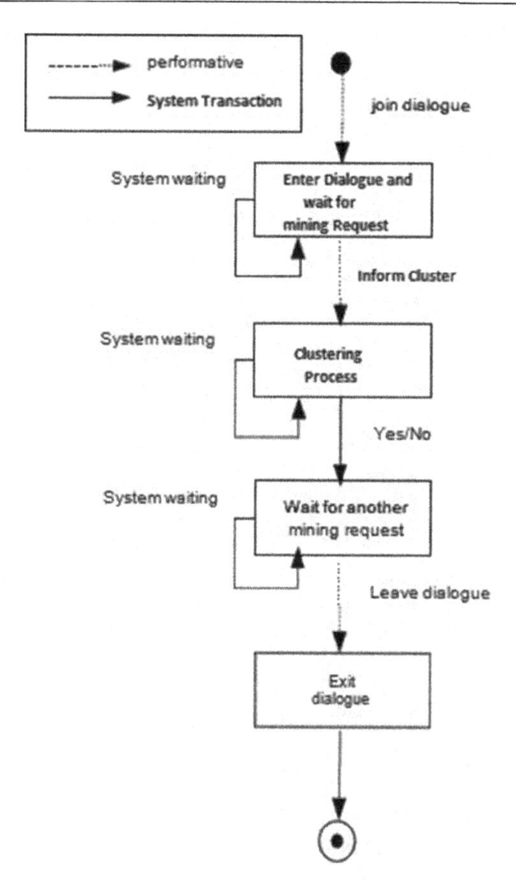

FIGURE 4.4 State transition diagram for classification agent.

Table 4.3 along with their values. The model was evaluated for performance measures such as accuracy, mean absolute error %, weighted average ROC, precision, recall and F-measure. The proposed model is compared for its performance with Naïve Bayes, multilayer perception and SMO. The experimentation results are given below.

Table 4.4 and Figure 4.1 show the accuracy comparison of the proposed agent-based model with existing Naïve Bayes, multilayer perception and SMO methods. It is noticed that the accuracy of the proposed agent-based model improves when compared to the existing models. The employment of agents has feasibly improved the accuracy and its consistent improvement is noticed for all distribution of training dataset.

Table 4.5 shows the mean absolute error % comparison of the proposed agent-based model with existing Naïve Bayes, multilayer perception and SMO methods. It is noticed that the error rate of the proposed agent-based model is decreased when compared to the existing models. The employment of agents has feasibly improved in its performance and its consistent improvement is noticed for all distribution of training dataset.

Table 4.6 shows the weighted average ROC comparison of the proposed agent-based model with existing Naïve Bayes, multilayer perception and SMO methods.

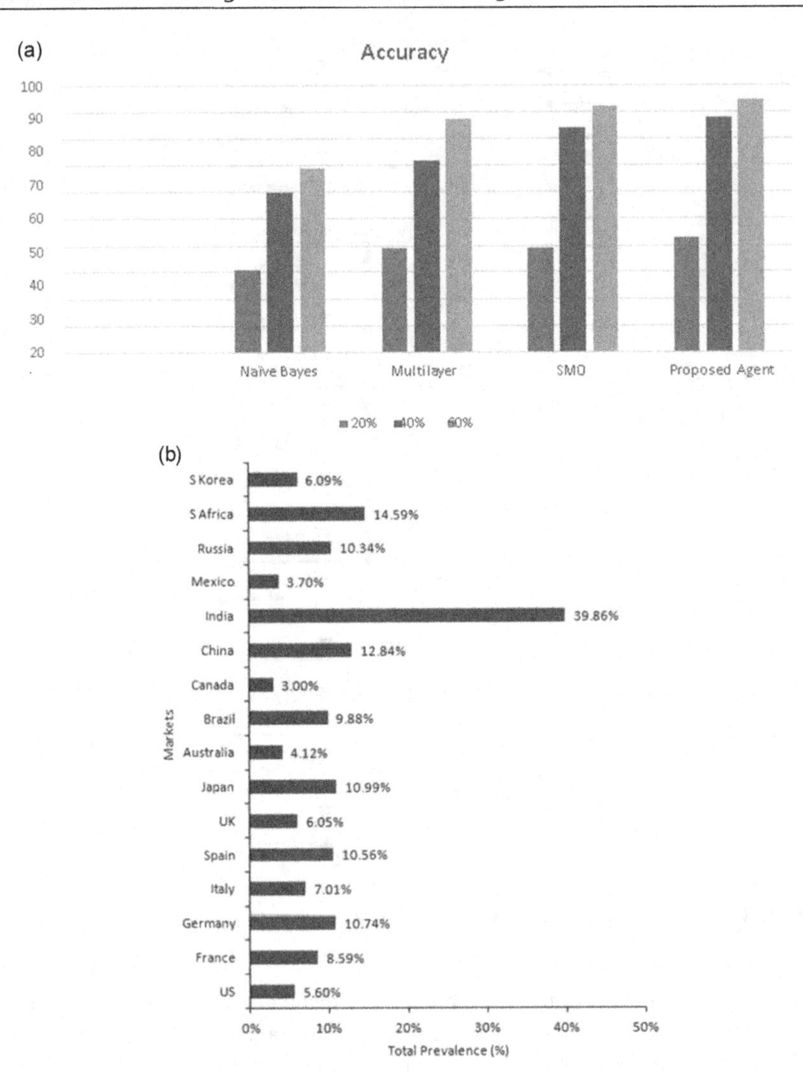

FIGURE 4.5 (a) Comparison of accuracy and (b) Total prevalence of anaemia in various countries.

Table 4.7 and Figure 4.2 show the weighted average ROC comparison of the proposed agent-based model with existing Naïve Bayes, multilayer perception and SMO methods. The results show that for 102 instances, the values are improved in the range of 0.01, and when the size of the dataset increases, definitely there would be an increase in the performance. The characteristics of agents play a vital role in the enhanced performance of the proposed model.

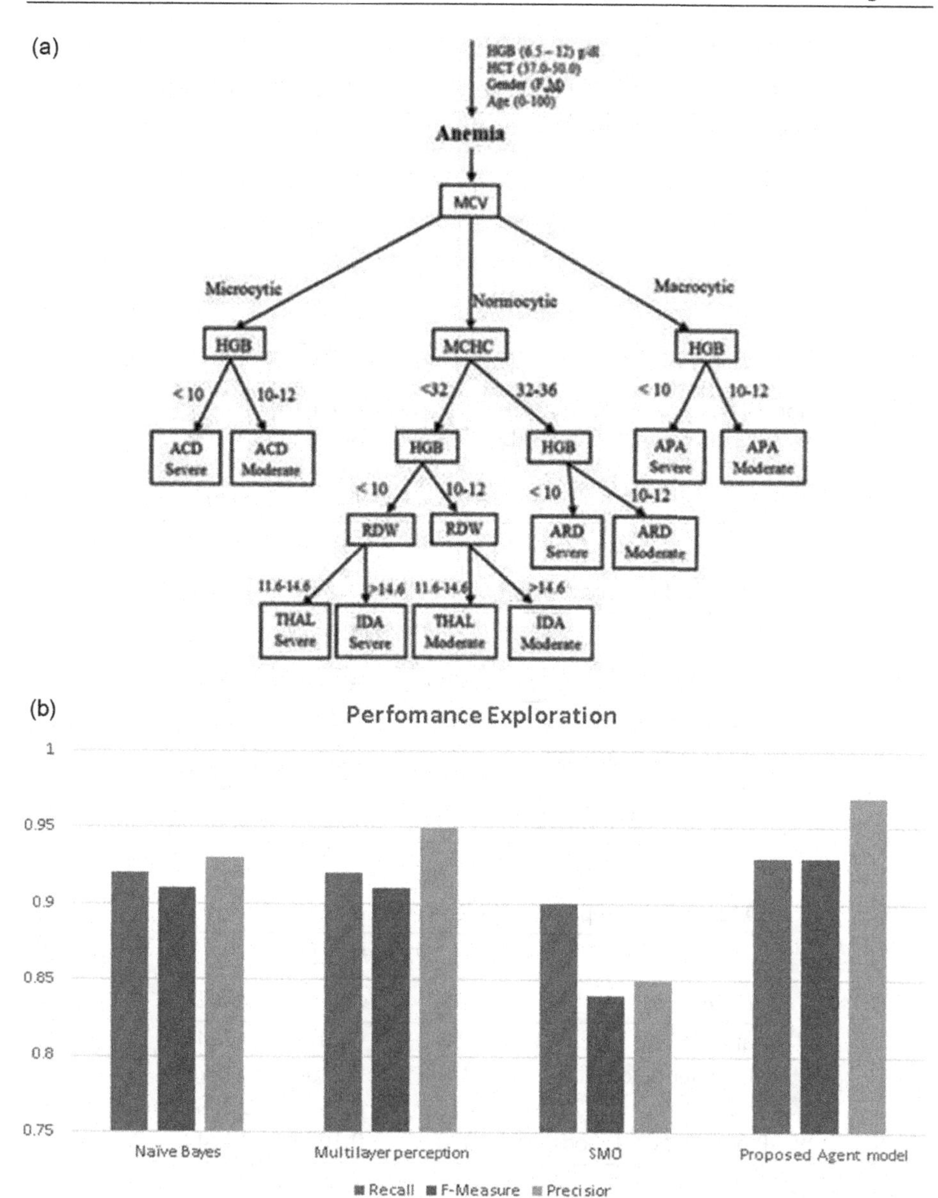

FIGURE 4.6 (a) Anaemia type classification and (b) Comparison of precision, recall and F-measure.

TABLE 4.3 Anaemia Dataset Description

ATTRIBUTE	*ATTRIBUTE VALUE*	*ATTRIBUTE CATEGORY*
Age	0–12	Child
	>12	Adult
MCV	<80	Microcytic
	80–100	Normocytic
	>100	Macrocytic
HCT	<37	Low
	37.0–50.0	Normal
MCHC	<32	Hypochromic
	32–36	Normochromic
Normal	<10	Severe
HGB	10–12	Moderate
RDW	>14.6	High
	11.6–14.6	Normal

TABLE 4.4 Comparison of Accuracy

	TRAINING SET		
ALGORITHM	*20%*	*40%*	*60%*
Naïve Bayes	31	60	68.75
Multilayer perception	39	72	87.5
SMO	39	84	92.47
Proposed agent model	43	88	94.78

TABLE 4.5 Comparison of Mean Absolute Error %

	TRAINING SET		
ALGORITHM	*20%*	*40%*	*60%*
Naïve Bayes	0.458	0.3372	0.2645
Multilayer perception	0.3744	0.2198	0.1372
SMO	0.4126	0.2547	0.2361
Proposed agent model	0.3207	0.1689	0.1743

TABLE 4.6 Comparison of Weighted Average ROC

	TRAINING SET		
ALGORITHM	*20%*	*40%*	*60%*
Naïve Bayes	0.50	0.70	0.825
Multilayer perception	0.77	0.852	0.921
SMO	0.855	0.868	0.95
Proposed agent model	0.677	0.902	0.96

TABLE 4.7 Comparison of Precision, Recall and F-Measure

ALGORITHM	*RECALL*	*F-MEASURE*	*PRECISION*
Naïve Bayes	0.92	0.91	0.93
Multilayer perception	0.92	0.91	0.95
SMO	0.90	0.84	0.85
Proposed agent model	0.93	0.93	0.97

4.8 CONCLUSION AND FUTURE WORK

In this work, AI technologies and agent framework has attempted to assist in the diagnosis of the disease in question. In this work, a novel approach is proposed to develop a CDSS for classifying the severity of anaemia disease diagnosis using data-mining and AI techniques. The major goal is to build an expert system for diagnosing the presence of anaemic disease with an integrated automated classifier using AI techniques. Real-time dataset is employed in the work with major ranking attributes MCV, MCHC and RBC.

The dataset is pre-processed and tests were run using Weka classifiers available in Weka 3.8. The proposed algorithm formalises the treatment of vagueness in decision support architecture which also evaluates the performance measures among them. Experimental results demonstrate the effectiveness of the proposed CDSS in anaemia diagnosis. The purpose of this work is to develop a cost-effective analysis using JADE and J48 Decision Tree Data-Mining technology for facilitating database of the decision support system. Experimental analysis is done to identify the effectiveness of the proposed methodology. The performance exploration is carried out in terms of accuracy, mean absolute error %, weighted average ROC, precision, recall and F-measure. The proposed model is compared for its performance with Naïve Bayes, multilayer perception and SMO. It is evident from the experimental analysis that the proposed agent model outperforms the existing models. Further enhancements to the proposed model can be incorporated using multi-agent systems (MAS) where the classification task could be evenly shared among agents for minimalising the time consumption of the classification process and hence improving the performance and helping the hospitals in identifying the disease in a concise period of time.

BIBLIOGRAPHY

Ahmad, A., Mustapha, A., Zahadi, E. D., Masah, N., & Yahaya, N. Y. 2011. Comparison between neural networks against decision tree in improving prediction accuracy for diabetes mellitus. In *International Conference on Digital Information Processing and Communications*, pp. 537–545. Springer, Berlin, Heidelberg.

Amin, M. N., & Habib, M. A. 2015. Comparison of different classification techniques using WEKA for hematological data. *American Journal of Engineering Research* 4(3): 55–61.

Bellifemine, F., Bergenti, F., Caire, G., & Poggi, A. 2005. JADE—A java agent development framework Multi-Agent Programming. Springer: 125–147.

Bellifemine, F., Caire, G., Poggi, A., & Rimassa, G. 2008. JADE: A software framework for developing multi-agent applications. Lessons learned. *Information and Software Technology* 50 (1): 10–21.

Bellifemine, F. L., Caire, G., & Greenwood, D. 2007. *Developing Multi-Agent Systems with JADE* (Vol. 7). John Wiley & Sons.

Dogan, S., & Turkoglu, I. 2008. Iron-deficiency anemia detection from hematology parameters by using decision trees. *International Journal of Science & Technology* 3(1): 85–92.

Elshami, E. H., & Alhalees, A. M. 2012. Automated diagnosis of thalassemia based on data mining classifiers. *Paper Presented at the International Conference on Informatics and Applications (ICIA2012).*

Garner, S. R. 1995. Weka: The Waikato environment for knowledge analysis. *Paper Presented at the New Zealand Computer Science Research Students Conference.*

Green, R. 2012. Anemias beyond B12 and iron deficiency: The buzz about other B's, elementary, and nonelementary problems. *ASH Education Program Book* 2012(1): 492–498.

Huang, F., Wang, S., & Chan, C.-C. 2012. Predicting disease by using data mining based on healthcare information system. *Paper Presented at the Granular Computing (GrC) IEEE.*

Kaur, P., Singh, M., & Josan, G. S. 2015. Classification and prediction based data mining algorithms to predict slow learners in education sector. *Procedia Computer Science* 57: 500–508.

Kesavaraj, G., & Sukumaran, S. 2013. A study on classification techniques in data mining. *Paper Presented at the Computing, Communications and Networking Technologies (ICCCNT).*

Kishore, C. R., Rao, K. P., & Murthy, G. 2013. Performance evaluation of entropy and Gini using threaded and non threaded ID3 on anaemia dataset. *Life* 6(10): 10–12.

Prakash, V. A., Ashoka, D., & Aradya, V. M. 2015. Application of data mining techniques for defect detection and classification. *Paper Presented at the Proceedings of the 3rd International Conference on Frontiers of Intelligent Computing: Theory and Applications (FICTA)* 2014.

Sabitha, R. & Karthik, S. 2013, Performance assessment of feature selection methods using K-means on adult dataset. *Research Journal of Computer Systems Engineering* 04: 606–612.

Sanap, S. A., Nagori, M., & Kshirsagar, V. 2011. Classification of anemia using data mining techniques Swarm, evolutionary, and memetic computing. In *International Conference on Swarm, Evolutionary, and Memetic Computing*, pp. 113–121, Springer, Berlin, Heidelberg.

Shashidhara, M. *Classification of Women Health Disease (Fibroid) Using Decision Tree Algorithm.*

Shouval, R., Bondi, O., Mishan, H., Shimoni, A., Unger, R., & Nagler, A. 2014. Application of machine learning algorithms for clinical predictive modeling: A data-mining approach in SCT. *Bone Marrow Transplantation* 49(3): 332–337.

Siadaty, M. S., & Knaus, W. A. 2006. Locating previously unknown patterns in data-mining results: A dual data-and knowledge-mining method. *BMC Medical Informatics and Decision Making* 6(1): 13.

Vijayarani, S., & Muthulakshmi, M. 2013. Comparative analysis of Bayes and lazy classification algorithms. *International Journal of Advanced Research in Computer and Communication Engineering* 2(8): 3118–3124.

Yilmaz, A., Dagli, M., & Allahverdi, N. 2013. A fuzzy expert system design for iron deficiency anemia. *Paper Presented at the Application of Information and Communication Technologies (AICT).*

Self-Diagnosis in Healthcare Systems Using AI Chatbots

Bhawna Nigam and Naman Mehra
Devi Ahilya Vishwavidyalaya

M. Niranjanamurthy
M.S. Ramaiah Institute of Technology

Contents

5.1 Introduction 80
5.2 Healthcare Chatbots 81
5.3 Healthcare Chatbots in Use 83
5.4 Developing Healthcare Chatbots 84
5.4.1 Data Pre-Processing 85
5.4.2 Model: Training 85
5.4.2.1 Custom Models 87
5.4.2.2 Deep Learning 87
5.4.2.3 NLP 87
5.5 Need for Chatbots 88
5.6 Research Works 89
5.7 Limitations 89
5.8 Conclusions 90
Bibliography 90

DOI: 10.1201/9781003051022-5

ABSTRACT

Self-diagnosis is the process of diagnosing, or identifying, medical conditions in oneself. It may be assisted by medical dictionaries, books, resources on the Internet, past personal experiences, or recognising symptoms or medical signs of a condition that a family member previously had. A chatbot is a computer program that simulates human conversation through voice commands or text chats or both. Chatbot, short for chatterbot, is an artificial intelligence (AI) feature that can be embedded and used through any major messaging application. Identify the right opportunity to develop an AI-driven chatbot.

You can build a chatbot using established frameworks or development (non-coding) platforms depending on the features your start-up requires. Understand the goals of your business and your customer early on in the process. An AI chatbot makes a decision by leveraging pre-existing knowledge and one that it acquires continuously. To maximise the ability of AI chatbots to improve service, save money, and increase engagement, businesses and organisations need to understand how these programs work and what they can do. In this guide, you'll get a crash course on talking bots, including the technology behind them, how they have transformed marketing and customer service, and how you can start putting them to work. Users can assuredly benefit from chatbots if they can diagnose several kinds of illness and render the required data. Text diagnosis bot enables sufferers to join in analyses of their medicinal matters and present a personalised analysis report with reference to the symptoms. This chapter represents the self-diagnosis in healthcare system using AI chatbots.

5.1 INTRODUCTION

The healthcare industry is one of the world's fastest-growing sectors. The advancement in deep learning has brought various changes, especially in the field of healthcare. Artificial intelligence (AI)-based systems are used for the diagnosis of various lung diseases as well as for identifying tumours in the brain and prognosis for identifying any chances of risk related to a patient's health in future events using predictive modelling. Using the lab result, we can estimate the risk of an event such as a heart attack or some life-threatening disease. The field of natural language processing (NLP) has advanced tremendously in recent years. Personal assists are used in every industry – for example, manufacturing, educational institutes, IT and healthcare. The use of chatbots or conversational bots in the field of healthcare is advancing. Chatbots are smart computer programs that conduct a conversation via audio or text methods. These chatbots are becoming more advanced and are used widely in various industries. Chatbots can be integrated with any mobile application and website. AI is giving personalised medical treatment to patients. Using AI to estimate the effect of treatment and conversational bots to help assist patients in answering their questions. The use of AI, along with IoT, can reduce the dependency on doctors at the time of basic needs, such as giving medicine or monitoring patients continuously for better treatment. These bots can extract information from the patient's reports and can effectively assist doctors and patients in treatment as well as answering questions. Healthcare chatbots provide information to patients related to their disease and symptoms and also offer solutions for the diagnosis. Widely, healthcare chatbots are used

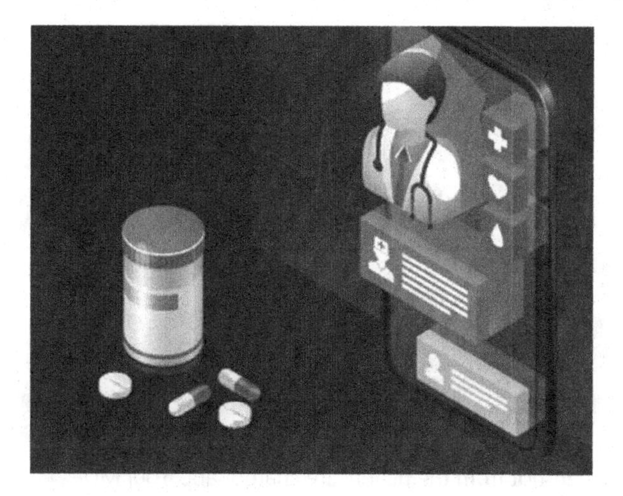

FIGURE 5.1 Healthcare Chatbots.

for symptom checking. They are using the patient's or user's information to guide them. These chatbots also provide useful health tips (Demirkan, 2013; Wickramasinghe et al., 2007; Patel et al., 2019) (Figure 5.1).

Various NLP models are used for developing chatbots for medical question answering systems. Using these chatbots, a doctor can stay in touch with patients even at home by using their smartphones. These chatbots will monitor the patient's health and analyse the data collected using the queries given by patients to send reports to the doctor or assist the patient. These chatbots can be used in hospitals too, which can be embedded in IoT-based applications such as robots. These chatbots can often be used with different devices.

From the healthcare industry perspective, chatbots can be of three types, namely

- Assistants – These chatbots provide predefined information to the users.
- Concierges – These chatbots provide contextualised information and can facilitate a service such as a text doctor where you can schedule appointments.
- Advisors – These chatbots can handle complex requests to perform actions such as scheduling appointments or texting a doctor.

5.2 HEALTHCARE CHATBOTS

The chatbots are used in the medical industry for a variety of purposes such as personalised medical follow-up, advice to patients, preliminary diagnosis, assistance during medication, monitoring, communication, and delivering test results. The appropriate rules and regulations should be applied depending on the level of risk associated while developing healthcare chatbots. Let's see how chatbots are used in the healthcare industry (Prasad & Ranjith 2020; Kao et al., 2019; Albayrak et al., 2018; Srivastava & Singh 2020) (Figure 5.2).

FIGURE 5.2 Screenshot from the healthcare chatbot app Your.MD.

- Chatbots in consultants – Healthcare chatbots are available as apps that can answer various health-related queries of their users and can provide essential information in general as well as based on symptoms as given by users. They issue reminders, schedule appointments, provide dietary information, and can initiate medical refill.
- Chatbots as hospital administration – It can take a patient's medical history, schedule appointments and doctor visits, and provide information to patients according to their queries and also helps doctors in assessing the patients. It can also monitor the patients in the hospital to help doctors so that they can help other patients as well. It helps healthcare personnel to spend more time effectively with patients and improve their routines. They also track the patient's health and take feedback from patients to help hospitals.
- Chatbots as self-care coach – Chatbots can track users routinely and keep their exercise records. They also count calorie information and monitor the amount of water and food the users consume. They can also assist them in yoga and meditation. They help to overcome stress and depression and provides emotional support. They can also boost the user's moods by providing them with motivational quotes and can also suggest a healthy diet based on the user's dietary information. These are often used for improving lifestyle and giving nutritional guidance to the users (Ravi, 2018; Zhang et al., 2016; Fiddin et al., 2020; Belfin et al., 2019).
- Chatbots as an elderly care provider – Voice-based personal assistants can look after the needs of older people in hospitals as well as in nursing homes to assist them with their health conditions. These chatbots can also help patients with visual and mobility disabilities.
- Chatbots as dispensers of drug information – Some chatbots can provide drug-related information to the patients and can also suggest alternate medication just in case. They can help chemists and can also provide more relevant information to doctors regarding the drug they prescribe to the patients (Nagarhalli et al., 2020).
- Chatbots in research and treatment – With the vast amount of data available, machines can perform better than humans in predicting disease. Chatbots

can help researchers in clinics in analysing the amount of enormous data and assist researchers in using powerful voice and speech recognition algorithms to communicate.

5.3 HEALTHCARE CHATBOTS IN USE

We will see some of the best-in-class healthcare chatbots being used in the industry today. At the heart of these chatbots is AI. These chatbots use advanced NLP algorithms.

- Youper – It uses personalised conversation to engage with users using psychological techniques. The central concept of Youper is to understand human emotion and help people to take care of their emotional health. Generally, it works as your psychological assistant. It is made by a professional team of doctors and engineers to help people cope with their stress, anxiety, and depression. As users engage more with the bot, it will learn more about them and adjust itself so that it can be more like human. The app also features personalised meditation and also tracks the mood of the user and monitors their emotional health.
- Mediktar – It is an AI-based medical assistant for pre-diagnosis and decision-making support. The user states their symptoms, and its AI system asks them questions, and according to those, it suggests the users about possible conditions and recommends on what to do next. It has a great NLP capability.
- Babylon health – It is the UK-based online health service and medical consultation provider. The healthcare service provider now values about $2 billion. The company offers live video consultations with the doctor as well as provides consultations based on recent medical history. It uses advanced speech recognition in order to assist the user. It is used by the National Health Service (NHS) of the UK. It also offers personalised healthcare assessments and treatment advice and face-to-face appointments with doctors 24×7.
- Your.Md – It offers accurate sources based on actionable health information and allows the user to make the best choices for their health. It checks symptoms and AI powers it. It is available in mobile apps, Facebook messenger, and browsers. It also helps to find nearby pharmacies, clinics, and doctors.
- Sensely – Sensely is a character-based virtual assistant, which can help the patients to assess symptoms using speech, images, videos, and text. Users can communicate using address and text. It recommends diagnosis based on user data and information fed to its algorithm. It also provides health assessments using which it provides information to the user to visit a doctor or to consider self-care. The character-based assistant is more engaging and gives a lifelike experience to the users. It uses an advanced text-to-speech and speech recognition system underneath to provide its service (Figure 5.3).

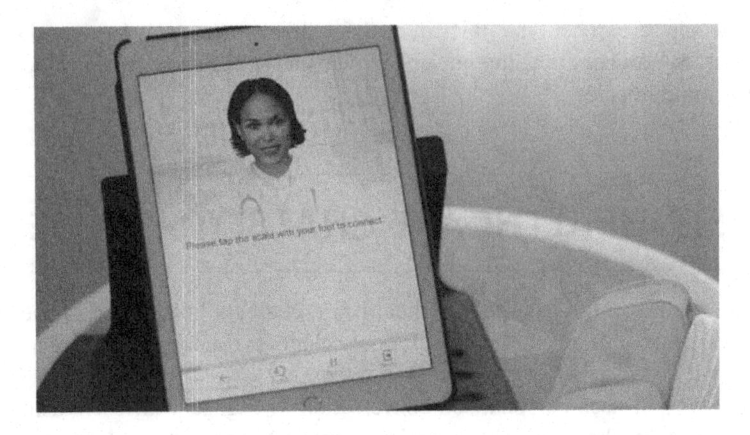

FIGURE 5.3 Screenshot from the app Sensely. Sensely is a character-based virtual assistant.

- Safedrugbot – This is a chat messaging service that helps the doctor as an assistant, which provides information about breastfeeding. Also, it provides related to the medication and their alternatives.
- One remission – The application empowers them by providing a list of diet exercises and post-cancer practices, curated by anti-medicine experts. So, they don't need to constantly rely on the doctor.
- Infermedica – It assesses the user's health status, and based on the symptoms, it assesses the user's health status, and based on the status, it sets up the possible diagnosis and gives an actionable recommendation.
- Gyant – It takes the information from the patients and sends it to the doctor who provides diagnosis and prefers medicines in real time.

And many more.

5.4 DEVELOPING HEALTHCARE CHATBOTS

Building healthcare chatbots requires medical data and expertise in NLP. They are depending upon the requirements, such as delivering personalised information to patients for identifying the disease for symptoms, scheduling appointments, monitoring health status, and much more. Most traditional chatbots are rule-based, or we can say, script-based. They cannot easily define symptoms, semantics, or context.

While conversational chatbots are driven by deep learning more specifically by natural language understanding (NLU), these bots use speech recognition, text-to-speech, and many other language technologies to contextualise written and spoken words. The job of the developer is to teach chatbots to extract information from speech or typed text and transform that information into structured data.

The healthcare chatbot's NLP engine has the following elements to process the queries of patients or clients:

- Intent – The user's intention is called Intent. It is the task a user wants to achieve.
- Utterance – Anything the user or patient says or gives as input.
- Context – This involves breaking down sentences into n-grams to understand the meaning of a given query.
- Entity – The details of the user's intent are called entities. An entity is also called slots.
- Session – A starting and end of the user's conversation are called session.

5.4.1 Data Pre-Processing

NLU requires a high-level data preparation. The chatbots learn from user interactions. The chatbot's performance is directly affected by the quality of data given by the user. The data that we use to train our model should be taken from valid medical resources since building a healthcare chatbot involves risk when we want to deploy it in production. Also, the chatbots often learn from the patient's historical data depending on the functionality and particular use case. The data which we use to feed to the model should be in JSON format. This JSON file is a python dictionary that consists of patterns (input) and responses (output). We define patterns and reactions in this JSON file along with context.

The first step is to stem the words. Stemming means to bring the terms to the root word. For example, "What's up?" After arising we will get the root word which is "What". The reason for arising is to eliminate extra words like **stopwords** and only include a single word for training our model. This will make our model more accurate. In NLP terms we call the process of stemming **Tokenization**.

For this purpose of data pre-processing, we can use various python libraries such as NLTK, spaCy, and Gensim.

The next step is to create a bag of words. This bag of words will represent each word as one-hot. This one-hot vector for one particular word will be equal to the total words in our bag. For example, if we will train our healthcare chatbots using 10k words, we will have to represent one word using a vector that will have 10k elements, where the place where the word exists will have one, while other values in the vector will have zero.

5.4.2 Model: Training

For training our model for healthcare chatbot, we can use Tensorflow, Keras, Pytorch, or Fastai. These are the popular open-source deep learning libraries used in industries by developers as well as by researchers.

There are several ways in which we can develop user-friendly chatbots, which can understand a person's problem and can give a solution.

There are two methods through which one can implement chatbots:

- Pre-trained
- Custom trained

Figure 5.4 gives a better understanding of the types of chatbot.

Pre-trained – Several companies provide you with the option to use their chatbot services/APIs.

Chatbot APIs help combine NLP to build an efficient chatbot—they take in the request, extract the intent or meaning of the message, and deliver the response.

Let's see how it works.

Chatbot APIs use user-defined data to train themselves.

For example, you are a renowned medical shop owner, and you need to start online delivery services. You don't have much knowledge of machine learning or NLP, so you decided to use chatbot API/services. You will just need to add your data which includes

- Greetings – Hello, Hi, how can I help?
- Intents – Name of medicines you want and enter the delivery address
- Entities – Weekdays, names, etc.

You will get to know more about these terms in further readings.

These are some examples of data you can feed to your chatbot.

FIGURE 5.4 Screenshot from the app Gyant. It has empathetic AI technology that provides a better experience to users.

5.4.2.1 Custom Models

The two main approaches for building your custom chatbot are

- NLP
- Deep learning

5.4.2.2 Deep Learning

We have a whole bunch of libraries like NLTK and spaCy in python from which we can train our deep learning algorithm.

Let's have a deep dive into the steps used for training a deep learning model.

- Cleaning of text for training the data with the help of libraries like NLTK and spaCy.
- Preparing data into JSON format.
- Now, we will take the word list, lemmatise (the process of grouping together the different inflected forms of a word so they can be analysed as a single item. For example, walking or walked all they are converted into one word, i.e. walk) and lowercase them. This will reduce our time and avoid unnecessary errors.
- Building a training model, we can use the sequential model in Keras, and they save the model in.h5 format.

5.4.2.3 NLP

First, NLP is a tool for computers to analyse, comprehend, and derive meaning from natural language in an intelligent and useful way. In order to understand the human language from unstructured data to structured data that computers can interpret, different techniques are required. When a user sends a message, it needs to use the algorithm to get meaning and context from every sentence to collect data from them. This process is called natural language understanding (NLU), and it's a subset of NLP. It consists of interpreting the user's message by extracting the user's details from it.

A way to extract the essential parts of a sentence is to differentiate between the entities and the intent. An intent of a sentence is the goal of the statement. What does the user actually want to achieve? For instance, if the message was, "When the Quality drug store closes" the intent of the message is to know when the medical store closes. An entity of a sentence is something that modifies or supports the intent. For instance, the entities of the question, "What are your closing hours on Sunday?" are Sunday and closing hours. An entity is basically anything that can be named (like the place, person, name, or object).

The chatbot basically needs to recognise the entities and intents of the user's messages. In order to do that, we need to build an NLP model for every entity for an intent. For example, we can build an NLP intent model for the chatbot to recognise when a user wants to know the opening hours of a place. We can build an NLP entity model for the chatbot to recognise locations and directions. We can then use these NLP models for the chatbot to offer the opening hours of any place, based on the user's location.

The NLP process is a core part of the chatbot architecture and process, since it is the foundation for translating the natural human language to structured data.

Let's see the step-wise process of how a chatbot works:

- Let's say, you want to purchase medicine and you decided to order from a chatbot. You put in a request.
- When you send a message to the chatbot, the chatbot sends the plain message to the NLP engine.
- The NLP engine, which uses NLP and NLU, converts the text message into structured data for itself. This is where the different NLP models come into play for extracting the intents and entities of the message.
- The chatbot moves the data that was collected (the intents and entities) to the decision-making engine.
- The decision-making model derives a solid decision based on previous actions and results are taken.
- This is where the chatbot converts the decision data to text. Natural language generation (NLG) consists of converting data into plain text. Using NLG, the message generator outputs the message. This message is presented to the user in the form of a text message or voice.

5.5 NEED FOR CHATBOTS

Let's take a real-life example – you visited the nearby medical store and bought some medicine. But when you returned home, you realised that you forgot to buy one more medicine. In that case, you need to go back again and buy it.

Also, if we take another scenario where you wanted to buy medicine, but it is not available in your area.

So, here such Chabot serves as a link for you to deliver your medicine request to nearby medical shops.

- Many a time, patients want more information. Medical chatbots have emerged as useful tools to provide additional information to patients.
- Healthcare industry lacks medical professionals; therefore, it takes time for the patients to receive one-to-one consultations. Medical chatbots can provide one-to-one assistance to patients.
- Chatbots can gather feedback from patients about the organisation's website, and this helps the organisations to provide their websites.
- HealthBots can remind about their medicines; moreover, these chatbots can monitor their health parameter.

- Medical chatbots can fix appointments with, doctors, and also they can find the nearest hospitals, pharmacies, and clinics.

5.6 RESEARCH WORKS

There are many research papers available on healthcare chatbots that can enhance your understanding of chatbots. Table 5.1 gives the research paper and its details.

5.7 LIMITATIONS

User privacy is a critical issue in the use of AI chatbots. Although chatbot builders do their best to implement better safety measures, somewhere it lags in security. Also,

TABLE 5.1 Research Papers and Their Contribution in Healthcare Chatbots

PAPERS	*DESCRIPTION*
Chatbot for healthcare system using artificial intelligence	In this paper keyword, ranking, and sentence similarity calculation is done using n-gram, TF-IDF, and cosine similarity. From the given input sentence, the score will be obtained for each sentence, and more similar sentences are obtained for the given query
Text-based healthcare chatbots for patients and health professional teams	A text-based healthcare chatbot (THCB) system that was designed to support patients and health professionals likewise
Pharmabot	A paediatric generic medicine consultant chatbot
Intelligent HealthBot for transforming healthcare	A support from healthcare industry. The goal is to introduce HealthBot, a system designed to improve the eHealth paradigm by using a chatbot to simulate human interaction in medical contexts. Based on machine learning
Healthcare assisting chatbot	Chatbot for symptoms and dosage prediction using IoT
Emergency patient care system using a chatbot	A chatbot can provide healthcare at low costs and improved treatment if the doctors and the patient keep in touch after their consultation. To answer the questions of the user, a chatbot is used. There is less number of chatbots in the medical field

sometimes chatbot is not accurate, for example, if a chatbot prescribed medicine to a patient and that medicine is not suitable for diabetic patients and the patient is suffering from diabetes, then in this situation chatbot cannot make its own decision for prescribing medicines.

5.8 CONCLUSIONS

Through chatbots, one can communicate with text or voice interface and get a reply through AI. Typically, a chatbot will communicate with a real person. Chatbots are programs built to automatically engage with received messages. Chatbots can be programmed to respond the same way each time, to respond differently to messages containing certain keywords, and even to use machine learning to adapt their responses to fit the situation. A developing number of hospitals, nursing homes, and even private centres presently utilise online chatbots for human services on their sites. These bots connect with potential patients visiting the site, helping them discover specialists, booking their appointments, and getting them access to the correct treatment. In any case, the utilisation of AI in an industry, where individuals' lives could be in question, still starts misgivings in individuals. It brings up issues about whether the task mentioned above ought to be assigned to human staff. This healthcare chatbot system will help hospitals to provide healthcare support online 24×7 – it answers deep as well as general questions.

BIBLIOGRAPHY

N. Albayrak, A. Özdemir and E. Zeydan, "An overview of artificial intelligence based chatbots and an example chatbot application," *2018 26th Signal Processing and Communications Applications Conference (SIU)*, Izmir, 2018, pp. 1–4, Doi: 10.1109/SIU.2018.8404430.

R. V. Belfin, A. J. Shobana, M. Manilal, A. A. Mathew and B. Babu, "A graph based chatbot for cancer patients," *2019 5th International Conference on Advanced Computing & Communication Systems (ICACCS)*, Coimbatore, 2019, pp. 717–721, Doi: 10.1109/ICACCS.2019.8728499.

H. Demirkan, "A smart healthcare systems framework," *IT Professional*, vol. 15, no. 5, pp. 38–45, Sept.–Oct. 2013, Doi: 10.1109/MITP.2013.35.

M. T. Fiddin Al Islami, A. Ridho Barakbah and T. Harsono, "Interactive applied graph chatbot with semantic recognition," *2020 International Electronics Symposium (IES)*, Surabaya, 2020, pp. 557–564, Doi: 10.1109/IES50839.2020.9231678.

C. Kao, C. Chen and Y. Tsai, "Model of multi-turn dialogue in emotional chatbot," *2019 International Conference on Technologies and Applications of Artificial Intelligence (TAAI)*, Kaohsiung, 2019, pp. 1–5, Doi: 10.1109/TAAI48200.2019.8959855.

T. P. Nagarhalli, V. Vaze and N. K. Rana, "A review of current trends in the development of chatbot systems," *2020 6th International Conference on Advanced Computing and Communication Systems (ICACCS)*, Coimbatore, 2020, pp. 706–710, Doi: 10.1109/ICACCS48705.2020.9074420.

F. Patel, R. Thakore, I. Nandwani and S. K. Bharti, "Combating depression in students using an intelligent ChatBot: A cognitive behavioral therapy," *2019 IEEE 16th India Council International Conference (INDICON)*, Rajkot, 2019, pp. 1–4, Doi: 10.1109/INDICON47234.2019.9030346.

V. A. Prasad and R. Ranjith, "Intelligent chatbot for lab security and automation," *2020 11th International Conference on Computing, Communication and Networking Technologies (ICCCNT)*, Kharagpur, 2020, pp. 1–4, Doi: 10.1109/ICCCNT49239.2020.9225641.

R. Ravi, "Intelligent chatbot for easy web-analytics insights," *2018 International Conference on Advances in Computing, Communications and Informatics (ICACCI)*, Bangalore, 2018, pp. 2193–2195, Doi: 10.1109/ICACCI.2018.8554577.

P. Srivastava and N. Singh, "Automatized Medical Chatbot (Medibot)," *2020 International Conference on Power Electronics & IoT Applications in Renewable Energy and its Control (PARC)*, Mathura, 2020, pp. 351–354, Doi: 10.1109/PARC49193.2020.236624.

N. Wickramasinghe, S. Chalasani, R. V. Boppana and A. M. Madni, "Healthcare system of systems," *2007 IEEE International Conference on System of Systems Engineering*, San Antonio, TX, 2007, pp. 1–6, Doi: 10.1109/SYSOSE.2007.4304283.

W. Zhang, H. Wang, K. Ren and J. Song, "Chinese sentence based lexical similarity measure for artificial intelligence chatbot," *2016 8th International Conference on Electronics, Computers and Artificial Intelligence (ECAI)*, Ploiesti, 2016, pp. 1–4, Doi: 10.1109/ECAI.2016.7861160.

Digital Water

New Approach to Build Efficient Water Management Systems

6

Priyanka Verma and Anubha Jain
IIS (deemed to be University)

Contents

6.1 Introduction 94
6.2 Artificial Intelligence 95
6.3 Applications of AI 95
6.3.1 Categories of AI 96
6.4 Considerations While Using AI 97
6.5 Water Resource Management 97
6.6 Digital Water 98
6.7 What AI Requires 100
6.8 Technologies Used by AI for Effective Water Management 100
6.9 Benefits of Working with AI 103
6.10 Conclusion 104
Bibliography 105

Abstract

Water is the most essential resource for life on this planet. In recent years, however, we have taken this resource for granted. The sudden rise in the occurrences of droughts and water scarcity situations in different parts of the world are making scientists look for better, more robust ways to implement water resource management, they have

DOI: 10.1201/9781003051022-6

realized that they need to opt for a different approach for maintaining the existing water management systems. One such approach is artificial intelligence (AI), by applying AI in water resource management, scientists are now building sustainable water supplies for people around the world. AI models enable water management companies and cities to build efficient water re-use systems. It has the potential to bring significant improvement in terms of cutting wastage of water, increasing the efficiency of treatment plants, and keeping overall water infrastructure healthy. AI is expected to bring in the trend of "digital water." AI can continuously work with large amounts of data sets, tirelessly. The technology also learns on the go, improving itself with each set of data that it processes. This adaptability makes AI the best option to deal with an everyday altering entity, like water. Digital water, in simple terms, is the water managed by data analytics, regression models, and probabilistic algorithms. It will enable water managers and government bodies to build efficient water systems that will help to solve water shortage problems.

6.1 INTRODUCTION

Water whose chemical denotation is **H_2O is a vital substance** and is signified by a chemical bond that is formed between two hydrogen elements and one oxygen element. It exists in three forms in nature: solid, liquid, and gas. It is also said to be among the most crucial resources for all known life forms. Water covers 70% of our planet, and it is easy for us to think that it will always be plentiful, but the reality is that approximately 1.1 billion people worldwide lack access to water, and 2.7 billion experience water scarcity for at least a period of one month of the year.

Clean water is one of the cornerstones of our living, and therefore we need to always thrive for the best and highest quality water. However, freshwater – the stuff we drink, bathe in and irrigate our farm fields with—is incredibly rare. According to facts given by WWF (World Wide Fund for Nature), which is an international non-governmental organization, only 3% of the world's water is freshwater. Moreover, two-thirds of it is laid away in frozen glaciers or otherwise unavailable for our use. As a result, water is indispensable for life. For thousands of years, human settlement and advancement have been dictated by a reliable supply of clean, safe water.

The need to effectively and efficiently manage water has not changed over time, though the management tools available in the market most certainly have. Water utilities are now equipped with data-driven technologies that allow them to extract previously unattainable information about water supply and demand, empowering water managers to build a dependable clean water future. Data-driven water management is now employing artificial intelligence (AI) techniques to fundamentally change the way of ensuring clean water for all.

AI refers to the simulation of human intelligence in machines that are programmed to think like humans and mimic their actions. The term may also be applied to any machine that exhibits traits associated with a human mind such as learning and problem-solving. Algorithms form the basis of AI. AI has been extensively used in many areas such as computer science, robotics, engineering, medicine, translation, economics, business, and psychology and is continuously evolving to benefit different industries.

The topmost priority is water controlling, tracing, and conservation to reiterate the importance of water and ensure its sustainability. It would further lead to a world having clean water for future generations to come. Thus, each country needs to exercise some regulations on water resources and to protect its own environment. In this chapter, we will discuss how AI and its techniques can help in managing water effectively to save it for future generations.

6.2 ARTIFICIAL INTELLIGENCE

AI, sometimes called machine intelligence, involves the development of computer systems that are able to perform tasks that generally require human intelligence, such as visual perception, speech recognition, decision-making, and translation between languages. It is based on the principle that human intelligence can be defined in a way such that a machine can easily reproduce it and execute tasks, from the most simple to those that are even more complex. The goals of AI include learning, reasoning, and perception. The ideal characteristic of AI is its ability to rationalize and take actions that have the best chance of achieving a specific goal.

6.3 APPLICATIONS OF AI

The applications for AI are endless. The technology can be applied to many different sectors and industries. The following are a few prevalent applications:

- Healthcare industry for dosing drugs and different treatments in patients, and for surgical procedures in the operating room.
- Gaming industry for designing the game, developing the characters, and also framing the story to a certain extent. It aims at creating AI-powered programs that can compete and win against human players.
- Personalized online shopping.
- E-marketing.
- Agriculture – AI embedded in smartphones helps farmers find more efficient ways to protect their crops from various elements such as weather, weeds, market consumption rates, and much more.

- Chatbots that are able to extract information from the site and present it to us on request.
- Social media for face verification and to detect facial features.
- Travel Industry uses AI tools for determining prices for various locations.
- Self-driving cars, a real-world application of AI wherein the computer system must account for all external data and compute it to act in a way that prevents a collision.
- Finance industry, where it is used to detect and flag activity in banking and finance such as unusual debit card usage and large account deposits to identify fraud.
- Autopilot in commercial flights.
- Voice-to-text conversion.
- Trading industry by making supply, demand, and pricing of securities easier to estimate.

These are fields where AI has established its prominence. In this chapter, we propose the use of AI for effective water management

6.3.1 Categories of AI

AI can be divided into two different categories: *weak and strong.*

Weak AI, or Narrow AI, embodies a system designed to carry out one particular job. They have specific intelligence. Weak AI systems include **video games** such as chess and **personal assistants** such as **Amazon's Alexa and Apple's Siri**. You ask a question to the assistant, it answers for you (Figure 6.1).

Strong AI systems, also known as artificial general intelligence, are systems that carry on the tasks considered to be human like. These are with generalized human cognitive abilities. These tend to be more complex and complicated systems. They are programmed to handle situations in which they may be required to solve a problem without

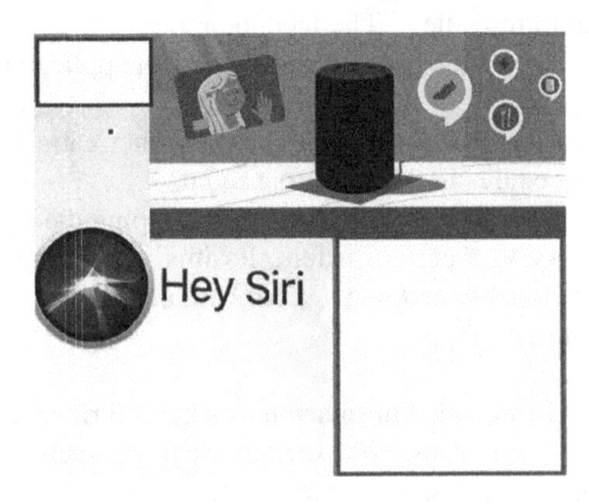

FIGURE 6.1 Weak Artificial Intelligence systems.

FIGURE 6.2 Strong Artificial Intelligence systems.

having a person to intervene. These kinds of systems can be found in applications such as **self-driving cars or in hospital operating rooms** (Figure 6.2).

6.4 CONSIDERATIONS WHILE USING AI

There are a few considerations to be kept in mind while using AI for any of the applications stated above.

- Machines can become highly developed; they may take off on their own, redesigning themselves at an exponential rate that humans will not be able to meet up.
- Machines can hack into people's privacy.
- Machines cannot have the same rights as humans.
- Machines can impact human employment. With many industries looking to automate certain jobs through the use of intelligent machinery, there is a concern that people would be pushed out of the workforce. For example, self-driving cars may remove the need for taxis and car-share programs, while manufacturers may easily replace human labor with machines, making people's skills more obsolete.

6.5 WATER RESOURCE MANAGEMENT

Water is essential for life on this planet. In recent years, however, we have taken the resource for granted. Concerned by the sudden peak in the occurrences of droughts, drop in groundwater levels and other water scarcity situations in different parts of the world, scientists are looking for better, more robust ways to implement water resource management.

Though the need to effectively manage water and its resources hasn't changed over the years, scientists have realized that they need to opt for a different approach for designing and maintaining the existing water management systems.

Water managing bodies have two main tasks before them. The first task is to seek new sources of water due to the volatility of the resource, and the second is to manage the existing water reserves and systems sustainably.

Researchers are trying algorithms and programs to build water plants that can not only give updated statistics about the current status of a resource but also help in building models for the future. **By applying technologies like AI in water resource management, scientists are now building sustainable water supplies for people around the world.**

6.6 DIGITAL WATER

Digital water, in simple terms, is the water managed by data analytics, regression models, and probabilistic algorithms which enables water managers and government bodies to build efficient water systems.

AI is making its mark on the water industry. It is powering intelligent operations using machine learning (ML) to optimize resource usage and operational budgets for organizations, as well as delivering truly intelligent built water systems. AI programming replicates the way humans learn in many ways. During the "learning phase" of AI programming, input data is correlated to known outputs to allow the algorithms to learn over time. Then, in the "operational phase," the program begins to make sense of patterns as new data is introduced. Because of AI's ability to constantly adapt and process large amounts of data in real time, **it is an ideal tool for managing water resources in an ever-changing environment, and the business of water, allowing water utility managers to maximize current revenue and effectively plan for the years ahead.**

The following are the ways in which AI can help in water conservation:

- Creation of software-powered programs with sensors and neural networks to strategize water operations dynamically and intelligently, which includes prediction models for future use of water, robotic sensors in water disposal plants, and blockchains for operating the financial transactions can all be made possible with the appropriate application of AI techniques. AI uses pattern recognition applications that employ a set of inputs, weighting factors, summation, and transfer functions that, by definition, are autonomously and dynamically updated as new information is presented; so as and when new data is provided, output also changes.
- AI can also be used to automate the inspection of sewer systems or the accuracy of meters through the use of hydraulic models used for several purposes:

(i) explanation and prediction tools, for what happened in one location at one time in the water distribution network without any instrumental and sensor data; (ii) forecasting tools, for the "what if" scenario for planning and operation of the water distribution network; and (iii) prescriptive tools, for decision support platforms, which are becoming popular by advising on the best options available to solve a particular problem or constraint and, in some cases, automate the decision.

- AI will drive investment in water and wastewater operations.

  Water and wastewater operations are investing in AI. Recent market research forecasts huge investment in AI solutions by the year 2030. This investment is a part of a growing trend for the water industry to "go digital" with smart infrastructure solutions.
- AI will deliver significant operational expenditure savings in water and wastewater operations.

  AI can save 20%–30% on operational expenditures by reducing energy costs, optimizing chemical use for treatment, and enabling proactive asset maintenance.
- AI will predict emergency events and learn from them at an accelerated rate.

  Water main breaks are costly for utilities – in both financial and social capital. AI and ML can predict the data patterns that indicate a break event and learn from these patterns so that alerts can be generated.
- AI will optimize energy use for water and wastewater operations.

  Energy consumption makes up 25%–30% of total operation and maintenance (O&M) costs. AI can optimize pump runtimes so that they are only using energy when they need to. This is a quick cost-reduction win for many early adopters of AI.
- AI will keep the water clean, cost-effectively.

  Meeting effluent compliance standards is a requirement for all organizations. AI learns from the unique characteristics of the site to ensure that effluence standards are met and that compliance fees are avoided.
- AI will simplify data integrity.

  An explosion of available data for water operations managers has brought along a challenge of data management. AI can take heterogeneous data and process it so that it is clean, useful, secure, and drives high-fidelity recommendations.
- AI will retain institutional knowledge.

  AI-powered dashboards will keep institutional knowledge documented and standardized so that all information is always readily available.
- AI will accelerate the move to value-based asset maintenance.

  Time-based maintenance is easy to manage but results in unnecessary uptime and deterioration, through the AI team can be told what assets need to be serviced, and when.
- AI will power truly smart water systems.

  The journey to AI adoption empowers organizations to pursue data-driven, intelligent management of water systems. The result is resilient, sustainable, and cost-effective water management for years to come.

Utilities today have two clear choices: seek new supply to deal with resource volatility or optimize existing use within the available supply. **By using AI and new software-as-a-service platforms, as well as low-cost sensors and affordable communication networks, managers can dynamically create strategic, tactical, and financial operations for their utilities. Municipalities can also better plan and execute capital project requirements, better understand real-time water loss, efficiently operate distribution networks, and ensure maximum revenue capture, all while meeting ever-increasing customer demands.**

The power of AI unleashes the imagination of our water professionals. For example, AI-driven planning can combine growth projections with future trends in water availability and infrastructure condition assessment to maximize the impact of investment in infrastructure.

The benefits go beyond the utility. AI-enabled online platforms offer customers a personalized, engaging, and informative way to view real-time water consumption, pay bills and access information about dynamic water resource conditions. These online portals also use big data to educate consumers about conservation and push them along a path where small changes in behavior add up to substantial cost savings for them while enabling large-scale water resource savings for the community. AI, however, is only as good as its data and the understanding of the output.

6.7 WHAT AI REQUIRES

For optimal results, AI requires the availability of an adequately sized, validates data set that includes the information that can characterize the problem, and trains and tests the network. Equally important is a fundamental understanding of the problem. AI can handle massive data sets and computations. Quality of data is one of the most important factors in creating a working AI system.

6.8 TECHNOLOGIES USED BY AI FOR EFFECTIVE WATER MANAGEMENT

- Internet of Things

  The Internet of Things (IoT) is the first pillar of effective water management through AI (Figure 6.3). The number of connected IoT devices in the world is constantly growing, exceeding expectations year after year. The integrated water cycle is experiencing a similar situation. Technologies enable device connectivity and data capture in real time. Water utilities these days are equipped with plenty of IoT sensors and other data-driven technologies that constantly collect data on different phases in the water supply and

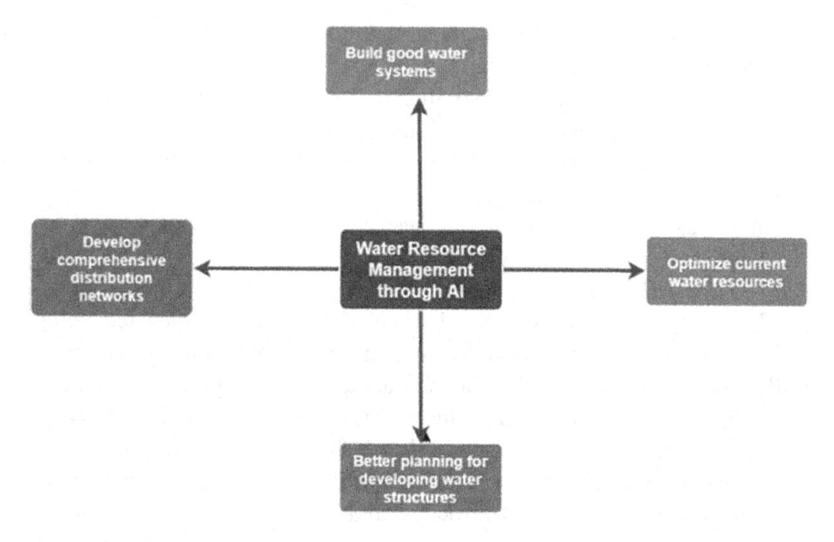

FIGURE 6.3 Water resource management through Artificial Intelligence (AI).

demand. These sensitive instruments, when used correctly, are able to draw previously unattainable information on the treatment of water and empower water utilities to predict their water operations; thus, the quality of the measurements is higher, and information management is more efficient.

Smart sensors, control technology and the capacity to interact in real time enable more efficient and effective operations with equipment and processes. The implementation of IoT solutions helps to improve qualitatively the entire management process. A good example is remote metering solutions (smart metering).

The evolution of communication technologies means that management can be closer to the citizen. Thanks to more frequent and higher quality data capture, companies can offer new and enhanced **services to citizens, such as communications about any issues, warnings about unusual consumption, and recommendations for responsible consumption**. Aside from a qualitative improvement in direct citizen services, remote metering offers the possibility of enhanced smart network management, early detection of leaks, enhanced energy efficiency, and ultimately, the optimization of processes and efficient integrated water cycle management.

LPWAN (Low Power Wide Area Network) is one of the communication technologies relevant for remote metering. NarrowBand-Internet of Things (Nb-IoT) and LoRaWAN are other technologies for qualitative and quantitative improvement in remote measurements from smart meters. These technologies are now part of the range of available remote metering technologies, providing new solutions for wireless data transmission with better results between devices separated by long distances; the penetration is higher in hard-to-access locations, and they have low battery consumption. With the new technologies, we can move from individual entities working in isolation to globally connected systems.

The use of 5G technologies will also result in a quantitative increase in transmission speed and bandwidth, also with lower energy consumption and massive device connectivity. 5G will enable enhanced communication and processing of images and sounds; the use of sensors will also evolve in the coming years, and so will integrate water cycle management.

- Big data and cloud computing

  Traditionally, business knowledge used to originate in independent and isolated information silos such as ERP and CRM. With the new technologies, there has been a shift from individual entities working in isolation, to globally connected systems. The aim is to achieve a better global understanding, and thus, more efficient management. Also, with IoT, an increasingly larger number of devices can be connected, thereby increasing the volume of data being captured exponentially. This data needs to be processed, first to turn it into information, then into knowledge, and finally into business intelligence.

  In this scenario, computing capacity is essential. Real-time connectivity to information silos and IoT devices is useless unless you can process the data and transform it quickly enough to enable more efficient management.

  To address these issues big data and cloud computing emerged. These technologies have boosted the traditional computing capacity, enabling the processing of large volumes of information in a shorter time. Furthermore, with cloud-based processing, solutions can be better scaled, to increase capacities in a flexible way when demand increases. Thanks to big data and cloud computing we can integrate and process information from traditional information, and integrate the IoT world, thereby obtaining a global perspective of the integrated water cycle for overall efficiency.

- AI and ML (AI/ML)

  AI and ML entail techniques and mathematical algorithms to solve complex problems. Although they are not new concepts, they have become more important because of the benefits generated by them, which include the need within those ecosystems, and secondly, the technological capacity. The current computing capacity allows processing a huge volume of data, turning them into business insight, but the transformation of that insight into action within industrial processes is more and more complex, because more decisions are needed, given the hyper-connectivity and hyper-integration of our systems. It is here where, thanks to AI and ML techniques, management and decision-making can evolve, shifting from traditional analytics, based on descriptive reporting and assessment analytics, to prescriptive and cognitive analytics to automate processes.

- Cybersecurity

  The privacy of citizens must be key in the service provided, and, therefore, a basic element of the corporate strategy. To ensure the security of clients' information, a security approach can be introduced based on intelligence to prevent, detect, analyses, and respond to threats in an ever-changing technological environment. Apart from client's security which is important, it is also necessary to prevent cyber-attacks on network information provided by devices. The analytical needs linking IT and OT require the adoption of security measures based on safe segmentation. Nevertheless, corporate awareness is more important than any technological solution.

Water analytics is an analytics platform for smart water cycle management. The platform captures information from IoT water network devices to turn it into knowledge and business intelligence, using technologies such as big data, cloud computing, ML, and AI.

The platform encompasses the whole data cycle, from capture by IoT devices to real-time processing, enrichment, and transformation, to the generation of business intelligence, allowing process automation and integration with technological solutions.

By using all these technologies, a good technological environment can be designed to build solutions that can adapt according to existing needs, which will take into account a plural ecosystem of connected devices, multiple communication technologies, and multiple technological field solutions. This can be achieved by following strict cyber security policies and employing AI/ML techniques, within a safe framework.

Three types of analysis can be performed:

Geospatial analytics – The user can perform analysis through geolocation in maps that include navigation features and interactive data view.

Comparative analytics – The user can generate custom queries and perform analyses using interactive graphs.

Advanced analytics – Through the use of AI/ML, prediction, clustering, and prescription tools are provided.

These three types of analysis will help in demand prediction, early detection of leaks and fraud, automatic generation of communications and work orders, etc., which summarizes all the tools needed for smart and efficient management.

6.9 BENEFITS OF WORKING WITH AI

AI is not, then, a fire-and-forget solution. **A key feature of a successful application is a necessary detailed understanding of the environment to both ask the right questions and evaluate the applicability of the response.** Critically interpreting the outcome becomes increasingly important, interpreting the output is going to be fundamentally important. Machines will most certainly perform this work more quickly, with fewer errors and with greater precision, water utility staff will continue to be a critical and essential element of delivering safe, reliable water to utility customers.

To access the value delivered by AI systems, water utilities need to create the processes around which data is collected, accessed, and analyzed. **This requires the creation of an ecosystem in which there is unfettered access to primary data and other supporting data sets, as well as a means of normalizing and storing this data combined with a team of skilled professionals to make sense of it all.**

Setting the smart monitoring system up and maintaining it can cost up to hundreds of thousands of euros. However, often this data is not analyzed until some trouble is already up. Typically, when a problem arrives, data collected by these sensors and water sample laboratory results are sent to an external analyst, who then tries to research the correlation between different parameters and understand what went wrong. **This process of analyzing raw data is quite difficult for humans, and it requires a lot of time.** What if the analysis process could be done constantly, drawing predictions on the performance of the water facility on a daily basis, even before any disturbance in the system?

AI, and ML, in particular, can be leveraged to create more effective water treatment processes, make sure problems can be recognized ahead of time, and help direct efforts in those areas early enough. ML, which is often used for predicting changes, can also create new insights that can be used as information for future investments and planning the usage of the water utilities.

The AI solution predicts the quality of the water leaving from the water utilities. The quality of the water is analyzed in the context of environmental permissions and terms.

With the help of ML model water treatment process is monitored constantly, shifting the focus from troubleshooting to predictive risk assessment and dynamic optimization of the facilities. With ML, it is possible to use the already invested smart systems and their measurements more efficiently, and identify risks in advance. Our AI system also contributes to finding out new factors that affect the performance of the water utilities.

The data analyzed consisted of measurement data from the laboratory and reporting data produced by the water utility's IoT sensor data. Also, other sources of public data such as weather data and network information could be used to form more accurate predictions.

Machine learning environments can be used to create more efficient and accurate support for clients. The algorithms can be leveraged to build cost-effective AI-driven systems on top of the existing IoT infrastructure, that improve the day-to-day operations at water utilities, optimizing the current usage of water facilities.

Water management through AI improves human–machine collaboration. Using AI in water resource management has proven to make way for economic water distribution and water-saving systems (Figure 6.4).

6.10 CONCLUSION

AI refers to the simulation of human intelligence in machines. The goals of AI include learning, reasoning, and perception. It is being used across different industries including

FIGURE 6.4 Technologies used in water resource management.

finance and healthcare. It can continuously work with large amounts of data sets, tirelessly. The technology also learns on the go, improving itself with each set of data that it processes. This adaptability makes AI the best option to deal with an everyday altering entity, such as water.

AI can bring in the trend of "**digital water**." It can fundamentally change the way water utilities operate and manage water resources in an increasingly volatile environment. The water sector's immediate challenge is to structure data and smart water management services to maximize AI's potential. Once this is addressed, the industry can then focus on developing dynamic models that run continuously to optimize utility operations, increase customer delight, and best protect our shared clean water resources.

This revolution means a disruption with regard to traditional management, transforming how companies operate, and leading to new challenges and opportunities. New technologies enable enhanced connectivity, and therefore, enhanced access to relevant data (IoT), turning information into knowledge much faster (big data and cloud computing); they also help with decision making, management, and process monitoring (AI/ML). In essence, new technologies allow enhanced efficiency, more sustainable management, and offer a better service to citizens in terms of effective water management.

BIBLIOGRAPHY

Ay, M., & Özyıldırım, S. "Artificial intelligence (AI) studies in water resources." *Natural and Engineering Sciences*, 2018, 3: 187–195.

Hatler, D. Industry 4.0 & the water sector. August 2020. https://waterfm.com/industry-4-0-the-water-sector/.

Hill, T. How artificial intelligence is reshaping the water sector. March 2018. https://waterfm.com/artificial-intelligence-reshaping-water-sector/.

Jenny, H., Alonso, E.G., Wang, Y., & Minguez, R. "Using artificial intelligence for smart water management systems." *ADB Briefs* No. 143, 2020: 1–10.

Joshi, N. 4 ways AI is helping with water management. October 2018. https://www.allerin.com/blog/4-ways-ai-is-helping-with-water-management.

Pinto, A., Fernandes, A., Vicente, H., & Neves, J. "Optimizing water treatment systems using artificial intelligence based tools." *WIT Transactions on Ecology and the Environment*, 2009, 125: 185–194.

Tang, H., Lei, Y., Lin, B., Zhou, Y., & Gu, Z. "Artificial intelligence model for water resources management." *Journal of Water Management*, 2010, 163: 175–187.

Online Recommendation Using Machine Learning (ML) and NLP

7

Mohnish Vidyarthi, Parth Vidyarthi and Rohit Maheshwari
Career Point University

Contents

7.1 Introduction 108
7.2 Content-Base Methods 110
7.3 Collaborative Filtering 111
7.4 Knowledge-Based 111
7.5 Hybrid Recommendation System 112
7.6 Deep Learning Models for Recommendation Systems 112
7.7 Recommendation System Pitfalls 114
7.8 NLP-Based RS without User Preferences 116
7.8.1 Practical Aspect: The Data 116
7.9 Conclusion 117
Bibliography 117

DOI: 10.1201/9781003051022-7

7.1 INTRODUCTION

When you watch a video on YouTube and you see a list of suggested videos to watch next, that list is being built by a recommendation machine-learning model, often called a recommendation engine. But just a machine-learning model is not enough; it needs a data pipeline that collects whatever input data the model needs – inputs like the last five videos the user watched – and that is done by a recommendation system. Recommendation systems are not just about suggesting products to the customers, sometimes they can be for suggesting uses for products. Like in marketing applications, if you may have new promotion and you want to find the thousand most relevant customers, that's called targeting and that is also done by recommendation systems. Many times recommendation systems are not about products. When Google maps suggest the shortest path and when Gmail suggests possible quick replies, that's also done by recommendation systems. Recommendation systems are about personalization. It's about taking your product that works for everyone and personalizing it for individual users. So let's say you want to recommend movies to users.

You can do this in several ways. In a context-based recommendation system, you can use the metadata about your products. For example, you know which movies are cartoons and which are sci-fi. Now, suppose a user who has seen and rated a few movies. Some that she liked and gave a thumbs up and some that she did not. We would like to know which movie in our database we need to recommend next. So, remember that we have metadata about the movies and we know that this particular user likes sci-fi and doesn't like cartoons. So, we might use that information to recommend popular sci-fi dramas to this user. So, perhaps we recommend *The Dark Knight Rises*.

Notice that this recommendation is based on knowing something about the content. You're simply recommending the most popular items in a category that the user likes. Maybe you don't even have the individual user's preferences. All you might have is market segmentation. Which movies are liked by users in which regions of the country, and that is enough to build a content-based recommendation system. Arguably, there is no machine learning here, it's a simple rule that relies on the builder of the recommendation system to assign proper tags to items and users. That's a content-based recommendation system.

In collaborative filtering, you don't have any metadata about the products, instead, you learn about item similarity and user similarity from the rating data itself. We might store our user movie data in a matrix like this (Figure 7.1).

Which user liked which movie and how many stars he/she gave to that particular movie? This matrix is very large. Since you might have millions to billions of users and hundreds to millions of movies, an individual will tend to have watched only a handful of these movies. So, most of this matrix is not only large but also sparse. The idea behind collaborative filtering is that this very large, very sparse user-by-item matrix can be approximated by the product of two smaller matrices called user factors and item factors. Then, if we need to find whether a particular user will like a particular movie, it's as simple as taking the row corresponding to the user and the column corresponding

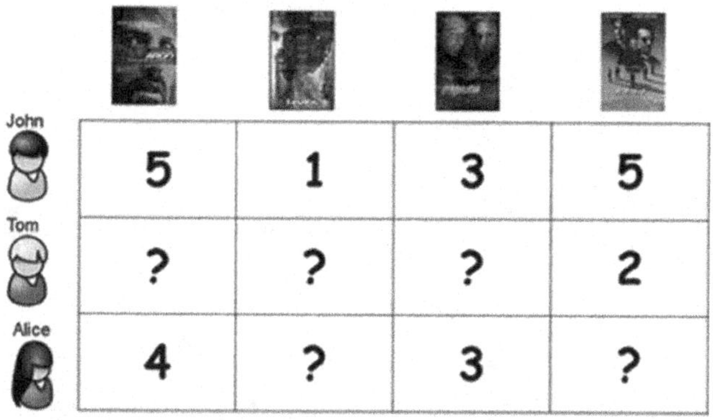

John	5	1	3	5
Tom	?	?	?	2
Alice	4	?	3	?

FIGURE 7.1 Data Matrix for user movie data.

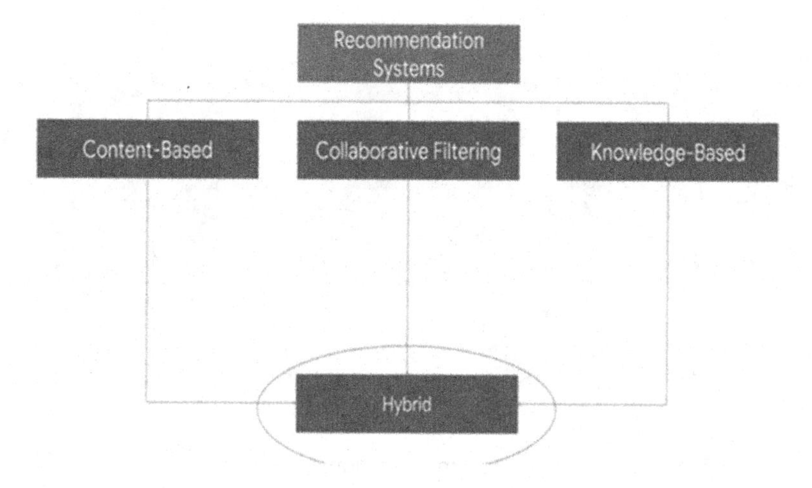

FIGURE 7.2 Real World recommendation systems are a hybrid of three broad theoretical approaches.

to the movie and multiplying them to get the predicted rating. To recommend movies to users, we recommend the movies that we predict they will rate the highest. One of the really cool things about collaborative filtering is that you don't need to know any metadata about your items. You also don't need to do market segmentation of your users as long as you have an interaction matrix, you're ready to go. But if you do have metadata about users and items, you can use content-based recommendation systems. But what if you have both? You have metadata and you have an interaction matrix. If you have both, you can use neural networks to combine all of the advantages and eliminate the disadvantages of all three types of recommendation systems (Figure 7.2).

Well, there's a third type of recommendation system called a knowledge-based recommendation system that can be used to provide business impact inputs to systems.

The following picture represents how YouTube's recommendations work (Figure 7.3).

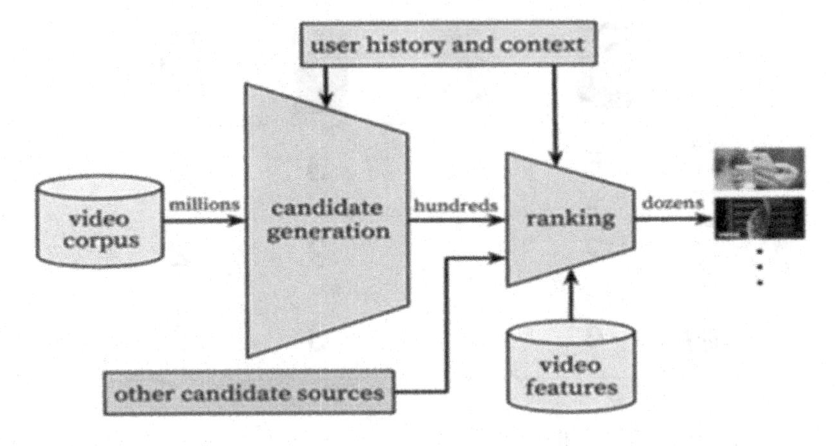

FIGURE 7.3 YouTube video recommendations.

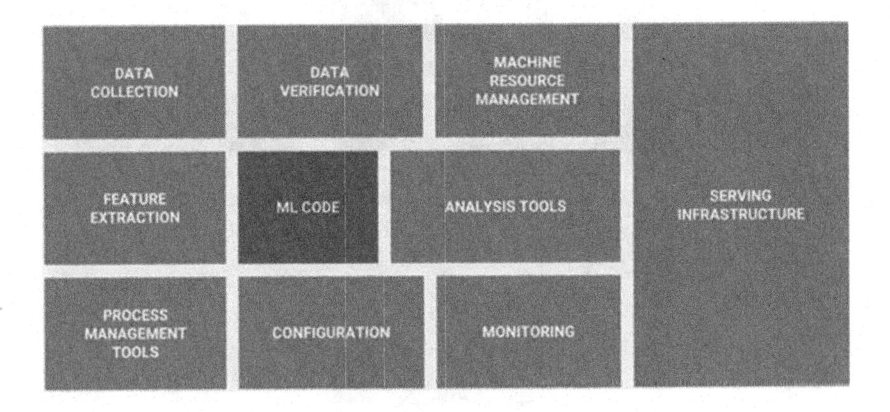

FIGURE 7.4 Overall small parts of recommendation system.

If we think of machine learning, well that's just a small part of the overall recommendation system (Figure 7.4).

Recommendation systems are often the machine-learning systems that you encounter the most often in enterprise settings. From a business impact standpoint, recommendation systems can allow you to sell products you could never sell before to users you could never reach before (Figure 7.5).

7.2 CONTENT-BASE METHODS

A content-based method uses attributes of the items to recommend new items to a user. It doesn't take into account the behaviour or ratings of other users. For example, if a user has rented and liked a lot of vacation homes on the beach, this method will suggest

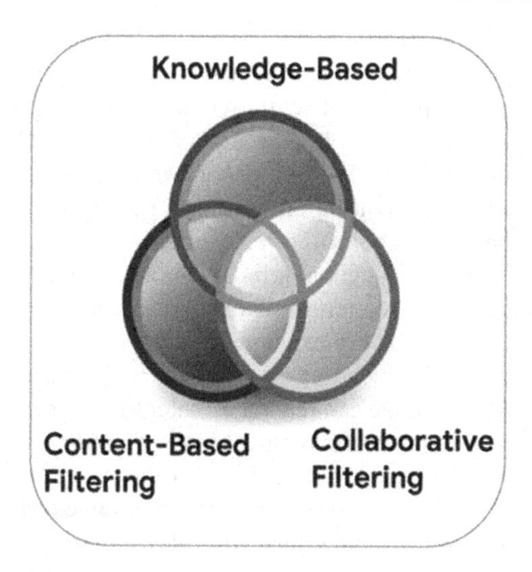

FIGURE 7.5 Different types of recommendation systems.

other similar homes that are also on the beach. This is often done by hand engineering features for the items and learning how much a single user aligns with each of those features. Using that feature, representation of the user is impossible to extrapolate how a given user would rank unseen items.

7.3 COLLABORATIVE FILTERING

A collaborative filtering model works with the entire user-item interaction matrix. They consider all users, all items, and all user-item ratings. Loosely speaking, they work with the idea that similar users will like similar items. That is, they use similarities between the users and the items simultaneously to provide recommendations.

This can allow for seemingly serendipitous recommendations. Meaning, they can recommend an item to user A based only on the interests of a similar user B. Another useful advantage is that the feature representations can be learned automatically. So, you don't have to rely on hand engineering specific features as you would for a content-based filtering method. This process often involves matrix factorization and behaves similarly to a content-based approach but does not rely on previously constructed features.

7.4 KNOWLEDGE-BASED

Knowledge-based recommender systems (RSs) are based on obvious information about the user's likings, things, and or recommendation measures. They're especially useful when alternative approaches such as collaborative filtering or content-based methods cannot be applied. This occurs in situations where items are not purchased very often. For example, if instead of renting a vacation house, suppose we want to build a recommendation engine for buying a vacation house. Because most people don't buy houses often, we probably wouldn't have enough previous house buying information to use either a content-based or collaborative filtering approach. In this scenario, knowledge-based systems will often explicitly ask users for their preferences and then use that information to begin making recommendations. Often, there is value in combining different types of recommendation models into a single hybrid approach. This can be done in several ways. For example, we could develop a few recommenders and then use one or the other depending on the scenario. If a user has already rated a large number of items, perhaps we can rely on a content-based method. However, if the user has rated only a few items, we may instead prefer to use a collaborative filtering approach. This way, we can fully leverage the information we have about other users and their interactions with items in our database, to gain some insight into what we can recommend. Of course, if we have no information about a user's previous item interactions or we like any information about a given user, we may instead want to rely on a knowledge-based approach and ask the user directly for their preferences via a survey before making any recommendation.

7.5 HYBRID RECOMMENDATION SYSTEM

A hybrid model can be created by simply combining the outcomes of more than one of these models. The multiple outcomes could then form the input to a more sophisticated model that makes the final recommendation that we then serve to the uscr. The idea is that the more sophisticated the model, we'll learn a more nuanced relationship between the query and the various model outcomes, and we'll have a much better recommendation. In fact, some research suggests that a hybrid approach combining multiple outcomes like this can provide more accurate recommendations in a single approach on its own.

7.6 DEEP LEARNING MODELS FOR RECOMMENDATION SYSTEMS

In addition to content-based, collaborative filtering and knowledge-based approaches, deep learning models can also be used when building a recommendation system. Deep neural networks work well because they are flexible and can be trained to have varying outcomes, such as predicting ratings, interactions, or even next items. For example, suppose we wanted to recommend videos to our users, we could approach this from a deep learning point of view by taking attributes of the user's behaviour input, for example, a sequence of their previously watched videos embedded into some latent space, combined with video attributes, either genre or artists information for a given video. These user and item attributes are combined into a single dense layer of a neural network, and then again to another fully-connected layer until a single value is ultimately produced, with the objective function comparing the difference with the user's rating for the given video. At inference time, we can apply this model to rate previously unseen videos and recommend to the user the video with the highest score. Deep learning models like this are flexible enough to easily incorporate all kinds of query features and item features into the input layer of the network to help capture the specific interests of a user or to improve the relevance of the recommendations (Figure 7.6).

Example: Suppose that we have built a recommendation engine and suggest new Apps from our App store. Our model recommends a Hiking App to a user because they recently installed a running App and I have been using it a lot. What kind of recommendation approach is this an example of? Would it be content-based filtering, collaborative filtering, a deep neural network approach, or perhaps a hybrid of more than one of these? Well, this would be an example of content-based filtering. We're told that the Hiking App suggestion is a result of the user recently installing and using a similar App for tracking their runs. Because the content of these two Apps is similar, we expect that this user will appreciate both, hence the name content-based filtering. This wouldn't be collaborative filtering because the recommendation does not rely on the behaviours and item interactions of other users. Its recommendation is based only on that user's previous behaviour. This is what sets content-based filtering apart from collaborative filtering. And although it's possible that this recommendation came from a neural network or a hybrid approach, it doesn't sound like it from the question. The recommendation was made based on the similarity of features and content of the two Apps.

In fact, the features related to Apps and the user are likely hand-engineered by some software engineers, and they aren't learned solely for the data.

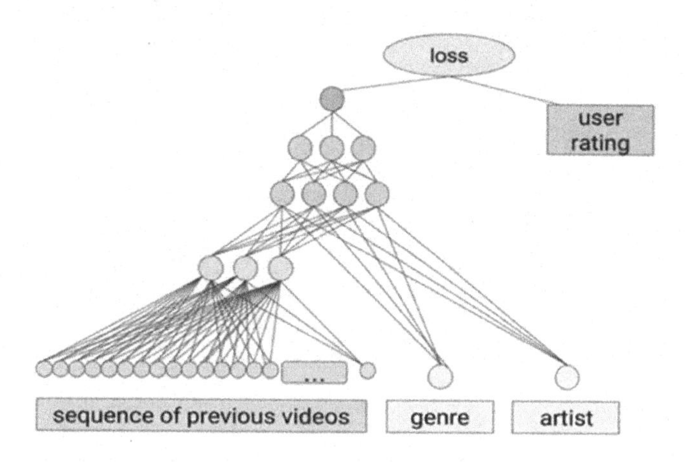

FIGURE 7.6 Recommended videos using deep learning.

7.7 RECOMMENDATION SYSTEM PITFALLS

When you're developing any recommendation system, some common difficulties are helpful to keep in mind, and some care must be taken to address them. For example, looking again at the user-item interaction matrix, know that the user space and the product space are sparse and skewed (Figure 7.7).

We say the user-item matrix is sparse because there are potentially few interactions within the entire user-item space. There could be billions of users and millions of items, and most of the entries in the matrix are zero. Sparse matrices are problematic because they take up a lot of memory and are slow to perform computations, even though most of the computations are simply adding or multiplying by zero. A typical person probably interacts with far less than one percent of your products. On the other hand, most products are in the long tail of usage and are probably rated by less than 0.1% of all users. In addition, the matrix is skewed. Some properties might be very popular. Maybe that property is a resort with 1,000 cabins, or there could be a few users that are very prolific. Maybe they're motorcycling around the country and staying in rental homes every day of the year, or maybe some users just like everything. If you naively take all the ratings for all users, you risk overemphasizing certain users or products. In addition, a cold start can occur for both products and users. This happens when there aren't enough interactions or information for users or items, for example, when a new user is added to a system. For a time, the recommender has no past interactions with which to make new recommendations. Consider when a new item is added to a catalogue. Because collaborative filtering relies on user-item interactions, without this information, reliable recommendations for users aren't generated. In this situation, a content-based approach would be better. Another problem to keep in mind is a lack of explicit user feedback in

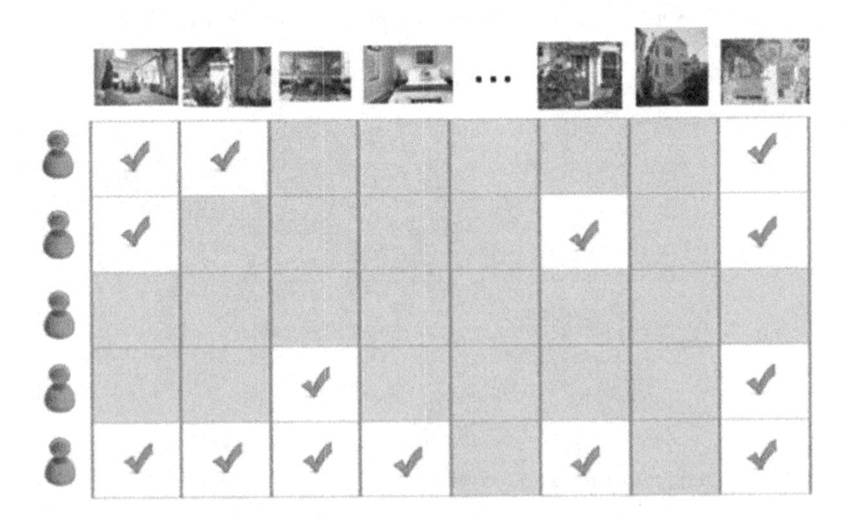

FIGURE 7.7 Recommended system user-item matrix.

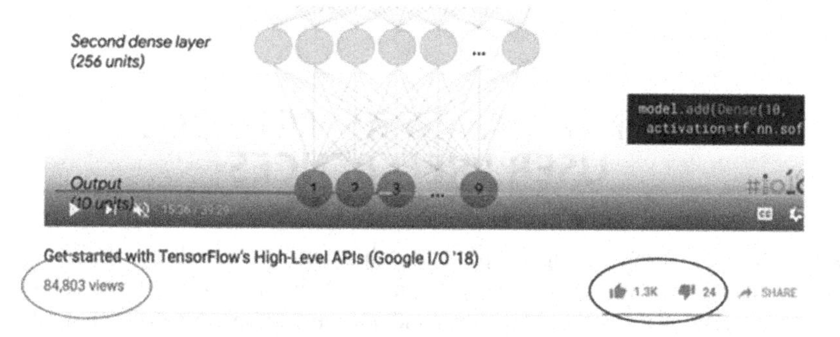

FIGURE 7.8 YouTube video views and ratings.

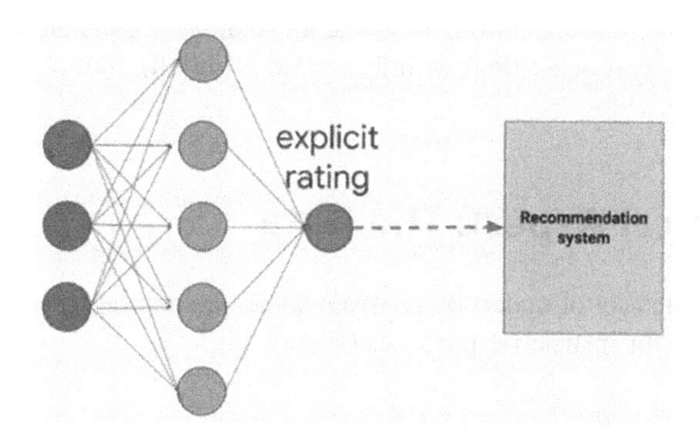

FIGURE 7.9 Explicit rating in recommendation system.

the form of ratings or thumbs up or down. Often we don't have explicit measures to feed to a RS. In such situations, it is necessary to rely on implicit ratings (Figure 7.8).

For example, consider a YouTube video like this one. Of the tens of thousands of views, there're substantially fewer upvotes or downvotes. Instead, implicit measures must be used.

Some other similar examples are like the portion of the video watched or the number of clicks, play counts, or other information about the user interaction like site navigation, or time spent on the page. In practical scenarios, implicit feedback like this is much more readily available, even though explicit user feedback is ground truth and preferred.

With enough data, you can train an initial model that uses implicit user feedback to learn an explicit rating (Figure 7.9).

Essentially, the input layer takes implicit ratings, and we use the explicit ratings as a label.

This explicit rating is what we want to use as our input when developing the recommendation engine. At inference time, in the absence of an explicit rating, we use the trained initial model to infer an explicit rating that we can then feed to the recommendation engine. As we collect more data, these signal scores can be improved.

7.8 NLP-BASED RS WITHOUT USER PREFERENCES

RSs have advanced into a principal apparatus for helping clients settle on educated choices and decisions, particularly in the period of enormous information wherein clients need to settle on decisions from countless items and administrations. A ton of RS models and methods have been proposed and a large portion of them have made extraordinary progress. Among them, the substance-based RS and cooperative separating RS are two the agents. Their adequacy has been exhibited by both examination and industry networks.

This methodology is exceptionally valuable when we are managing items that are included by a portrayal or a title (text information all in all), like news, employments, books and so forth.

7.8.1 Practical Aspect: The Data

We will utilize a dataset of undertakings from some outsourcing sites that were parsed from RSS channel for instructive purposes (Figure 7.10).

```
import pandas as pd
projects = pd.read_csv("proj.csv")
projects_done= pd.read_csv("projects_history.csv")
projects.head()
```

The history table includes two columns, user_id and project_id.

Presently, how about we set up the information and clean it?

```
projects = projects.drop_duplicates(subset="Title", keep='first',
inplace=False)
projects["Category"] = projects["Category"].fillna('')
```

	Category	Description	SubCategory	Tag	Title
0	Machine Learning	We are a learning startup company and we are l...	Data Science & Analytics	Big Data	Writing in Deep Learning and Artificial Intell...
1	Data Extraction / ETL	Hi, I wanted a program that: Pulls the most ...	Data Science & Analytics	Big Data	Coinbase Pro API & MySQL Integration
2	Web Development	You will be working on a current project, taki...	Web, Mobile & Software Dev	Big Data	UI/UX Frontend Web Developer Needed
3	Data Extraction / ETL	We are looking for Indian company who can work...	Data Science & Analytics	Big Data	Looking for company who does big data implemen...
4	Technical Writing	We are looking for an experienced technical co...	Writing	Big Data	Need Technical Content Writer for Big Data &am...

FIGURE 7.10 Dateframe of projects.

```
projects["Description"] = projects["Description"].fillna('')
projects["SubCategory"] = projects["SubCategory"].fillna('')
projects['Title']= projects['Title'].fillna('')
```

We should join the entire columns as one line for each archive.

```
projects["text"] = projects["Category"] + projects["Description"]
+ projects["SubCategory"] + projects["Title"]
```

Presently, we will figure TF-IDF "Term Frequency – Inverse Data Frequency"

```
from sklearn.feature_extraction.text import TfidfVectorizertf =
TfidfVectorizer(analyzer='word',ngram_range=(1, 2),min_df=0,
stop_words='english') tfidf_matrix =
tf.fit_transform(projects['text'])
```

It's an opportunity to produce the cosine sim framework.

```
from sklearn.metrics.pairwise import linear_kernel, cosine_
similaritycosine_sim = linear_kernel(tfidf_matrix, tfidf_matrix)
```

7.9 CONCLUSION

We have fabricated a substance-based recommender utilizing an Natural Language Processing (NLP) approach that is valuable with regard to items with text information.

Without the requirement for client's inclinations, we can suggest quality proposals for them, just from their finished undertakings.

This methodology is additionally having geniuses, which is the reason it is constantly prescribed to utilize various methodologies and consolidate them for precise outcomes.

BIBLIOGRAPHY

A. Louly "NLP based recommender system without user preferences", https://towardsdatascience.com/nlp-based-recommender-system-without-user-preferences–7077f4474107.

8

Natural Language Processing and Translation Using Machine Learning

Garima Tyagi
Career Point University

Contents

8.1 Introduction to Natural Language Processing 120
8.1.1 Examples of NLP 121
8.1.2 Stages of NLP 121
8.1.2.1 Lexical Analysis and Morphological 121
8.1.2.2 Syntactic Analysis (Parsing) 121
8.1.2.3 Semantic Analysis 121
8.1.2.4 Discourse Integration 121
8.1.2.5 Pragmatic Analysis 121
8.2 Machine Translation 122
8.3 Machine Learning for Natural Language Processing 122
8.3.1 Supervised Learning 123
8.3.2 Unsupervised Learning 123
8.3.3 Semi-Supervised Learning/Reinforced Learning 124
8.4 Machine Learning and Natural Language Processing 124
8.4.1 Supervised Machine Learning for NLP and Text Analytics 124

DOI: 10.1201/9781003051022-8

8.4.1.1 Tokenization 125
8.4.1.2 Part-of-Speech Tagging 125
8.4.1.3 Named Entity Recognition 125
8.4.1.4 Sentiment Analysis 126
8.4.1.5 Categorization and Classification 126
8.4.2 Unsupervised Machine Learning for Natural Language Processing and Text Analytics 126
8.4.3 Using Machine Learning on Natural Language Sentences 127
8.4.4 Hybrid Machine Learning Systems for NLP 128
8.5 Machine Translation 129
8.5.1 Neural MT's Evolution 129
8.5.2 Replacing an Algorithm with a System 131
8.5.3 MT with Neural Networks 131
8.5.3.1 Google Translate 131
8.5.3.2 Translator by Microsoft 131
8.5.3.3 Facebook Translator 132
8.6 Conclusion 132
Bibliography 132

Abstract

Natural language processing is a field of computer science that deals with the development of computer systems that can be used to interpret text and speeches. If NLP is used to analyze words and to deal with text then certain aspects of language analysis and understand the use of machine learning. When machine learning is used for natural language processing and text analytics, it involves the machine-learning algorithm and artificial intelligence to understand the meaning of the documents and the text written. As per the current needs, it seems that machine learning is an essential tool for natural language processing associated with the text classification and word sense analysis.

8.1 INTRODUCTION TO NATURAL LANGUAGE PROCESSING

Natural language processing is a field of artificial intelligence, which is used to analyze that text very quickly and easily so that the auto-correction can be done most easily. Various methods are used in Natural Language Processing (NLP) such that the machine operates in such a way that the commands can be comprehended accordingly. The field that is essentially used to determine how machines can be used for various purposes to understand natural languages is natural language processing. The different techniques are used by NLP researchers to figure out how a human being can understand and use language, and hence various tools and techniques can be developed in NLP.

8.1.1 Examples of NLP

- Voice/speech recognition systems, as well as query systems such as Siri, process the question and provide a response. You speak to a machine, and it knows what you're doing.
- Computer programs that read plain English financial reports and generate numbers.
- Employment platform retrieves applicant information and produces a resume and application for the job based on the candidate's skills.
- Google Translate parses the text in the input string and translates it into the target language.
- After you type a word of the topic into the search window, Google-like search engines return your documents.

8.1.2 Stages of NLP

NLP has traditionally been considered to be divided into the following stages for both spoken and written language:

8.1.2.1 Lexical Analysis and Morphological

The Lexical Analysis is the first step in NLP. The source code is scanned as a stream of characters and converted into meaningful lexemes in this process. The entire text is divided into paragraphs, sentences, and phrases.

8.1.2.2 Syntactic Analysis (Parsing)

Syntactic Analysis is used to check grammar and word arrangements and to describe the relationship among the words.

Example: "Jaipur goes to the Prakash"

In the real world, "Jaipur goes to the Prakash", there is no meaning of such sentence, so this sentence is rejected by the Syntactic analyzer.

8.1.2.3 Semantic Analysis

Semantic analysis is concerned with the representation of meaning in a sentence. It mainly focuses on the literal meaning of words, phrases, and sentences.

8.1.2.4 Discourse Integration

Discourse Integration is influenced by the sentences that come before it, as well as the context of the sentences that come after it.

8.1.2.5 Pragmatic Analysis

Pragmatic is the fifth and last phase of NLP. It helps you to discover the intended effect by applying a set of rules that characterize cooperative dialogues.

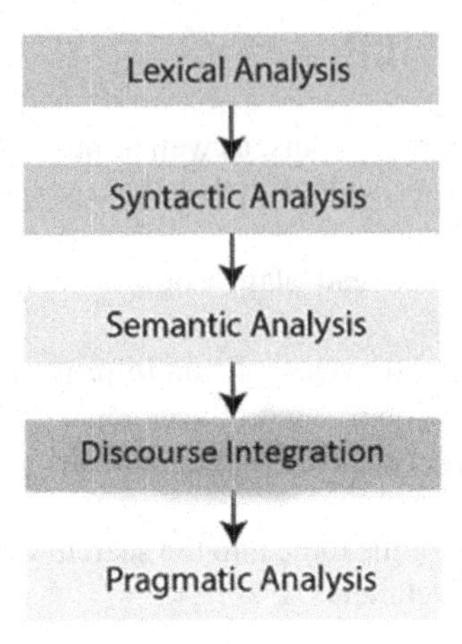

FIGURE 8.1 Stages of NLP (NLP Tutorial, n.d).

For example: “Open the window” is interpreted as a request instead of an order (Figure 8.1).

8.2 MACHINE TRANSLATION

Machine translation (MT) is a branch of computational linguistics that studies how software can be used to translate text or speech from one language to another. Computer translation essentially replaces words in one language with words in another, but this does not guarantee correct translation. MT is the process where the text present in the source language can be converted into the required target language. There is no human intervention in this text translation from one language to another; instead, the computer performs the conversion (Figure 8.2).

8.3 MACHINE LEARNING FOR NATURAL LANGUAGE PROCESSING

Machine learning can be defined as the process of training a machine. We know that the machine must learn, so we build a learning framework to offer specific data for it to learn from. When a machine-learning model is built, the training data may be included, and the model may evolve, as more learning is required. In algorithmic programming, we use predetermined data, whereas in machine learning, we use undefined data. It is necessary to define it since it is used to generate, generalize, and deal with real-time data.

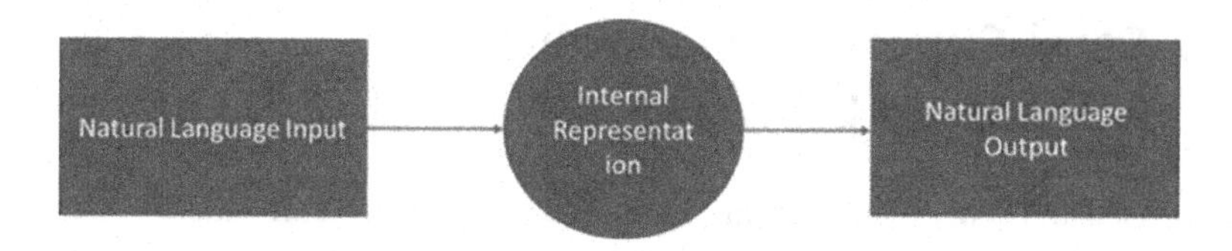

FIGURE 8.2 Machine translation.

When machine learning is utilized for NLP and text analytics, a set of statistical approaches must be applied for the various NLP operations, which include identifying the portion of the speech, sentiment, and other elements of the text. In the case of natural language processing and text analytics, artificial intelligence and machine learning can be applied to improve text analytics algorithms to interpret unstructured text or any useless data. When we build a model that contains unexpected approaches and apply them to other text, we call it supervised machine learning; however, when data is extracted and functions are performed, we call it unsupervised machine learning. Both types of systems can be employed when dealing with NLP. When dealing with text data, a different approach to machine learning is required because text data contains a group of dimensions that may have different types of functions. For example, in the case of English or Hindi, we may have more than lakhs of words that are commonly used in our general language, and they may differ from one another when used in a specific context.

Machine learning is divided into three categories: supervised learning, unsupervised learning, and semi-supervised learning. Lexalytics use all three, depending on the problem at hand.

8.3.1 Supervised Learning

The term "supervised learning" refers to giving a dataset to a machine-learning model that has been annotated in some way. For example, we might collect 10,000 customer support comments and categorize them according to whether they are connected to software or hardware. We're showing the machine what data it needs to assess each remark this way.

This is the most straightforward method of instructing a model on what you want it to accomplish. It's also the most time-consuming. We employ supervised learning at Lexalytics for NLP tasks such as sentiment analysis and specific categorization approaches.

Because the sentiment analysis perspective might change depending on the context, we train sentiment analysis models on hand-scored instances.

8.3.2 Unsupervised Learning

Unsupervised learning is when we give a machine a large amount of data and tell it to detect patterns in it.

8.3.3 Semi-Supervised Learning/ Reinforced Learning

Unsupervised and supervised learning techniques are used in semi-supervised learning. Both marked-up supervised content and unmarked data are used with this method. The marked-up information is used by the machine-learning model to generalize and make assumptions about the data.

8.4 MACHINE LEARNING AND NATURAL LANGUAGE PROCESSING

There are many applications in which machine-learning techniques have been used (Emms & Luz, 2007), including

- Speech recognition
- Document categorization
- Document segmentation
- Part-of-speech tagging, word-sense disambiguation
- Named entity recognition (selecting and classifying multi-word sequences as instances of semantic categories)
- Parsing
- MT

8.4.1 Supervised Machine Learning for NLP and Text Analytics

Single-task machine-learning models are excellent at determining the sentiment polarity of a document or the part of speech for a specific word. Models, on the other hand, are ineffective at jobs that involve multiple layers of interpretation.

In the case of supervised machine learning, a batch of text documents is utilized, and the machine must examine and interpret the results. Please note that comments are mostly used to train a specific machine model, which is a statistical model that is given a specific text you wish to analyze.

For developing any supervised NLP machine-learning model, we may use different algorithms, which are

- Support vector mechanics
- Bayesian network
- Maximum entropy

- Conditional random fields
- Neural networks for deep learning

Supervised machine learning is used to improve text analytics' fundamental operations as well as other NLP characteristics, which may include the following:

- Tokenization
- Part-of-speech tagging
- Named entity recognition
- Sentiment analysis
- Categorization and classification

8.4.1.1 Tokenization

Tokenization is the process of breaking down a written document into machine-readable chunks, such as words. You're probably quite skilled at determining what's a word and what's not. English is particularly simple. Look at how much white space there is between the letters and paragraphs. This makes tokenization a breeze. As a result, NLP principles are sufficient for tokenization in English.

But how does one teach a machine-learning algorithm to recognize a word? What if you're not working with documents written in English? Whitespace is absent in graphical languages such as Mandarin Chinese.

This is where machine learning comes into play for tokenization. Chinese, like English, has rules and patterns that we can teach a machine-learning model to recognize and interpret.

8.4.1.2 Part-of-Speech Tagging

Part-of-Speech Tagging (PoS tagging) is the process of recognizing and tagging each token's part of speech (noun, adverb, adjective, etc.). A lot of significant Natural Language Processing tasks rely on PoS labeling. To distinguish entities, extract themes, and process sentiment, we must correctly identify parts of speech. Even for short, nasty social media messages.

8.4.1.3 Named Entity Recognition

Named entities are people, places, and objects (products) described in a text document at their most basic level. Unfortunately, hashtags, emails, mailing addresses, phone numbers, and Twitter handles can all be considered entities. In fact, if you look at it the correct way, almost anything can be considered an entity. Don't get us started on off-topic references.

We've trained supervised machine-learning models on a huge number of pre-tagged things at Lexalytics. This method allows us to improve accuracy and flexibility. NLP algorithms have also been trained to recognize non-standard things (such as species of tree or types of cancer).

It's also worth noting that Named Entity Recognition models rely on precise PoS tagging from those models to function properly.

8.4.1.4 Sentiment Analysis

The act of detecting if a piece of writing is positive, negative, or neutral and then assigning a weighted sentiment score to each entity, subject, topic, and category within the document is known as sentiment analysis. This is a really difficult task that changes greatly depending on the situation. Take the phrase "CEO," for example. This can be used as the CEO of a company or in case you are the best at what you are doing.

It would be difficult to create a set of NLP rules that would account for every feasible sentiment score for every possible phrase in every possible circumstance. However, by using pre-scored data to train a machine-learning model, it may learn to distinguish between what "CEO" means in the domain of the CEO of a company and what it means in the context of talking to someone about doing the best in any case. Each language, predictably, requires its own sentiment categorization model.

8.4.1.5 Categorization and Classification

Categorization is the process of grouping stuff into groups in order to gain a rapid overview of what's in the data. Analysts employ pre-sorted content to train a text classification model and carefully nurture it until it achieves the desired degree of accuracy. As a result, text document categorization is accurate and trustworthy, and it requires significantly less time and effort than human analysis.

Supervised learning has been used for a range of natural language processing applications in addition to text categorization. Re-ranking probabilistic parser output (Collins & Koo, 2005), authorship identification, language identification, keyword extraction, and a variety of additional applications are among them. In order to address the issue of training data availability, which is a common problem in natural language applications, supervised learning versions that can learn with unlabeled input in addition to modest quantities of labeled data have been developed. These include algorithms such as expectation maximization (EM) and Baum–Welch algorithms, which are used in semi-supervised learning.

8.4.2 Unsupervised Machine Learning for Natural Language Processing and Text Analytics

Training a model without pre-tagging or labeling is known as unsupervised machine learning. Some of these methods are surprisingly simple to grasp.

Clustering is the process of combining comparable documents into groups or sets. The importance and relevance of these clusters are then sorted (hierarchical clustering).

Latent semantic indexing is another sort of unsupervised learning (LSI). This method discovers words and phrases that are commonly used together. For faceted searches or for returning search results that aren't the exact search word, data analysts utilize LSI.

For example, the terms "gold" and "jewelry" are closely related to each other. So, when you Google "gold" you get results that also contain "jewelry."

Another technique for unsupervised NLP machine learning is matrix factorization. This method uses "latent factors" to split down a huge matrix into two smaller matrices. Similarities between the items are known as latent factors.

Consider the phrase, "I threw the ball over the mountain." The word "threw" is more commonly connected with the word "ball" than with the word "mountain."

Humans, in fact, have a built-in ability to recognize the variables that make anything throwable. However, this distinction must be taught to a machine-learning NLP system.

Unsupervised learning is difficult, but it requires significantly less time and data than supervised learning. Lexalytics employs unsupervised learning algorithms to generate a "basic grasp" of how language functions. To help our models comprehend the most likely interpretation, we extract some significant patterns from massive volumes of text documents.

The Concept Matrix, which is based on unsupervised learning, is the foundation of our understanding of semantic information.

The Syntax Matrix is a tool that helps us comprehend the most likely parsing of a sentence, and it is the foundation of our knowledge of syntax.

8.4.3 Using Machine Learning on Natural Language Sentences

Let's have a look at the sentence "Rohan hits a ball over the wall," this will provide three types of information:

- Semantic information: person – the act of striking an object with another object – spherical play item – place people live
- Syntax information: subject – action – direct object – indirect object
- Context information: this sentence is about a child playing with a ball

They give a general concept of what the statement is about, but full comprehension necessitates the successful integration of all three elements.

This can be done in a variety of methods, including using machine-learning models or entering rules for a computer to follow while evaluating text. These strategies, on their own, aren't very effective.

Machine-learning models excel in recognizing entities and overall sentiment in documents, but they struggle to extract themes and topics and aren't very strong at matching sentiment to individual entities or themes.

A system can be taught to identify the basic rules and patterns of language. Unfortunately, recording and implementing language rules takes a lot of time. What's more, NLP rules can't keep up with the evolution of language. The Internet has butchered traditional conventions of the English language. And no static NLP codebase can possibly encompass every inconsistency and meme-ified misspelling on social media.

Very early text mining systems were entirely based on rules and patterns. Over time, as natural language processing and machine-learning techniques have evolved, an increasing number of companies offer products that rely exclusively on machine learning. But as we just explained, both approaches have major drawbacks.

That's why at Lexalytics, we utilize a hybrid approach. We've trained a range of supervised and unsupervised models that work in tandem with rules and patterns that we've been refining for over a decade.

8.4.4 Hybrid Machine Learning Systems for NLP

Our text analysis functions are based on patterns and rules. Each time we add a new language, we begin by coding in the patterns and rules that the language follows. Then our supervised and unsupervised machine-learning models keep those rules in mind when developing their classifiers. We apply variations on this system for low-, mid-, and high-level text functions (Figure 8.3).

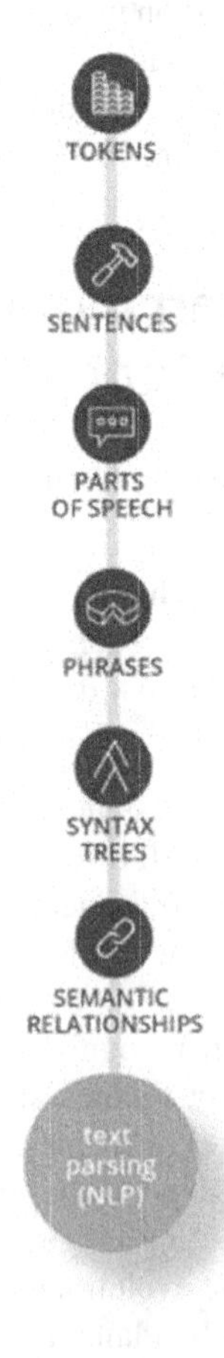

FIGURE 8.3 Steps for parsing a text.

Low-level text functions are the initial processes through which you run any text input. These functions are the first step in turning unstructured text into structured data. They form the base layer of information that our mid-level functions draw on. Mid-level text analytics functions involve extracting the real content of a document of text. This means who is speaking, what they are saying, and what they are talking about.

The high-level function of sentiment analysis is the last step, determining and applying sentiment on the entity, theme, and document levels.

Low Level

Tokenization: ML+Rules
PoS Tagging: Machine Learning
Chunking: Rules
Sentence Boundaries: ML+Rules
Syntax Analysis: ML+Rules

Mid-Level

Entities: ML+Rules to determine "Who, What, Where"
Themes: Rules "What's the buzz?"
Topics: ML+Rules "About this?"
Summaries: Rules "Make it short"
Intentions: ML+Rules "What are you going to do?"

High Level

Apply Sentiment: ML+Rules "How do you feel about that?"

You can see how this system pans out in the chart in Figure 8.4.

8.5 MACHINE TRANSLATION

Given the flexibility of human language, automatic translation or MT is likely one of the most difficult artificial intelligence projects. Historically, rule-based systems were employed for this activity, but statistical methods were introduced in the 1990s to replace them. Deep neural network models have recently achieved state-of-the-art outcomes in the field of neural MT (Brownlee, 2017).

8.5.1 Neural MT's Evolution

MT had been a skeptic for William Mamane, Head of Digital Marketing at Tomedes, a professional language services organization. However, the quality of MT has been steadily improving. MT does not yet match the accuracy of a professional mother tongue linguist, but AI and MT definitely have a position in the translation services value chain."

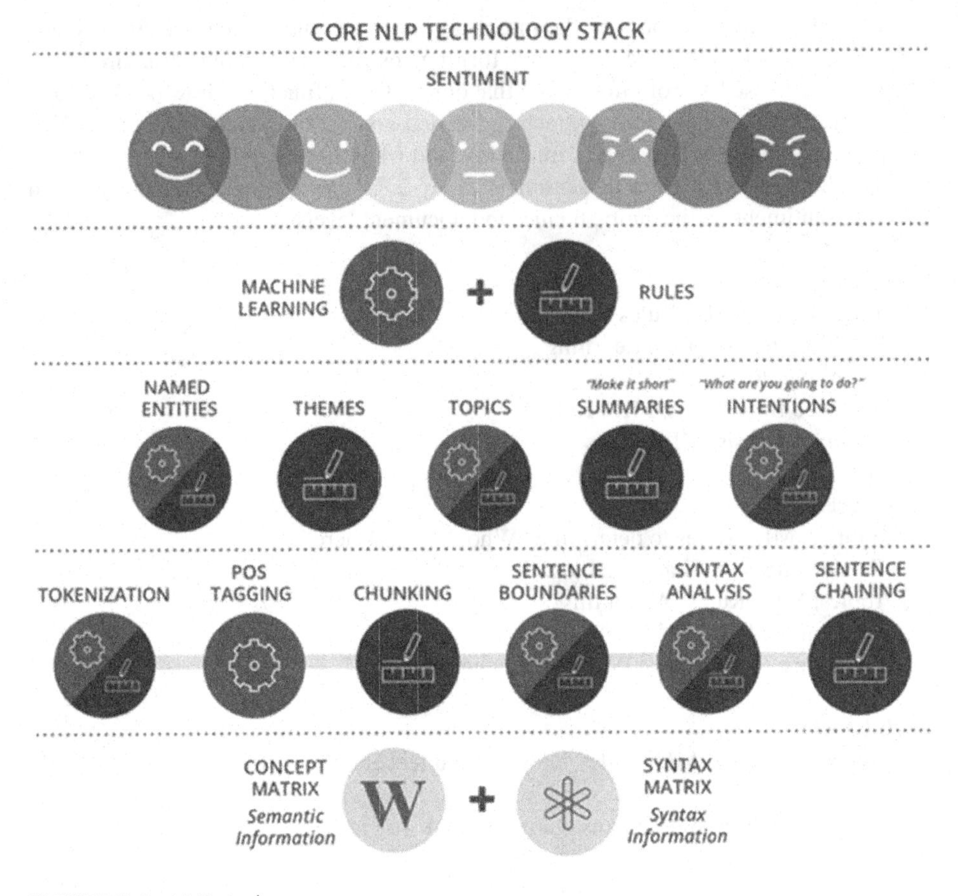

FIGURE 8.4 NLP stack.

Beginnings of AI relate to MT to follow this growth. At its most basic level, MT employs algorithms to replace words in one language with those in another. This is insufficient for a proper translation.

Both the source and target languages require a thorough understanding of entire sentences. MT is the process of decoding a source language and encoding its meaning in a destination language.

There are several techniques to tackling this problem, one of which is to use statistics to select the best translation for a particular sentence. Others use established rules to choose the most likely interpretation. However, even the greatest MT engines do not sound natural in complicated linguistic forms like fiction or other genres of literature.

For some purposes, machines do better with structured language. Weather data, financial reports, government regulations, legal paperwork, and sports results are just a few examples. In these situations, idioms and language are limited. There are linguistic structures and formats that are formulaic.

8.5.2 Replacing an Algorithm with a System

MT is already in use on a regular basis in. There were some systems that were able to recognize proper names, resolve ambiguities, and understand idioms. The supervisory, editorial, or auditing position, on the other hand, is less hard and time-consuming than translation (Koret, 2019).

8.5.3 MT with Neural Networks

A neural network created artificially is used in neural machine translation (NMT). When translating, this deep learning algorithm considers entire sentences rather than just individual words. The memory used by neural networks is a fraction of that required by statistical approaches. They are much more efficient.

The NMT approaches that are defined as bidirectional recurrent neural network, or RNN, are used in the most recent NMT techniques. These networks include an encoder that generates a source sentence for a decoder, which is a second RNN. The words that should appear in the target language are predicted by a decoder. This method is used by Google in the NMT that powers Google Translate. Microsoft's Microsoft Translator and Skype Translator both employ RNN. Harvard's NLP department recently launched OpenNMT, an open-source neural MT system, with the goal of realizing the long-held ambition of simultaneous translation. Facebook is conducting substantial research with open-source NMT, learning from its users' words.

8.5.3.1 Google Translate

Google Translate is a text translation tool provided by Google that is free and multilingual. It has a website interface as well as Android and iOS mobile apps. Developers can use its API to create browser extensions and software apps. Over 100 languages, both living and dead, are supported by Google Translate. As of May 2017, it was serving over 500 million individuals every day. It was translating more than 100 billion words per day as of 2018.

Google Translate is coming closer to being as reliable as human translation. Last year, Google conducted a survey in which native speakers of each language were asked to score Google Translate's translation on a scale of 0–6: it received a 5.43 on a scale of one to ten. Languages have different levels of performance. When English is the target language and the source language is European, for example, Google Translate performs best.

8.5.3.2 Translator by Microsoft

Microsoft Translator is a multilingual MT cloud service that works with a variety of consumer, developer, and enterprise applications. A free tier of the Translator Text API allows for two million characters each month. It includes paid tiers that allow billions of characters to be translated per month.

The duration of the original audio stream is used to measure the duration of speech translation via Microsoft Speech services. When it comes to live translation, Microsoft adopts a fresh approach. Conversation starters are given a code to pass on to other participants.

8.5.3.3 Facebook Translator

As of 2017, Facebook used neural networks to perform roughly 4.5 billion automatic translations each day. For its 2 billion-plus users, it ditched phrase-based MT. Convolutional neural networks (CNNs) were chosen above RNNs by Facebook. It was believed that the localized/translated text would resemble natural English more closely.

CNNs approach information in a hierarchy, whereas RNNs handle data in a linear and methodical manner. Non-linear data relationships can be recognized thanks to the hierarchy. This method has already been proven to be useful in machine vision, which is superior to human eyesight. It understands context better when translating.

The Facebook Translator's multi-hop attention capability is a huge benefit. The way humans translate is modeled by attention. Here they don't break out a statement and then translate it all at once. Instead, they double-check and triple-check its meaning. This is a procedure that CNN imitates. It evaluates the sentence several times before deciding what to translate first.

8.6 CONCLUSION

Language is messy and complex. The meaning varies from speaker to speaker and listener to listener. Machine learning can be a good solution for analyzing text data. In fact, it's vital – purely rules-based text analytics is a dead-end. But it's not enough to use a single type of machine-learning model. Certain aspects of machine learning are very subjective. You need to tune or train your system to match your perspective.

The best way to do machine learning for NLP is a hybrid approach: many types of machine-learning work in tandem with pure NLP code. With the rapid pace of change of NMT, one might think so. However, the reality is different. Translators of Facebook, Microsoft, and Google do well at producing approximate meanings. NMTs can assist in translating while skilled linguists can finish and polish the translation output. Future translators will be more often working with artificial intelligence rather than against it.

BIBLIOGRAPHY

Brownlee, J. (2017, December 29). A gentle introduction to neural machine translation. Retrieved from Machine Learning Mastery: https://machinelearningmastery.com/introduction-neural-machine-translation/.

Collins, M., & Koo, T. (2005). Discriminative reranking for natural language parsing. *Computational Linguistics* 31(1): 25–69.
Emms, M., & Luz, S. (2007). *Machine Learning for Natural Language Processing*.
Koret, R. (2019, November 02). Machine learning for translation: What's the state of the language art? Retrieved from ReadWrite.

Text and Multimedia Mining through Machine Learning

9

T. Kohilakanagalakshmi
Dayananda Sagar Institutions

K.R. Radhakrishnan
St. Joseph's College of Engineering

Salini Suresh
Dayananda Sagar Institutions

Ruchi Nanda
IIS (deemed to be University)

Contents

9.1 Introduction 136
9.1.1 About Text Mining 136
9.1.2 About Multimedia Mining 137
9.1.3 What Exactly Is Machine Learning 138
9.2 Text Mining and Machine Learning 138
9.2.1 Text Mining Fundamental Principles 138
9.2.2 Text Mining Architecture and Its Process 139

DOI: 10.1201/9781003051022-9

9.2.2.1 Information Retrieval 140
9.2.2.2 Information Extraction 140
9.2.2.3 Choosing ML Algorithms 141
9.2.3 Text Mining Techniques 141
9.2.3.1 Word Frequency Analysis 141
9.2.3.2 Collocation Analysis 141
9.2.3.3 Concordance Analysis 142
9.2.4 Feature Selection Using Machine Learning 142
9.2.4.1 Multivariate Relative Discrimination Criterion 142
9.2.4.2 Minimal Redundancy-Maximal New Classification Information 142
9.2.5 Feature Extraction Using Machine Learning 142
9.2.5.1 Bag of Words (BOW) 142
9.2.5.2 TF-IDF 143
9.2.5.3 Word2Vec 143
9.2.6 Machine Learning Algorithms for Text Mining 144
9.2.7 Accuracy, Precision, Recall, F1 Score, and Cross-Validation 145
9.2.8 Challenges of ML Text Analysis 147
9.3 Multimedia Mining and Machine Learning 147
9.3.1 Multimedia Mining Process 147
9.3.2 Machine Learning Algorithms for Multimedia Mining 149
9.4 Conclusion 153
Bibliography 153

Abstract

The process of mining in a certain field of study was taken into consideration of deep analysis in the field of multimedia, which comprises audio, video, text, images, and so on. This chapter helps to understand the concepts of text mining, video, and audio mining with the perspectives of the working principles and its extension in the work relation with machine learning. Many works have been done in the form of research by the industries in order of providing products to the society. We have listed the core concern of the algorithm of machine learning to be implemented in this field of mining.

9.1 INTRODUCTION

9.1.1 About Text Mining

Generally utilized in information-driven associations, text mining is the way toward inspecting enormous assortments of archives to find new data or help answer explicit examination questions.

Text mining distinguishes realities, connections, and affirmations that would in some way or another stay covered in the mass of printed large information. When separated, this data is changed over into an organized structure that can be additionally examined or introduced straightforwardly utilizing bunched HTML tables, mind maps, graphs, and so on. Text mining utilizes an assortment of philosophies to handle the content, perhaps the most significant of these being Natural Language Processing (NLP).

The organized information made by text mining can be incorporated into data sets, information stockrooms, or business knowledge dashboards and utilized for enlightening, prescriptive, or prescient investigation.

9.1.2 About Multimedia Mining

Interactive media information mining is the disclosure of intriguing examples from sight and sound data sets that store and oversee huge assortments of interactive media objects, including picture information, video information, sound information, just as grouping information and hypertext information containing text, text markups, and linkages. Mixed media information mining is an interdisciplinary field that coordinates picture preparing and understanding, PC vision, information mining, and example acknowledgment.

Issues in interactive media information mining incorporate substance-based recovery and comparability search, speculation, and multidimensional examination. Media information 3D shapes contain extra measurements and measures for sight and sound data. Different points in media mining incorporate order and forecast investigation, mining affiliations, and video and sound information mining.

Picture mining is the procedure to identify irregular examples and concentrate verifiable and helpful information from pictures that are stored in huge data sets. It manages to make the relationship between various pictures from huge picture databases. The applications territories of Image mining are clinical determination, distant detecting, agribusiness, businesses, and space research and furthermore take care of hyper otherworldly pictures.

Video structure mining. The primary goal of video structure mining is the identification of the substance design and examples to carry out the quick arbitrary access of the video data set. Video structure mining is defined as the process of finding the central rationale structure from the preprocessed video program embracing information mining methods such as classification, grouping, and affiliation rule.

It is fundamental to examine video content semantically and use multimodality data to overcome any issues between human semantic ideas and PC low-level highlights from both the video successions and sound transfers. Video structure mining is executed in the accompanying advances:

1. video shot detection,
2. scene identification,
3. scene grouping, and
4. occasion mining.

The momentum explores its center around mining object semantic data and occasion location.

9.1.3 What Exactly Is Machine Learning

Artificial intelligence is a subfield of electronic thinking (AI). The target of AI overall is to appreciate the development of data and fit that data into models that can be seen and utilized by people.

Disregarding the way that AI is a field inside computer programming, it fluctuates from standard computational philosophies. In standard handling, computations are sets of unequivocally adjusted rules used by PCs to figure or issue tackle. Computer-based intelligence estimations rather contemplate PCs to get ready on data information sources and use quantifiable assessment to yield regards that fall inside a specific reach. Thus, AI supports PCs in building models from test data to automate dynamic cycles subject to data inputs.

Any development customer today has benefitted from AI. Facial affirmation advancement grants electronic media stages to help customers tag and offer photos of buddies. Optical character affirmation (OCR) advancement changes over pictures of text into compact sort. Proposition engines, constrained by AI, prescribe what movies or organization shows to watch next subject to customer tendencies. Self-driving vehicles that rely upon AI to investigate may after a short time be available to clients.

Computer-based intelligence is a field that is making progress continually. Thus, there are a couple of examinations to recollect as you work with AI frameworks or explore the impact of AI measures.

9.2 TEXT MINING AND MACHINE LEARNING

9.2.1 Text Mining Fundamental Principles

Text mining, otherwise called text information mining, is the way toward changing unstructured content into an organized configuration to distinguish important examples and new experiences. By applying progressed scientific procedures, for example, Naïve Bayes, Support Vector Machines (SVM), and other profound learning calculations, organizations can investigate and find shrouded connections inside their unstructured information.

Since 80% of information on the planet dwells in an unstructured configuration, text mining is an incredibly significant practice inside associations. Text mining devices and common language handling (NLP) strategies, similar to data extraction (PDF, 127.9 Kb) (connect live outside of IBM), permit us to change unstructured records into an organized organization to empower investigation and the age of great experiences. This, thus, improves the dynamic of associations, prompting better business results.

Text is perhaps the most well-known information type inside data sets. Contingent upon the data set, this information can be coordinated as follows:

Organized information – This information is normalized into a plain organization with various lines and segments, making it simpler to store and measure for investigation and AI calculations. Organized information can incorporate data sources, for example, names, locations, and telephone numbers.

Unstructured information – This information doesn't have a predefined information design. It can incorporate content from sources, similar to web-based media or item surveys, or rich media designs like video and sound records.

Semi-organized information – As the name recommends, this information is a mix of organized and unstructured information designs. While it has some association, it needs more design to meet the prerequisites of a social data set. Instances of semi-organized information incorporate XML, JSON, and HTML records.

The five crucial advances associated with text mining are as follows:

1. Gathering unstructured information from different information sources such as plain content, site pages, pdf documents, messages, and online journals.
2. Detect and eliminate abnormalities from information by directing pre-handling and purging tasks. Information purifying permits you to separate and hold the significant data covered up inside the information and to help recognize the underlying foundations of explicit words. For this, we get various content mining devices and text mining applications.
3. Convert all the important data extricated from unstructured information into organized organizations.
4. Analyze the examples inside the information utilizing the Management Information System (MIS).
5. Store all the significant data into a safe data set to drive pattern investigation and improve the dynamic interaction of the association.

9.2.2 Text Mining Architecture and Its Process

The cycle of text mining includes a few exercises that empower you to conclude data from unstructured content information. Before you can apply distinctive content mining procedures, you should begin with text preprocessing, which is the act of cleaning and changing content information into a usable configuration. This training is a central part of regular language preparation (NLP), and it generally includes the utilization of strategies, for example, language distinguishing proof, tokenization, grammatical feature labeling, lumping, and linguistic structure parsing to organize information fittingly for investigation. At the point when text preprocessing is finished, you can apply text mining calculations to get bits of knowledge from the information. A portion of these regular content mining methods is included in Figure 9.1.

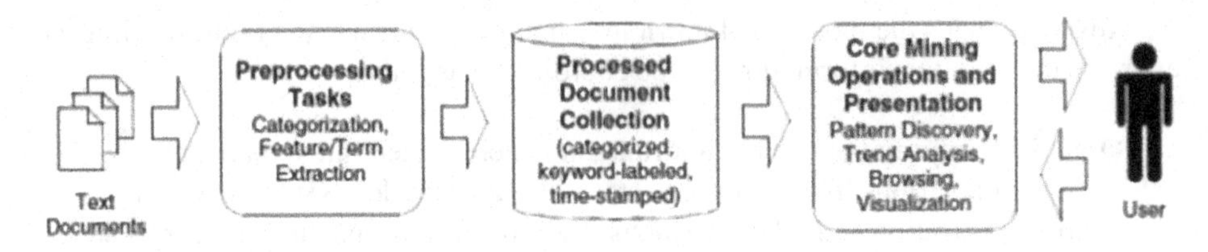

FIGURE 9.1 Basic working principle of text mining.

9.2.2.1 Information Retrieval

Data recovery (IR) returns important data or reports dependent on a pre-characterized set of inquiries or expressions. IR frameworks use calculations to follow client practices and recognize important information. Data recovery is normally utilized in library index frameworks and well-known web crawlers, similar to Google. Some normal IR sub-undertakings include the following:

- Tokenization – This is the way toward breaking out long-structure text into sentences and words called "tokens." These are, at that point, utilized in the models, similar to pack of words, for text grouping and archive coordinating assignments.
- Stemming –This alludes to the way toward isolating the prefixes and additions from words to determine the root word structure and importance. This method improves data recovery by lessening the size of ordering documents.
- Removing superfluous accentuation, labels.
- Removing stop words — continuous words, for example, "the," "is," and so on that don't have explicit semantics.

9.2.2.2 Information Extraction

Data extraction (IE) surfaces the important bits of information while looking through different records. It additionally centers around removing organized data from free content and putting away these elements, ascribes, and related data in a data set. Normal data extraction sub-assignments include:

- **Feature choice**, or characteristic determination, is the way toward choosing the significant highlights (measurements) to contribute the most to the yield of a prescient examination model.
- **Feature extraction** is the way toward choosing a subset of highlights to improve the exactness of an order task. This is especially significant for dimensionality decrease.
- **Named-entity recognition (NER)** otherwise called substance ID or element extraction plans to discover and sort explicit elements in the content, for example, names or areas. For instance, NER recognizes "California" as an area and "Mary" as a lady's name.

- **Sentiment investigation** Sentiment examination, or assessment mining, recognizes and contemplates feelings in the content. The feelings of the creator are significant for getting writings. SA permits to arrange assessment extremity about another item or survey a brand's standing. It can likewise be applied to audits, reviews, and online media posts. The confirmation SA is that it can viably break down even snide remarks.
- **Topic investigation** Topic displaying orders messages by subject and can make people's lives simpler in numerous areas. Discovering books in a library, merchandise in the store, and client service tickets in the CRM would be unimaginable without it. Text classifiers can be custom fitted to your requirements.
- **Content labeling** Students and educators, legal counselors, researchers, and lab collaborators would all be able to profit by the utilization of text order innovation. Since they are managing huge measures of unstructured information day by day, labeling and arranging writings into classifications would make their lives a lot simpler.

9.2.2.3 Choosing ML Algorithms

There are different ways to deal with building ML models for different content that put together applications depending on the issue space and information accessible.

Old-style ML approaches such as "Gullible Bayes" or "Backing Vector Machines" for spam sifting have been broadly utilized. Profound learning methods are giving better outcomes for NLP issues such as supposition examination and language interpretation. Profound learning models are delayed to prepare and it has been seen that for straightforward content arrangement issues old-style ML approaches also give comparable outcomes with faster preparing time.

9.2.3 Text Mining Techniques

9.2.3.1 Word Frequency Analysis

This procedure permits you to quantify how oftentimes words show up in the text. This is actually how people can recognize the subject of the content and lead estimation investigation. We realize that "fascinating" generally alludes to positive impressions. So in the event that you see this word in an audit, it implies that the customer is fulfilled. Nonetheless, this strategy isn't delicate to mockery, which may influence the overall consequences of your investigation.

9.2.3.2 Collocation Analysis

Two, three, or more words that are regularly utilized together in discourse are called collocations. A similar word in various collocations can have various implications. "Free" signifies "freed" as in "free soul." "Free" can likewise signify "gratis." "Free" is substantially more prone to show up on an online store's site along with "delivering," instead of with "soul" or even independently. Consideration makes the semantic investigation more exact.

9.2.3.3 Concordance Analysis

A concordance is a table that shows various implications of a similar word in various settings. Here is a model from a context-oriented word reference demonstrating how various individuals utilize "concordance."

Logical word references are useful for language students since they contain genuine models demonstrating various methods of utilizing a similar word. They are similarly as useful for machine interpretation and discourse age frameworks. Concordance and collocation investigation is helpful for watchword meaning disambiguation.

9.2.4 Feature Selection Using Machine Learning

9.2.4.1 Multivariate Relative Discrimination Criterion

While the reason for the highlighted choice is to choose a reduced element subset with maximal discriminative ability, which requires having a high significance to the class name and low excess inside the chose include subset. MRDC is proposed to consider both pertinence and repetition ideas in its assessment cycle. On how MRDC functions initially process the significance of each component utilizing relative discrimination criterion measure, and afterward, Pearson connection is utilized to register relationship esteems between highlights.

9.2.4.2 Minimal Redundancy-Maximal New Classification Information

There are two gatherings for highlight choice, one spotlight on limiting excess, and the other augmenting new arrangement data. The strategies that emphasize limiting element excess don't think about new characterization data and the other way around, along these lines bringing about chosen highlights with a lot of new grouping data yet high repetition, or highlights with low repetition minimal new order data.

Gao et al. proposed a crossbreed including determination technique that coordinates two gatherings of highlight choice strategies by thinking about two sorts of highlight repetition and conquers the constraints that are referenced previously.

MR-MNCI thinks about both new arrangement data and highlight repetition. Highlight excess can be separated into two classifications: class-subordinate element repetition and class-free component repetition. These two kinds of highlight excess are both critical for including determination.

9.2.5 Feature Extraction Using Machine Learning

9.2.5.1 Bag of Words (BOW)

In-text preparing, expressions of the content address discrete, straight out highlights. How would we encode such information in a manner that is fit to be utilized by the calculations? The planning from text-based information to genuine esteemed vectors is called include extraction. Perhaps the easiest method to mathematically address text is Bag of Words.

We make the rundown of one-of-a-kind words in the content corpus called jargon. At that point, we can address each sentence or report as a vector with each word addressed as 1 for present and 0 for missing from the jargon. Another portrayal can tally the occasions each word shows up in a report.

- The most well-known methodology is utilizing the Term Frequency-Inverse Document Frequency (TF-IDF) strategy.
- Term Frequency (TF)=(Number of times term t shows up in an archive)/(Number of terms in the report).
- Inverse Document Frequency (IDF)=log(N/n), where, N is the number of archives and n is the number of reports a term t has shown up in.
- The IDF of an uncommon word is high, though the IDF of a successive word is probably going to be less, consequently having the impact of featuring unmistakable words.
- We compute TF–IDF estimation of a term as=TF * IDF.

9.2.5.2 *TF-IDF*

An issue with the pack of words approach is that the words with higher recurrence become prevailing in the information. These words may not give a lot of data to the model. Also, because of this issue, area explicit words that don't have a bigger score might be disposed of or overlooked.

To determine this issue, the recurrence of the words is rescaled by thinking about how often the words happen on the whole archive. Because of this, the scores for successive words are likewise continuous among all the reports are decreased. This method of scoring is known as Term Frequency-Inverse Document Frequency.

- TF is the recurrence of the word in the current archive.
- Inverse Document Frequency (IDF) is the score of the words among all the records.

These scores can feature the remarkable words that address needful data in a predetermined record. Hence, the IDF of an inconsistent term is high, and the IDF of an incessant term is low.

9.2.5.3 *Word2Vec*

Word2Vec is utilized to develop word embeddings.

The models made by utilizing Word2Vec are shallow importance two-layer neural organizations. When prepared, they replicate the semantic settings of words. The model takes a gigantic corpus of text as information. It at that point makes a vector space which is a rule of many measurements.

- Each particular word in the corpus is permitted with a relating vector in the space. The words with regular settings are put close by in vector space.

- Word2Vec can utilize one of the two structures: persistent skip-gram or consistent bag of words (CBOW).
- In the persistent skip-gram, the current word is considered to foresee the adjoining window of setting words.
- In this engineering, the close-by setting words are viewed more vigorously than words with far offsetting.
- In the persistent sack of words engineering, the succession of setting words don't affect the forecast as it depends on the pack of words model.

9.2.6 Machine Learning Algorithms for Text Mining

There are many AI calculations utilized in the content order. The most habitually utilized are the Naïve Bayes (NB) group of calculations, SVM, and profound learning calculations.

The **NB** group of calculations depends on Bayes' Theorem and the restrictive probabilities of the event of the expressions of an example text inside the expressions of a bunch of writings that have a place with a given tag.

- Vectors that address messages encode data about how likely it is for the words in the content to happen in the writings of a given tag.
- With this data, the likelihood of a book having a place with some random tag in the model can be registered.
- Once the entirety of the probabilities has been figured for an information text, the characterization model will restore the tag with the most noteworthy likelihood as the yield for that input.
- One of the principle preferences of this calculation is that outcomes can be very acceptable regardless of whether there's hardly any preparation information.

SVM is an estimation that can detach a vector space of named compositions into two subspaces: one space that contains an enormous segment of the vectors that have a spot with a given tag and another subspace that contains most of the vectors that don't have a spot with that one.

- Classification models that utilize SVM at their center will change messages into vectors and will figure out what side of the limit isolates the vector space for a given label those vectors have a place with.
- Based on where they land, the model will know whether they have a place with a given tag or not.
- The most significant favorable position of utilizing SVM is that outcomes are generally better compared to those got with NB.
- However, more computational assets are required for SVM.

Deep Learning is a bunch of calculations and procedures that utilize "counterfeit neural organizations" to handle information much as the human cerebrum does. These calculations utilize immense measures of preparing information (a large number of guides) to create semantically rich portrayals of writings which would then be able to be taken care of into AI-based models of various types that will make considerably more precise expectations than conventional AI models:

Hybrid Systems: Mixture frameworks normally contain AI-based frameworks at their centers and rule-based frameworks to improve the forecasts.

Evaluation: Understanding what they mean will give you a more clear thought of how great your classifiers are at examining your writings. Classifier execution is typically assessed through standard measurements utilized in the AI field: exactness, accuracy, review, and F1 score.

It is additionally essential to comprehend that assessment can be performed over a fixed testing set (e.g., a bunch of writings for which we know the normal yield labels) or by utilizing cross-approval (e.g., a technique that parts your preparation information into various overlap so you can utilize a few subsets of your information for preparing purposes and some for testing purposes, see underneath).

9.2.7 Accuracy, Precision, Recall, F1 Score, and Cross-Validation

Accuracy is the number of right forecasts the classifier has made isolated by the all outnumber of expectations. As a rule, exactness alone is definitely not a decent marker of execution.

- For model, when classifications are imbalanced, that is, when there is one classification that contains a lot a bigger number of models than the entirety of the others, foreseeing all writings as having a place with that classification will restore high precision levels.
- This is known as the exactness oddity. To improve the thought of the exhibition of a classifier, you should consider exactness and review all things being equal.

Precision states the number of writings was anticipated effectively out of the ones that were anticipated as having a place with a given tag.

- In different words, accuracy takes the number of writings that were effectively anticipated as sure for a given tag and partitions it by the number of writings that were anticipated (accurately and inaccurately) as having a place with the tag.
- We need to remember that accuracy just gives data about the situations where the classifier predicts that the content has a place with a given tag.
- This may be especially significant, for instance, on the off chance that you might want to create robotized reactions for client messages.

- In this case, before you send a computerized reaction, you need to know without a doubt that you will send the correct reaction, isn't that so? All in all, if your classifier says the client message has a place with a particular kind of message, you might want the classifier to make the correct estimate. This implies you might want a high accuracy for that sort of message.

The recall states the number of writings was anticipated accurately out of the ones that ought to have been anticipated as having a place with a given tag.

- In different words, the review takes the number of writings that were accurately anticipated as sure for a given tag and partitions it by the number of writings that were either anticipated effectively as having a place with the tag or that were inaccurately anticipated as not having a place with the tag.
- The review may demonstrate to be valuable when directing help passes to the proper group, for instance.
- A robotized framework may want to identify whatever number of tickets could be expected under the circumstances for a basic tag (for instance, tickets about "Shocks/Downtime") to the detriment of making some wrong expectations en route.
- In this case, making an expectation will help play out the underlying steering and address the greater part of these basic issues ASAP.
- If the expectation is wrong, the ticket will get rerouted by an individual from the group. When handling a great many tickets each week, high review (with great degrees of exactness too, obviously) can save uphold groups a decent arrangement of time and empower them to settle basic issues quicker.

The **F1 score** is the symphonious method for accuracy and review. It reveals to you how well your classifier performs if equivalent significance is given to accuracy and review. All in all, the F1 score is a greatly improved marker of classifier execution than precision is.

Cross-approval is often used to assess the presence of text classifiers. The strategy is straightforward:

- First of all, the preparation data set is arbitrarily part into a few equivalent length subsets (e.g., four subsets with 25% of the first information each).
- Then, all the subsets aside from one are utilized to prepare a classifier (for this situation, three subsets with 75% of the first information), and this classifier is utilized to anticipate the writings in the leftover subset.
- Next, all the exhibition measurements are figured (e.g., exactness, accuracy, review, and F1).
- Finally, the interaction is rehashed with another testing fold until all the folds have been utilized for testing purposes.
- Once all folds have been utilized, the normal execution measurements are figured and the assessment cycle is done.

9.2.8 Challenges of ML Text Analysis

According to a recent study, about 80% of all data generated in enterprises is in the form of texts. A lot of insights can be drawn from it.

But ML textual analysis also presents some challenges:

Complexity – Transforming text into a format that can be processed by the computer requires several steps. For example, if we are solving a text classification problem, we need to collect the data, detect the keywords in it, define several classes, group the data according to these classes, and describe these processes in mathematical terms. It's challenging both intellectually and in terms of human/money/time resources.

Conceptual struggles – Computers don't understand concepts that are behind words, so working with homographs is difficult for them. Programmers have to come up with some effective tools for word meaning disambiguation to work with sentences such as "Will, will Will will Will Will's will?". Google Translate, for example, cannot cope with this sentence right now.

Understanding culture – Understanding human speech means understanding their emotions. One of the hardest emotions for a computer to grasp is sarcasm. Continuing the topic of disambiguation, the same meaning in different cultures can be expressed by different words such as slang or local variants. What is a "jumper" to a Brit is a "sweater" to an American. A computer program must have experience and cultural background to effectively communicate with speakers who use less conventional forms of language.

9.3 MULTIMEDIA MINING AND MACHINE LEARNING

9.3.1 Multimedia Mining Process

The common information mining measure mentioned in Figure 9.2 comprises a few phases and the general cycle is intrinsically intuitive and iterative. The fundamental phases of the information mining measure are: (i) area getting; (ii) information choice; (iii) cleaning and preprocessing; (iv) finding designs; (v) understanding; and (vi) revealing and utilizing found information.

- The area understanding stage requires figuring out how the aftereffects of information mining will be utilized to assemble all pertinent earlier information prior to mining.

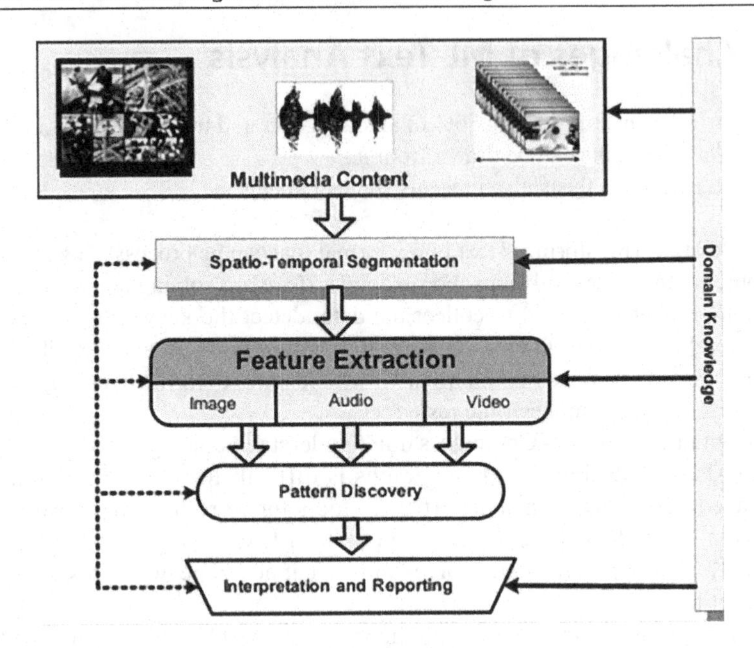

FIGURE 9.2 General working principles of multimedia mining.

- Blind use of information mining procedures without the essential area information regularly prompts the disclosure of unimportant or good for nothing designs.
- The information determination stage requires the client to focus on a data set or select a subset of fields or information records to be utilized for information mining.
- A legitimate space understanding at this stage helps in the recognizable proof of valuable information.
- This is the most tedious phase of the whole information-digging measure for business applications; information is never perfect and in the structure reasonable for information mining.

The next stage in a typical data mining process is the **preprocessing** step:

- This includes incorporating information from various sources and settling on decisions about addressing or coding certain information handle that fill in as contributions to the example disclosure stage.
- Such portrayal decisions are required on the grounds that specific fields may contain information at levels of subtleties not considered appropriate for the example revelation stage.

The **pattern-discovery** stage is the heart of the entire data mining process:

- It is the stage where the hidden patterns and trends in the data are actually uncovered.

- That includes coordinating information from various sources and settling on decisions about addressing or coding certain information handle that fill in as contributions to the example revelation stage.
- Such portrayal decisions are required on the grounds that specific fields may contain information at levels of subtleties not considered appropriate for the example disclosure stage.

The **interpretation stage** of the data mining process:

- It is utilized by the client to assess the nature of disclosure and its incentive to decide if past stages ought to be returned to or not.
- Proper space understanding is essential at this stage to put an incentive on found examples.
- The last phase of the information mining measure comprises detailing and utilizing the found information to create new activities or items and administrations or advertising procedures by and large.

The **spatio-temporal** segmentation step in the architecture is necessitated by the unstructured nature of multimedia data.

This step breaks interactive media information into parts that can be described as far as specific credits or highlights:

- Related to the component extraction step, this progression serves a capacity like that of the preprocessing stage in a commonplace information mining measure.
- In picture information mining, the spatio-fleeting advance just includes picture division.
- Both district and edge-based picture division techniques have been utilized at this stage in various applications.

9.3.2 Machine Learning Algorithms for Multimedia Mining

Regression Algorithms: Regression analysis is part of predictive analytics and exploits the co-relation between dependent (target) and independent variables.

The notable regression models are as follows:

- Linear regression,
- Logistic regression,
- Stepwise regression,
- Ordinary least squares regression (OLSR),
- Multivariate adaptive regression splines (MARS),
- Locally estimated scatterplot smoothing (LOESS), etc.

Simple linear regression is a statistical method that enables users to summarize and study relationships between two continuous (quantitative) variables.

- Linear relapse is a straight model wherein the model expects a direct connection between the info factors and the single yield variable.
- Here they can be determined from a direct blend of the information factors.
- When there is a solitary info variable, the strategy is called straightforward direct relapse.
- When there are different info factors, the strategy is alluded to as numerous straight relapses.

Linear regression is applicable in some of the most popular applications of the linear regression algorithm – financial portfolio prediction, salary forecasting, real estate predictions, and in traffic in arriving at ETAs.

Logistic Regression: Perhaps the most usually utilized relapse methods in the business, which are widely applied across extortion discovery, Visa scoring, and clinical preliminaries, any place the reaction is twofold have a significant preferred position.

One of the significant potential gains is of this mainstream calculation is that one can incorporate beyond what one ward variable which can be persistent or dichotomous.

The other significant favorable position of this administered AI calculation is that it gives a quantified incentive to quantify the strength of the relationship as per the remainder of the factors.

Despite its prominence, analysts have drawn out its impediments, referring to an absence of hearty procedure and furthermore an incredible model reliance.

Today endeavors send logistic regression to foresee house estimations in the land business, client lifetime esteem in the protection area and are utilized to deliver a nonstop result, for example, regardless of whether a client can purchase/will purchase situation.

Multivariate Regression Algorithm: This is a summed-up straight displaying method that might be utilized to demonstrate a solitary reaction variable that has been recorded on, in any event, a span scale.

The method might be applied to single or various informative factors and furthermore unmitigated illustrative factors that have been properly coded.

Ordinary least squares (OLS) relapse is one of the significant methods used to investigate information and structure the premise of numerous different strategies.

The convenience of the strategy can be significantly stretched out with the utilization of faker variable coding to incorporate assembled illustrative factors and information change strategies.

OLS relapse is especially amazing as it was moderately simple to likewise check the model suspicion, for example, linearity, consistent fluctuation, and the impact of exceptions utilizing straightforward graphical techniques.

Ordinary Least Squares Regression (OLSR): This is a summed-up straight displaying method that might be utilized to demonstrate a solitary reaction variable that has been recorded on, in any event, a span scale.

The method might be applied to single or various informative factors and furthermore unmitigated illustrative factors that have been properly coded.

OLS relapse is one of the significant methods used to investigate information and structure the premise of numerous different strategies.

The convenience of the strategy can be significantly stretched out with the utilization of faker variable coding to incorporate assembled illustrative factors and information change strategies.

OLS relapse is especially amazing as it was moderately simple to likewise check the model suspicion, for example, linearity, consistent fluctuation, and the impact of exceptions utilizing straightforward graphical techniques.

Instance-based learning creates classification forecasts utilizing just specific occurrences.

Occasion-based learning calculations don't keep a bunch of deliberations received from specific examples.

This methodology expands the closest neighbor calculation, which has enormous capacity necessities.

It portrays how stockpiling necessities can be significantly reduced with, probably, minor sacrifices in learning rate and classification precision.

While the capacity decreasing calculation performs well on a few certifiable data sets, its exhibition debases quickly with the degree of property commotion in preparing examples.

Accordingly, specialists broadened it with a significance test to recognize boisterous cases.

This all-encompassing calculation's exhibition debases effortlessly with expanding commotion levels and contrasts well and a clamor open-minded choice tree calculation.

The **k-closest neighbor** choice guideline has being regularly utilized in these example acknowledgment issues.

One of the difficulties that emerge while using this procedure is that each marked example is given equivalent significance in choosing the class participations of the example to be classified, paying little mind to their averageness.

Three techniques for appointing fluffy enrollments to the marked examples are proposed, and exploratory outcomes and correlations with the fresh form are introduced.

The k-nearest-neighbors technique for classification is perhaps the least complex strategy in AI and is an extraordinary method to acquaint yourself with AI and classification by and large.

At its most fundamental level, it is basically classification by finding the most comparative information focuses on the preparation information and making informed speculation dependent on their classifications.

Albeit easy to comprehend and actualize, this technique has seen wide application in numerous areas, for example, in proposal frameworks, semantic looking, and irregularity identification.

Regularization is basically the way toward neutralizing overfitting or subside the anomalies.

Regularization is only a straightforward yet incredible modification that is increased with other existing ML models, commonly regressive models.

It smoothes up the relapse line by censuring any bowed of the bend that attempts to coordinate the exceptions.

Models: Ridge regression, least absolute shrinkage and selection operator (LASSO), elastic net, least-angle regression (LARS), and so forth.

Ensemble methods are meta-calculations that consolidate a few AI methods into one prescient model to diminish fluctuation (stowing), inclination (boosting), or improve forecasts (stacking).

Gathering techniques can be partitioned into two gatherings:

1. Consecutive outfit strategies where the base students are created successively (e.g., AdaBoost). The essential inspiration of consecutive strategies is to abuse the reliance between the base students. The general presentation can be helped by weighing beforehand mislabeled models with higher weight.
2. Equal gathering techniques where the base students are created equal (e.g., Arbitrary Forest). The essential inspiration of equal strategies is to misuse autonomy between the base students since the blunder can be diminished significantly by averaging.

Most outfit techniques utilize a solitary base learning calculation to create homogeneous base students, for example, students of a similar sort, prompting homogeneous troupes.

A few strategies utilize heterogeneous students, i.e., learners of various kinds, prompting heterogeneous gatherings.

For gathering strategies to be more precise than any of its individual individuals, the base students must be just about as exact as could really be expected and as assorted as could be expected.

The principle motivation behind a group technique is to incorporate the projections of a few more fragile assessors that are independently prepared to help up or improve generalizability or vigor over a solitary assessor.

The sorts of students and the way to fuse them are deliberately picked to expand the precision.

Models: Boosting, Bootstrapped Aggregation (Bagging), AdaBoost, Stacked Generalization (mixing), Gradient Boosting Machines (GBM), Gradient Boosted Regression Trees (GBRT), Random Forest, Extremely Randomized Trees, and so on. Ensembling contains two primary methods:

Random Forest is a flexible, simple-to-utilize AI calculation that produces, even without hyper-boundary tuning, an extraordinary outcome more often than not.

- It is in like manner may be the most used estimations since it is clear and the way that it will, in general, be used for both classification and backslide endeavors.
- In this post, you will acknowledge, how the self-assertive forest count functions and a couple of other critical things about it.
- Irregular forest is a controlled learning computation.
- It makes boondocks and makes it somehow unpredictable.
- The boondocks it creates is a gathering of Decision Trees, usually set up with the stowing methodology.
- The by and large thought about the stowing methodology is that a mix of learning models constructs the overall result.

One major favorable position of arbitrary woods is that it tends to be utilized for both classification and relapse issues, which structure most of current AI frameworks.

Gradient Boosting Machines are a group of ground-breaking AI strategies that have demonstrated impressive achievement in a wide scope of useful applications.

They are exceptionally adjustable to the specific requirements of the application, such as being mastered concerning distinctive misfortune capacities.

The standard thought behind this calculation is to build the new base students to be maximally connected with the negative slope of the misfortune work, related to the entire group.

The misfortune capacities applied can be discretionary, yet to give a superior instinct, if the blunder work is the exemplary squared-mistake misfortune, the learning method would bring about successive mistake fitting.

As a rule, the decision of the misfortune work is up to the specialist, with both a rich assortment of misfortune capacities determined up until this point and with the chance of actualizing one's own undertaking specific misfortune.

9.4 CONCLUSION

This chapter is having the content of the working principles and the general structure of the text and multimedia mining and its strategies with machine learning – starting with a description of what is mining, what are text mining, and multimedia mining, the architecture, and the basic working principles of the text and multimedia mining. We have included the complete description of the algorithms that play a major role in this machine learning along with its applications. In the final part, we have stated how the growth is required and also the recent trends that were made and yet waiting for the growth of results in this field.

BIBLIOGRAPHY

Mukhopadhyay, K., Sil, J., and Banerjea, N.R. "A competency based management system for sustainable development by innovative exclude quotes on exclude bibliography on exclude matches off organizations: A proposal of method and tool", *Vision The Journal of Business Perspective*, 2011.

Salloum, S.A., Al-Emran, M., Monem, A.A. and Shaalan, K, "Using text mining techniques for extracting information from research articles", *Studies in Computational Intelligence*, January 2018.

Shrishrimal, P.P., Deshmukh, R.R. and Waghmare, V.B. Multimedia data mining: A review. *Conference: International Conference On Recent Trends and Challenges in Science and Technology at: Padmashri Vikhe Patil College of Arts, Science & Commerce Pravaranagar*, August 2014.

Suneetha, V., Suresh, S., and Jhananie, V. "A novel framework using apache spark for privacy preservation of healthcare big data", *2020 2nd International Conference on Innovative Mechanisms for Industry Applications (ICIMIA)*, 2020.

Talib, R., Hanif, M.K., Ayesha, S. and Fatima, F, "Text mining: Techniques, applications and issues, (IJACSA)", *International Journal of Advanced Computer Science and Applications*, Vol. 7 No. 11, 2016.

Uma, C., Krithika, S., and Kalaivani, C. "A survey paper on text mining techniques", *International Journal of Engineering Trends and Technology (IJETT)*, Vol. 40, No. 4, 225–229, October 2016. ISSN:2231-5381.

Vijayarani, S. and Sakila, A. "Multimedia mining research – An overview", *International Journal of Computer Graphics & Animation (IJCGA)* Vol. 5, No. 1, January 2015.

Yadav, P.K. and Rizvi, S. "An exhaustive study on data mining techniques in mining of multimedia database", *2014 International Conference on Issues and Challenges in Intelligent Computing Techniques (ICICT)*, 2014.

10 Application of IoT and Block Chaining for Business Analysis

K. Sheela, M. Mubina Begum and C. Priya
Vels Institute of Science, Technology and Advanced Studies (VISTAS)

Contents

10.1 Introduction 156
10.2 IoT 156
10.3 Introduction to Collaborating Technologies 157
10.4 Blockchain Technology 157
10.4.1 Blockchain Technology: Powering the Business of the Future 158
10.4.2 New Wave of Economic Opportunity and Digital Innovation 158
10.5 Advantages of Blockchain and IoT Collaboration 159
10.6 Business Analysis 162
10.7 Business Analyst 164
10.8 Application of IoT and Blockchain Technology for Business Analysis 164
10.8.1 Publicity 165
10.8.2 Decentralization 166
10.8.3 Resiliency 166
10.8.4 Security and Speed 166
10.8.5 Cost Saving and Immutability 166
10.8.6 Privacy 167
10.9 Conclusion 168
Bibliography 168

DOI: 10.1201/9781003051022-10

Abstract

Internet of Things (IoT) and blockchain technology have their own uniqueness in their methodology. When we consider the collaboration of technologies, IoT and blockchain play an iconic role in our environment. As we know, blockchain provides great security with its distributed and immutable nature. This integration in Business Process Management (BPM) satisfies many key requirements such as scalability, security and openness. IoT devices embedded in blockchain technology exhibits specified functionalities in various application areas such as business, healthcare, ambient assisted living, banking, agriculture and intelligent transportation. This in turn reflects a huge benefit in business analysis where the data can be stored and accessed securely without worrying about the privacy and integrity of the data. Automation can also be promoted with the integration of artificial intelligence with these technologies. Implementation of IoT faces certain challenges in ensuring privacy and security, which can be tackled by blockchain technology. This chapter briefs out the applications of IoT and blockchain for analysis in the business sector.

10.1 INTRODUCTION

The branch of economic activities such as business to consumer (B2C) and business to business (B2B) services are emerging with the technologies such as blockchain and artificial intelligence, along with harnessing data science capabilities for higher performance and speed from the traditional processes. These technologies renovate everyday experiences for people, making them more customized, classic, productive and impactful.

The Internet of Things (IoT) connects people, locations and productions. In doing the same, it provides opportunities for value creation and encapsulation. Enlightened chips, sensors and actuators are embedded into physical items, each disseminating data to the IoT network. The analytical capabilities of the IoT use this data to convert insights into action, impacting business processes and leading to new ways of working. However, there are still a few more technical and security concerns that remain uninvestigated.

10.2 IoT

IoT is a system of interrelated computing devices, mechanical and digital machines, objects, animals and peoples, which are provided with unique identifiers and the ability to transfer data over the network without requiring human-to-human or human-to-machine intervention. Whether you work in a large or small business, information technology is going to play a large role in your day-to-day tasks. The IoT platform enables the industry to have real-time tracking of assets and the environment. It also puts a critical impact on asset velocity, i.e., the assets linked to the inventory in the business world. The adoption of blockchain technology in the IoT industry is at its rise due to its provenance in security and tracking. Businesses tend to use IT in three main ways: to support basic information processing tasks, to help with decision-making, and to support innovation.

10.3 INTRODUCTION TO COLLABORATING TECHNOLOGIES

Converting information into digitalization is continuously growing and disrupting many aspects of our everyday life. It is becoming crucial to keep an eye on the animal's health and maintain batches with high proficiencies to increase the value of the product and reduce operating revenue. The demand for livestock identification and traceability increased the need for quality management and control of infectious diseases, medication and its effects on the environment and consumer health. For these to IoT helps the farmer. The farmer can receive the information anywhere through the web or mobile application. These data enable farmers to calculate the time and optimum insemination period after the estruses, which also prevents cattle from calving accidents. Furthermore, veterinary health services and owners are notified in real time with an alarm and message in case of any abnormality. The following are the fundamental characteristics of IoT:

Connectivity – Anything can be interlinked with the communication lines of IoT and transmission channels

Heterogeneity – The components and devices of IoT are heterogeneous since they work on different networks and platforms. They can also interact with other devices and services through different networks.

Tremendous scale – As the communication lines need to be managed, the order of magnitude must be higher than the existing one.

Dynamic changes – Swap between connections, disconnection, changes in speed and position may occur at any period.

Safety – IoT must be designed in such a way where the personal data, physical safety and privacy of the transactions are being preserved.

Small devices – These devices are encouraged to become smaller, cheaper, powerful and more reliable over time (Tzafestas, 2018). Wearable devices are also encouraged to improve the business sector along with collaborative technologies. Remote access control will be highly promoted (Sharma et al., 2019).

10.4 BLOCKCHAIN TECHNOLOGY

A blockchain is a chain of blocks that contain information. The data that is stored inside a block depends on the type of blockchain. A block contains confidential information about the sender and the receiver. The main benefit of the blockchain is its quality of preserving the data with full security. Blockchain is the future of modern data transfer technology. Most blockchains are entirely open-source software. This means that anyone and everyone can view its code. This also means that there is no real authority on who controls Bitcoin's code or how it is edited.

The major advantage of the blockchain is its quality of fixity, which makes it secure as well as easy to audit trials. The blockchain can be developed to record virtually anything this is expressible in coding. Many organizations are already adopting this technology, and others are moving toward this technology. In the manufacturing sector, the supply chain is the crucial factor. In a typical supply chain situation, multiple independent parties take part in moving payload from Point A to Point B; they should tract it to all destinations (Sheela & Priya, 2020). The grain supply chain usually transfers through various storages at different locations shipped by numerous logistics from farmers to end customers. A tamper-proof distributed ledger can track the travel of a particular batch of production that where, when and who shipped or stored it or if it needs to be shipped somewhere in a particular period.

10.4.1 Blockchain Technology: Powering the Business of the Future

Blockchain strategy is a result of continuously exploring and evaluating the latest technology innovations that demonstrate an opportunity to deliver more seamless, safe, efficient and impactful global experiences. This technology is a new, powerful tool that is already shaping the future of the Internet with simple, safe and secure transactions. It helps to achieve the vision to make the globe the happiest place for people.

Blockchain technology is an avenue worth exploring for improved business decentralization. However, this technology comes at the expenditure of added restraints. Table 10.1 defines the pros and cons triggered by the development of blockchains, which is highlighted in the literature.

10.4.2 New Wave of Economic Opportunity and Digital Innovation

The blockchain strategy will usher in economic opportunity for all sectors in the smart city. Globe has a long tradition of leading digital innovation. Now, for the first time, Dubai will be pioneering the application of new technology for cities and sharing it with

TABLE 10.1 Challenges of Technology Collaboration

CHALLENGES	*SUB-CHALLENGES*
Blockchain	Performance/trade offs
	Coding deployment
	Transaction stability
	Multi-business-to-customer integration
Business	User interaction/interface
	Privity and flexibility
IOT	Infrastructure
	Scalability

the world. When successful, Dubai will be the first blockchain-powered government, driving the future economy. Blockchain strategy is built on three pillars – government efficiency, industry creation and international leadership.

Participation from all city stakeholders – residents, visitors, business owners, parents and families – is a cornerstone of the strategy. This goal will be carried out by leveraging a wide range of technologies including blockchain, AI and IoT and by focusing on three strategic pillars: government efficiency, industry creation and international leadership. Collaborating with the private sector and government partners, to empower, deliver and promote an efficient, seamless, safe and impactful experience for peoples. To achieve its strategic pillars, organizations aim to introduce initiatives and develop partnerships to contribute to its smart economy, smart living, smart governance, smart environment, smart people and smart mobility dimensions.

10.5 ADVANTAGES OF BLOCKCHAIN AND IoT COLLABORATION

Security is a crucial concern with IoT-enabled devices that have obstructed its large-scale deployment. IoT devices often experience security vulnerabilities that make them an easy object for distributed denial of service (DDoS) attacks. In DDoS attacks, multiple understanding computer systems attack a target, i.e., a central server with a huge volume of concurrent data requests, thereby causing a denial of service for users of the targeted system. A number of DDoS attacks in recent years have caused disruption for organizations and individuals. Unsecured IoT devices provide an easy target for cybercriminals to exploit the weak security protection to hack them into launching DDoS attacks. Blockchain is the one solution for so many security problems in IoT environment because of its decentralized nature (Noor & Hassan, 2018).

The IoT network can manage data transactions through various devices that are powered and administered or monitored by various organizations. To reduce making it difficult to pinpoint the data leakages in case of an attack by cybercriminals. In addition to that, IoT produces a huge amount of data and with various stakeholders involved, the possession of data is not very clear that this data belongs to whom. IoT plays a crucial role in the economic and industrial growth of a developing region. Nevertheless, the security of data and information is an important concern, is highly desirable and is a great challenging issue to handle (Kumar et al., 2019). Smart contracts are simply a program that runs on the Ethereum blockchain. Smart contracts are a mutual agreement between two groups that is in the blockchain. Smart contracts help to exchange money, property or anything of value in a transparent, conflict free manner without involving the mediator. Smart contracts can also use to authorize payment automatically without human involvement.

Chain of Things (CoT) is a syndicate of technical people and prominent blockchain organizations. It examines the finest possible use cases where a mixture of blockchain and IoT can offer important advantages to industrial, environmental and humanitarian applications. Until now CoT has developed Maru, which incorporated blockchain and IoT hardware to resolve problems with identity and security. There are three separately developed use cases named Chain of Solar, Chain of Shipping and Chain of security.

IOTA is an open-source ledger that serves as a protocol for swift transaction settlement and data integrity, with a tangle ledger that eliminates the need for steep mining (validation of transactions). IOTA is a brilliant framework for IoT devices that need to process large amounts of microdata. This tangle ledger is a distributed ledger, facilitates machine-to-machine communication, fee-less micropayments, quantum-resistant data and so on. It has also built a sensor data marketplace and is entering the market for data-driven insights that were supported by more than 20 global corporations.

Modium.io unites IoT sensors with blockchain technology, supporting data integrity for transactions involving physical products. The Modium sensors record ecological conditions, such as warmth, that goods are subject to while in transfer. While the goods reach the next transfer point or end consumer, sensor data will be verified against predetermined conditions in a smart contract on the blockchain (https://www2.deloitte.com/ch/en/pages/innovation/articles/blockchain-accelerate-iot-adoption.html). The contract validates that all the requirements are met that are set out by the sender and the involving parties and it triggers various actions such as notifications to all involving parties.

Riddle code – Riddle code offers cryptographic identification solutions for blockchains in supply chain management and swift logistics. Riddle & code that operates on the convergence of IoT devices and distributed ledger networks provides a hybrid, original hardware and software solution that enables safe and trustworthy interaction with machines in the IoT era by giving machines and any physical system a "Trust factor." This technology bridges the gap between the physical and digital worlds to strike a balance between the need for paper documents and the benefits of blockchain technology.

As previously discussed, a fundamental flaw in today's IoT systems is their security infrastructure, which is based on a centralized client–server model operated by a central authority, making it vulnerable to a single point of failure (SPOF). Blockchain addresses this issue by decentralizing decision-making to a decentralized network of computers focused on consensus. However, there are three key tasks to remember when building the infrastructure for IoT devices in combination with a blockchain ledger:

Scalability – Scalability is an important consideration. One of the most significant challenges currently faced by IoT is scale: how to handle the massive volumes of data generated by a vast network of sensors without compromising transaction processing speeds or potentials. Defining a simple data model ahead of time will save time and avoid problems when it comes to putting the solution into production (Knirsch et al., 2019).

Network privacy and transaction confidentiality – Confidentiality of network transactions and network secrecy. On public blockchains, the secrecy of transaction history in a shared ledger for a network of IoT devices is difficult to guarantee. This is due to the fact that transaction pattern analysis can be used to infer the identities of users or devices hidden behind public keys. Organizations should check their privacy concerns and see whether hybrid or private blockchains will be a perfect match for them.

Sensors – Sensors are instruments that detect changes in the environment. Interfering with the correct calculation of the conditions that must be met to perform a transaction that could jeopardize the reliability of IoT sensors. To ensure a secure environment for data recording and transactions, steps to ensure the security of IoT devices so that they cannot be tampered with by external interventions are necessary.

Based on the analysis, Blockchain and IoT both are very leveraging developing technologies with great capability but still lack popularity or adoption in the companies due to technical and security concerns. Many organizations in the market are already working on use cases fusing these two technologies, as together they offer a way to reduce the security and accompanying business consequences. It's so uncommon to come across an outstanding combination such as blockchain and the IoT. "A kickass fusion" is the phrase that comes to mind when we believe about it. By offering a brief explanation of how blockchain and the IoT work so that we're all on the same page. Figure 10.1 depicts the workflow involved in IoT-enabled devices. The entire process will be maintained and managed by blockchain.

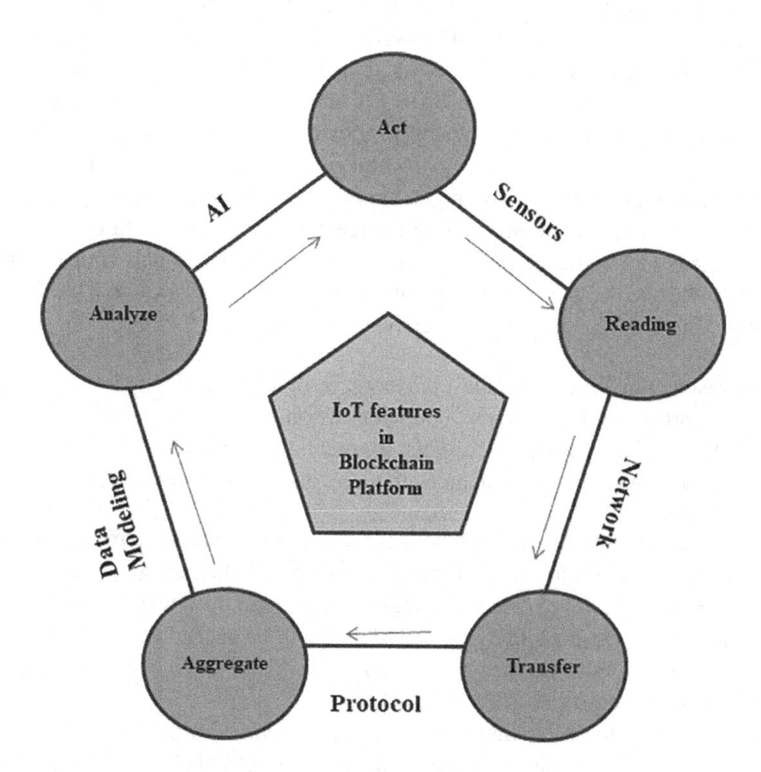

FIGURE 10.1 Operations involved in IoT-enabled process.

The idea of combining the two technologies is still foreign and intimidating to many. But slowly and steadily, things are changing, and it's only the beginning of what's possible. Blockchain is a sort of appropriated record for keeping up a changeless and carefully designed record of transactional data. Managed by computers belonging to a peer-to-peer (P2P) network, the tech successfully acts as a decentralized database. In a layman's language, each one of the computers in the disseminated network keeps up a duplicate of the record to keep a SPOF, and all copies are updated and approved at the same time. IoT, on the other hand, is all about associating any device with the internet and with other associated devices. These devices can be anything who are capable enough to represent themselves digitally, and they can be controlled from any place. Set up them together and in principle, you have a precise, secure and changeless technique for recording data prepared by "smart" machines in the IoT. As per the study, 0.022 MB is the highly utilized space (in 2019) in blockchain and 0.00007 MB is the smallest utilized space (Lo et al., 2019).

10.6 BUSINESS ANALYSIS

In general, business analysis includes the process of finding out solutions to the problems before or after taking place. These solutions may include growth in existing policy and planning, system components or software, procedural improvement and so on. Business analysis helps us to recognize the problems occurring at present in a target organization. This problem recognition stimulates us to improve our potentials and achieve desired goals. Hence, maximum delivery of orders to the stakeholders can be accomplished. Business analysis can be carried out by proposing a business model which includes detailed information about how a company can generate income and gain profit through the overall process, resources, customers, suppliers, capabilities, etc. The ultimate goal of business is to increase the revenue by decreasing the expenses (Ju et al., 2016). IoT business models can be developed using the business model canvas framework which is a tool to picturize the building blocks of a business from the basics. This framework commonly considers four major perspectives:

1. Infrastructure – key activities, partners, resources, etc.
2. Value proposition – reasons to be your customer.
3. Customer – relationship, channels, segmentation, etc.
4. Financial perspectives – includes income, expenditure and profit.

These analyses using IoT devices can be merged with blockchain technology to enhance its security level. Each data will be stored in a distributed manner, which makes the data immutable. Hence, the development strategies maintained by a business analyst will not be known to the competitors. The distributed data will be encrypted so that others could not view the original data without permission. Figure 10.2 depicts the challenges to be faced in imposing security on IoT.

Data collection for a business must be done from both primary and secondary sources in order to satisfy the target requirements (Chong et al., 2019). A business model essentially focuses on capturing and delivering the requirements of the customers.

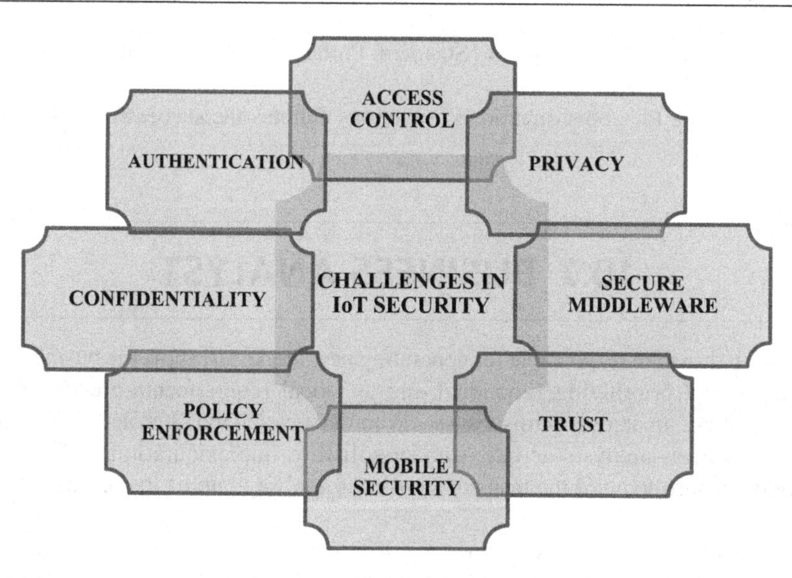

FIGURE 10.2 Predominant challenges of IoT security.

It also strives to work with 24/7-based connectivity in customer's environment to provide efficient results. IoT business models are mainly classified into seven types as follows:

1. Subscription model – This model enables IoT products to implement more benefits on software-oriented products. It considers SaaS (Software as a Service) as its basic platform and supports a subscription model with certain criteria like predictive maintenance as a service and monitoring as a service.
2. Outcome-based model – This model encourages the customers to pay only on the outcome (result) basis.
3. Asset sharing model – This model promotes the method of sharing IoT business models. They can be utilized by the people when they are required and they can be paid accordingly.
4. "Razor blade" model – Major goal of this model is to sell the product either for profit or loss with an intention to reproduce a new version of services.
5. Monetize your IoT data – It is essential to monitor the data collected through IoT devices. The goal of this model is to deploy more devices to collect data. IoT decision framework is recommended to monitor the impacts of new functionalities in our existing product.
6. Pay-per-usage – Monitorization of devices paves way for another model called pay-per-use. With this model, we can find the active interaction time between the product and the user. Hence, the customer can pay as per their utilization.
7. Offer a service – As the devices are getting upgraded, the customers can also receive extra services than expected service (https://www.linkedin.com/pulse/internet-things-business-analysis-product-design-mba-pmp-itil).

Analysis is the process of managing and delivering the proper service as per the requirements of the customer (https://danielelizalde.com/monetize-your-iot-product/). It can be represented as

$$\text{Value of a service} = \text{Function}\left(\text{Standard, Production, Affordability, Reliability}\right)$$

Hence, satisfying the above-mentioned functions denotes the successful completion of a task.

10.7 BUSINESS ANALYST

Business analysts are responsible for generating new models to uplift the business range. Identifying and prioritizing functional and technical requirements are the foremost responsibility of an analyst (https://www.cio.com/article/2436638/project-management-what-do-business-analysts-actually-do-for-software-implementation-projects.html). The following are some of the tasks of a business analyst to uplift the business:

Recognizing business requirements – It is necessary for an analyst to accept the stakeholders or the customers as the way they are without complaining. Observation has to be done to determine the exact requirement of a customer.

Knowing competitor's potentials – Also, analysts need to analyze his/her company along with competitor's company and their expertise. This helps them to improve themselves and provide better results to the stakeholders.

Furnishing ideas – Business analyst comes up with innovative ideas to satisfy the business requirements with respect to essential data and business insights. This makes the stakeholders go through the ideas and choose the right solution for implementation and development.

Operational efficiency – It is the responsibility of an analyst to define the communication protocol between the different role players in an IoT project team. This ensures the teammates to uniformly know the progress of the work.

Evaluation of result – The analyst has to manage the risk factors even at the development stage, functional strength of a project, problems with feasibility and so on. Also, it is mandatory to evaluate the outcome of a project to verify whether it suits the requirements of a customer (https://www.iiba.org/business-analysis-blogs/how-business-analysis-can-improve-iot-projects/).

10.8 APPLICATION OF IoT AND BLOCKCHAIN TECHNOLOGY FOR BUSINESS ANALYSIS

Supply Chain – According to IBM, 40 other organizations are trying to improve and speed up a very complicated and time-consuming process of international trade and shipping and bring transparency to its unavoidable bureaucracy. By merging these two technologies, temperature control to container weight, enables shippers, shipping lines, freight forwarders, port and terminal operators, inland transportation and customs authorities to interact more efficiently through real-time access to shipping data and shipping documents in a single shared view of a transaction without compromising details, privacy or confidentiality. In addition to this, all the parties involved in international trade can easily collaborate in cross-organizational business processes and information exchanges, all backed by a secure, non-reputable audit trail.

Logistics – Pharmaceuticals and many other industries are seen using these solutions that have the potential to generate value by leveraging IoT and blockchain in a relevant way. With several IoT sensors being verified against predetermined conditions, medicines can be given to specific conditions.

Insurance – claims management, fraud management, health insurance, and property and casualty insurance seems to have improved in many ways with smart contracts combined with IoT data from wearable personal technologies, sensors on objects (vehicles, shipping containers), location-based sensors (factories, warehouses, homes, alarms, cameras and industrial control systems). Numerous advantages of adopting Blockchain with IoT are shown in Figure 10.3. The following are the advantages of collaborating blockchain and IoT.

10.8.1 Publicity

Since each participant has their own ledger, all participants have access to all transactions and blocks. The transaction summary is shielded by the participant's private keys, so even though all participants can see it, it is encrypted. The IoT is a complex system

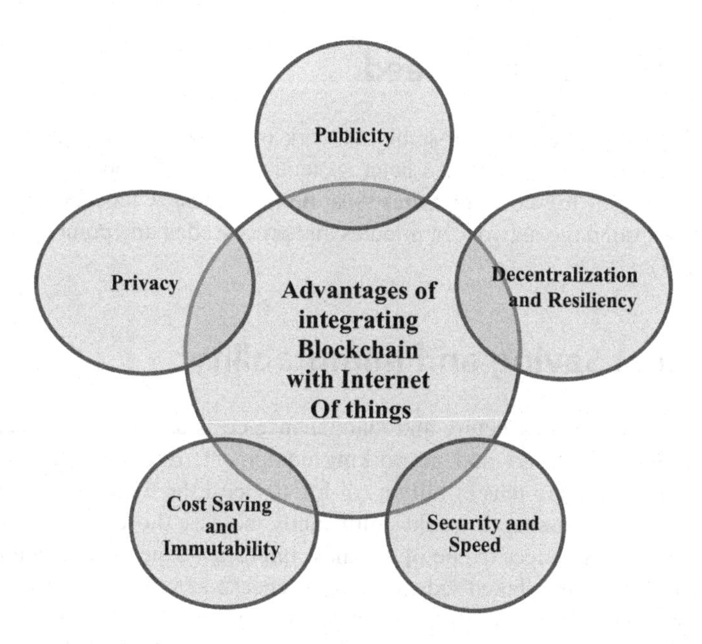

FIGURE 10.3 Advantages of integrating blockchain with AI.

in which all connected devices can exchange information while also protecting the privacy of all participants. The overwhelming amount of data produced by the network is beyond the reach of any single organization.

10.8.2 Decentralization

The majority of participants should verify the transaction for them to approve it and add it to the distributed ledger. There is no verification authority that can approve the transactions or set of specific rules to have the transactions approved and agreed. Since the majority of the network members must consent to validate transactions, there is a significant amount of confidence involved. As a result, blockchain can provide IoT devices with a stable network. More localized traffic flows and single points of failure in the centralized IoT infrastructure are being eliminated. In this blockchain technology, there is no need for a central authority due to this decentralized structure (Moin et al., 2019).

10.8.3 Resiliency

Each node has its own copy of ledger, which contains all of the network's previous transactions. As a result, the blockchain is more resistant to attacks. And if one node were to be hacked, blockchain would still be maintained by all other nodes. Having a copy of data at each node in the IoT would help with knowledge sharing. It does, however, incorporate new processing.

10.8.4 Security and Speed

Blockchain's ability to establish a stable network over untrusted parties is critical in the IoT, where there are various and heterogeneous devices. To put it another way, to initiate an attack, all IoT network nodes must be malicious. A blockchain transaction can be spread around the network in minutes and processed at any point during the day (Kshetri, 2017).

10.8.5 Cost Saving and Immutability

Because of the high infrastructure and maintenance costs associated with centralized architecture, massive server and networking equipment. Existing IoT solutions are costly. Although there are tens of billions of IoT devices, the total amount of communications that must be managed would significantly increase those costs. Immutability: Having an immutable ledger is one of the most important aspects of accounting. Any modifications to the distributed ledger must be checked by the majority of network nodes. As a result, the transactions are difficult to change or erase. Having an immutable ledger for IoT data will increase security and privacy which are the major challenges in this technology and all new technologies.

10.8.6 Privacy

Both the buyer and the seller use anonymous and exclusive address numbers to complete a deal, keeping their identities secret. This feature has been panned because it encourages the use of crypto-currencies in the black market. However, if used for other reasons, such as electoral voting systems, it may be seen as a benefit. It alleviates the security and scalability concerns associated with IoT through blockchain.

Blockchain offers transparency, by allowing anyone with permission to access the network to view and monitor previous transactions. This can be a reliable way to pinpoint the root of any data leaks and take immediate corrective steps. By providing a way to enable trust among the stakeholders, blockchain can allow IoT companies to reduce their costs by eliminating the processing overheads related to IoT gateways (e.g., traditional protocol, hardware or communication overhead costs).

Smart companies – The companies that use IoT and blockchain to improve their long-term and short-term success in operations. This can be applied on applications to monitor and collect actionable data on consumer usage, inputs, outputs, real-time behavior, acquisitions, etc. With all these observations, a company can be clear with its target and requirements. This can be deliberately achieved with blockchain and IoT.

Smart security – Beyond the general usage of devices at a single location, we can use blockchain and IoT to improve organizational operations. To enhance the quality of a business portfolio, remote controls for scan cards, on site areas, mobile apps and device upgrades can be introduced. These data are hashed using hashing algorithms of blockchain which guarantees data security.

Smart management – Blockchain and IoT are capable of revolutionizing the business industry with their unique features like communicating with devices in a distributed and secure manner. Every data stored in the network will be securely managed by the blockchain platform (https://www.moqdigital.com/insights/applications-of-iot-in-business). IoT is enabling successful digital transformation in today's world in collaboration with blockchain technology with the following necessary key factors.

Improved business insights and customer experience – Interconnected equipment in supply chain, aviation, manufacturing, agriculture, healthcare and other industries are creating greater insights into their business. They can clearly analyze the requirements and usage of products by their customers. It can also be enabled by cloud platforms provided by AWS, Microsoft Azure, Google, IBM and so on with the involvement of a third party. Blockchain overcomes this drawback by introducing the concept of decentralization.

Cost & downtime reduction – This technology helps to overcome the operational and system downtime in factories and other installations in industries. If a fault was identified in a system, it consumes hours or days to overcome. Hence, availing the support of readily available software helps to increase the scalability of a process.

Efficiency and productivity gains – Collaborating a business with blockchain and IoT increases the possibility to increase the productivity and efficiency of a service. Continuously tracking a process will always be encouraged in order to maintain customer's trust and fulfill the customer's experience.

Asset tracking and waste reduction – Usually, efficiency and productivity help in waste reduction. Supply-chain management with IoT and block-chain supports the business industry to maintain an error-free record where every transaction will be clearly maintained in a secure platform. These data can be viewed at any time by authorized users. New business models have also been generated to enhance the business sector with all the features of collaborating technologies (https://internetofbusiness.com/5-ways-the-internet-of-things-is-transforming-businesses-today/).

IoT security – IoT security creates a technical environment in which IoT-related objects and data can be protected. All objects with a unique identifier and data transmission capabilities over a network are linked via the internet with digital computers, devices for computing purposes on electronic, mechanical, people, livestock, houses and so on. If these IoT devices are not properly connected. Then there may be serious vulnerabilities (Sharma & Lal, 2020).

IoT security creates a technical environment in which IoT-related objects and data can be protected. All objects with a unique identifier and data transmission capabilities over a network are linked via the internet with digital computers, devices for computing purposes on electronic, mechanical, people, livestock, houses and so on. If these IoT devices are not properly connected.

10.9 CONCLUSION

Thus the adoption of IoT-blockchain technology is not prevalent due to the technical issues and operations challenges. Scalability and storage are the foremost challenges in a blockchain system, which are having a huge centralized ledger, saving and storing the ledger on ledger notes (https://isg-one.com/consulting/blockchain/articles/blockchain-in-iot-a-vital-transformation). With blockchain, it's impossible to overwrite the existing data because of its high level of encryption. So it is recommended to use blockchain to store IoT data, which will acknowledge as another layer of security that makes it difficult for a hacker to go through a network. Blockchain can enable the fast processing of transactions and coordination among billions of connected devices. As the number of interconnected devices grows, distributed ledger technology provides a viable solution to support the processing of a large number of transactions. The synergy of blockchain and IoT provides greater advantages in the field of business sector from data collection to storage.

BIBLIOGRAPHY

Chong, A.Y.L., Lim, E.T., Hua, X., Zheng, S. and Tan, C.W. Business on chain: A comparative case study of five blockchain-inspired business models. *Journal of the Association for Information Systems*, 20(9), 1310–1339, 2019, Doi: 10.17705/1jais.00568.

Ju, J., Kim, M.S. and Ahn, J.H. *Prototyping Business Models for IoT Service, Elsevier*, 2016.

Knirsch, F., Unterweger, A. and Engel, D. Implementing a blockchain from scratch: why, how, and what we learned. *EURASIP Journal on Information Security*, 2019(1), 1–14, 2019.

Kshetri, N. Can Blockchain Strengthen the IoT? *IEEE*, 68–71, 2017.

Kumar, S., Tiwari, P. and Zymbler, M. Internet of Things is a revolutionary approach for future technology enhancement: A review. *Journal of Big Data*, 6(1), 1–21 Springer, 2019.

Lo, S.K., Liu, Y., Chia, S.Y., Xu, X., Lu, Q., Zhu, L. and Ning, H. Analysis of blockchain solutions for IoT: A systematic literature review. *IEEE Access*, 7, 58822–58835, 2019.

Moin, S., Karim, A., Safdar, Z., Safdar, K., Ahmed, E. and Imran, M. Securing IoTs in distributed blockchain: Analysis, requirements and open issues, 2019, Doi: 10.1016/j.future.2019.05.023.

Noor, M.M. and Hassan, W.H. Current research on Internet of Things security: A survey. *Computer Networks*, 2018, Doi: 10.1016/j.comnet.2018.11.025.

Sharma, V. and Lal, N. A detail dominant approach for IoT and blockchain with their research challenges. *2020 International Conference on Emerging Trends in Communication, Control and Computing (ICONC3)*, 2020.

Sharma, V., Som, S. and Khatri, S.K. Future of wearable devices using IoT synergy in AI, IEEE Explore, 2019.

Sheela, K. and Priya, C. Enabling the efficiency of blockchain technology in tele-healthcare with enhanced EMR. IEEE, 2020, https://ieeexplore.ieee.org/document/9132922.

Tzafestas, S.G. Synergy of IoT and AI in modern society: The robotics and automation case. *Robotics, Automation & Engineering Journal*, 31(5), 1–15, 2018.

Available at https://www2.deloitte.com/ch/en/pages/innovation/articles/blockchain-accelerate-iot-adoption.html.

Available at https://www.linkedin.com/pulse/internet-things-business-analysis-product-design-mba-pmp-itil.

Available at https://danielelizalde.com/monetize-your-iot-product/.

Available at https://www.cio.com/article/2436638/project-management-what-do-business-analysts-actually-do-for-software-implementation-projects.html.

Available at https://www.iiba.org/business-analysis-blogs/how-business-analysis-can-improve-iot-projects/.

Available at https://www.moqdigital.com/insights/applications-of-iot-in-business.

Available at https://internetofbusiness.com/5-ways-the-internet-of-things-is-transforming-businesses-today/.

Available at https://isg-one.com/consulting/blockchain/articles/blockchain-in-iot-a-vital-transformation.

11 Applications of Body Sensor Network in Healthcare

K.R. Radhakrishnan
St. Joseph's College of Engineering

T. Kohilakanagalakshmi and Salini Suresh
Dayananda Sagar Institutions

T.M. Thiyagu
Anna University

Amita Sharma
IIS University

Contents

11.1 Introduction 172
11.1.1 Sensor Network 172
11.1.2 Wireless Sensor Networks 172
11.1.3 Body Sensor Network 173
11.2 Wireless BSN Architecture 174
11.3 Sensors Used for Treatment and Health Observing 176
11.3.1 An Introduction To Sensors in Healthcare 176
11.3.2 Non-Invasive Applications 177
11.3.2.1 Electrophysiological Measurement 178
11.3.2.2 Environmental, Biochemical and Biophysical Sensors 181

DOI: 10.1201/9781003051022-11

11.4 Future Scope in Healthcare 193
11.5 Future Trends 194
11.6 Conclusion 195
Bibliography 195

Abstract

The network field is extending its service from a wired to wireless network. As far as concern, using sensors in the network makes huge progress in various applications such as military forces, security violations, and healthcare. Improving technology growth for the care of people is the aim of any country. In this base healthcare needs more improvement in equipping its standards to frame a comfort zone for the people in the name of body sensor network (BSN). This acts as a bridge between the people and the infirmary. This platform provides amenities such as monitoring medical needs, accessing the data of patients, communicating during emergencies, enhancing the memory of the system, and so on. BSN in healthcare has many applications, these applications can be the type of sensors used in the detection or identification of the illness or the diseases they are getting affected. The goal of this chapter is to deliver complete knowledge of the BSN, the working principle of BSN, categories of the sensors used in BSN, its applications, and the challenges confronted with BSN.

11.1 INTRODUCTION

11.1.1 Sensor Network

It is always wondering to know the new concept such as what is this and that. Here it is mandatory to know what is sensor network. The sensor network has the following components that are explained in Figure 11.1

i. Sensor nodes – it is used for the monitoring purpose of transmitting the collected data from the sensor and passes it to the various sensor nodes.
ii. Actuator nodes – it is also called management nodes to which the processed data transferred from the sensor nodes through the gateway reaches this actuator node.
iii. Clients – the clients or users who manage the management node and collect the required data from the managed node and monitored node.

11.1.2 Wireless Sensor Networks

Implementing any application or improvement in research-based activities using a sensor as the major comfort to the device is the upcoming trend. From the starting stage of using the network for applications, it was designed and implemented using the wired network. This wired network also had many stages of extension or versions.

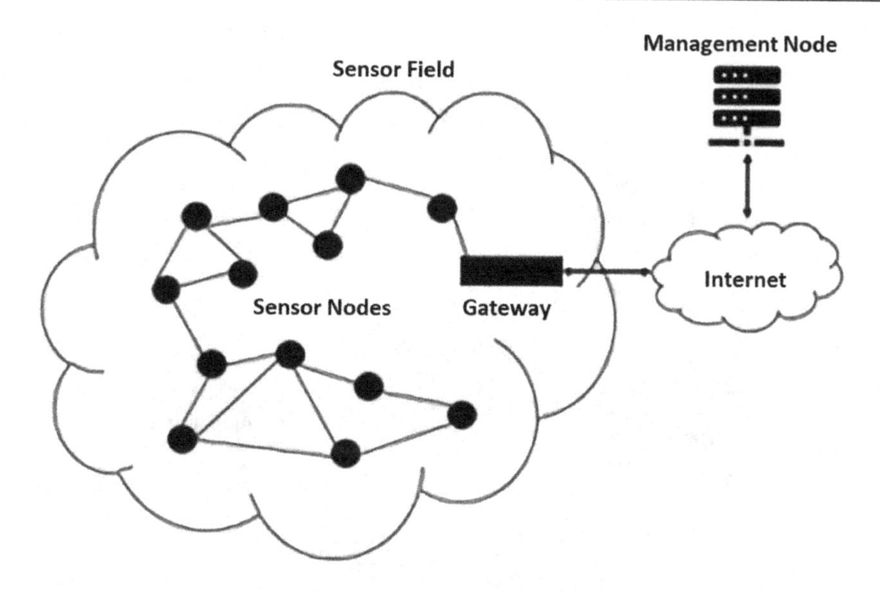

FIGURE 11.1 Sensor network components and working principle.

The extensions are produced to the cables by improving the quality and the security as the base standard. Little by little this made the growth for wireless from wired. The wireless was much more comfortable than the wired in many ways. The very common thing was portable. The usage of LAN to Wi-Fi, Bluetooth, cellular, and so on is the best practical way that we follow nowadays.

There are many advantages in using Wireless Sensor Networks (WSN), such as without moving the infrastructure arrangements of the network can be shifted. Suppose we need an extra workstation, it's very flexible, inexpensive in pricing, avoids manpower, makes the communication easier, acts as a centralized system, etc.

The WSN can be implemented in certain structural formats called topologies such as star, tree, mesh, and hybrid. Depending on the environment, the WSN can be deployed underground, underwater, and under the land. Since it is not having a specific infrastructure, it can be deployed with a larger number of wireless sensors in the form of an Ad hoc network also. The WSN is providing better performances in the fields of healthcare; it is also proven in the means of equipment getting used in the hospitals or clinics for the scanning and treatments. These kinds of improvements are possible by the hands of research facilitators working in giving with their competitions all around the world.

WSN can be used for wide applications and also it includes much research in the fields for better improvement as suggested in given in Figure 11.2; with this extension of WSN, we are going to see the next version called body sensor network (BSN).

11.1.3 Body Sensor Network

Very often we have many questions once a new technology emerges for an application that is very necessary to society's security. "What exactly is BSN? How does it work?

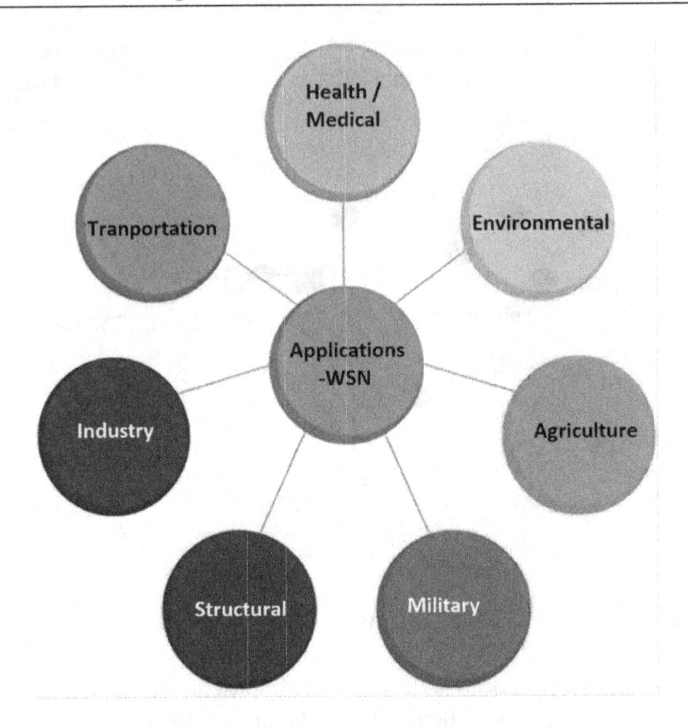

FIGURE 11.2 Applications of Wireless Sensor Networks.

What components does it use in the design? What are its principles and standards? Will it give any harm to society? Getting treatment through BSN is expensive or not? What are the applications of the sensors used in BSN?". These are very basic questions that have to be solved so that people will feel free to access the trends with no fear. That is because people are the primary source of the applications developed either by research teams of corporate or the government sectors. This session in the introduction will give the basic description of the BSN and further the applications of the sensors used in designing the BSN are discussed in the upcoming sections on wireless BSN in the same chapter.

11.2 WIRELESS BSN ARCHITECTURE

Nowadays, wireless body sensor network (WBSN) helps society in the need of better medical facilities and treatments at low cost by monitoring the human body either in the hospital or in the native living area as a real-time field. Besides the ability of inconspicuous, cost-effective, and unsupervised unremitting monitoring, WBSNs consume a varied choice of supervising applications such as sports activity, medical and healthcare, and rehabilitation systems. BSN works under the concept of the wireless network, so it can also be named WBSN. It uses unobtrusive sensor nodes along with the wireless network having small light. These are attached to the human body to record the essential data needed.

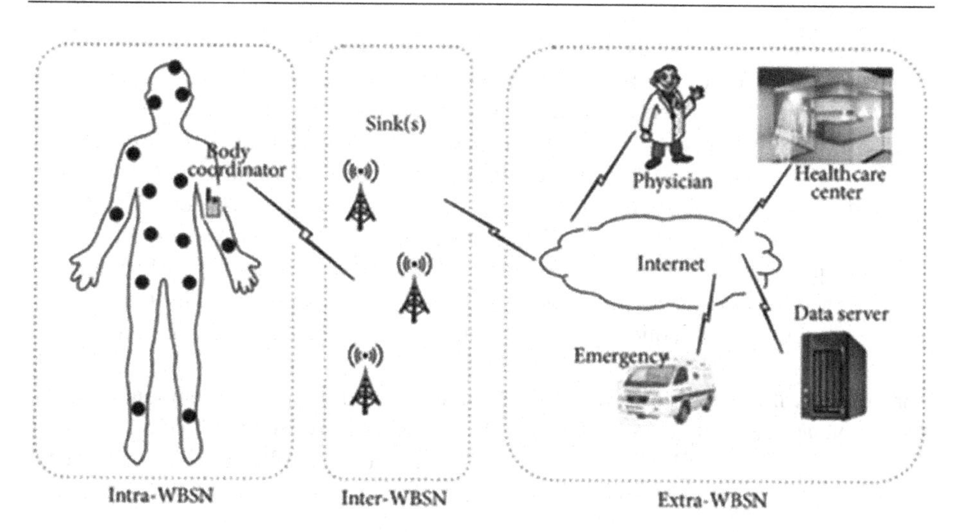

FIGURE 11.3 Categories of WBSN.

According to the architecture of the BSN described by Iqbal Bangash (Bangash et al., 2014) mentioned in Figure 11.3, BSN is categorized into three subdivisions: Extra WBSN, Intra WBSN, and Inter WBSN.

- **Intra – WBSN** has the short co-ordinate ranges fixed in the body by wireless monitoring networks deployed either in the human body or in covered clothes. These co-ordinate ranges are designed in such a way that they are easily recharged or replaced to provide comfortable health monitoring. The sensor nodes used in this field are blood pressure (BP), electroencephalography (EEG), electrocardiography (ECG), electromyography (EMG), body temperature, and various others that can also be equipped away on the wearers, well-positioned on the human body, or even embedded within the human body. These kinds of nodes used in Intra WBSN are also known as body sensor units (BSU). The sensors used in this tier are explained in detail in Section 11.2.
- The BSUs communicate with each other and perform Intra WBSN/tier-1 communication. After collecting the data from the sensor nodes, it is transferred to the body control unit so that from BCU the data is forwarded to many other devices such as laptops, mobile, etc., or to other networks. For this transformation, it is expected that the devices have to be user-friendly, and also software and hardware-based requirements can satisfy the data collection. This transformation of data from BCU to another node is known as **Inter WBSN / tier-2** communication. It provides communication between the PS to the internet or between PS and one or more RPs and then to the internet.
- Using the existing wireless methods such as ZigBee and Bluetooth, soon the data from the tier-2 devices is further forwarded to an access point through the gateway that makes use of connecting a huge network platform called **Extra WBSN**. This Extra WBSN includes network centers such as hospitals, medical centers, and telephones for communication purpose. The data

present in this WBSN or tier-3 device get access worldwide as if it is located on the internet. This means the data is stored in the database server or on the specific domain network so that it is accessible by the researchers in the field, doctors to diagnose the results, and patients to know their reports at any time.
- As if the data are stored on the internet and accessible worldwide for corrections and any sort of development, there is a possibility to get access by other users or clients in the form of hackers so it is better to have security as the important aspect in this regard to avoid the access of unauthorized persons. Another solvable method is that it can be stored as private data in its own accessories consequently whenever needed it can be taken out for access.

For example, it creates a bridge between the doctor and the patient communication in the remote access using the 4G/5G network. This kind of WBSN/tier-3 communication is known as **Beyond WBSN**. Some designs are essential to fulfill the data server concerning high capacity and processing.

11.3 SENSORS USED FOR TREATMENT AND HEALTH OBSERVING

11.3.1 An Introduction To Sensors in Healthcare

Most prominently used sensors and their performance for the people's personal health monitoring, home-based fitness systems, and continuous health monitoring care-based applications are presented. This theme concentrates on the development and usage of sensors such as wearable devices that help in monitoring the patient's details at home or for any outside preferences; sports persons can also use this in their daily occasions as a part of doing exercises.

Many new and old versions of the detecting devices are found in the consideration of two categories called wearable and non-wearable devices.

The wearable devices are used in the means of holding an accelerometer on the object or a subject that has to be detected in the cases like movement, increase in speed, fall in the detection of the subject will also be taken for the consideration. Non-wearable devices have equipment like sensors that may be normal or acoustic in nature, cameras, and pressure sensors that are leveled in the subject with the very normal environment later can be measured in various determination.

Healthcare and sensors are the built-in play in the current hospital management system. Many have produced their results in the form of products and even in the form of applications by the researchers and the industries that are top in the market for the welfare of the people as well as the growth in technology. To make the compactness of the devices in handheld manner. At the same time, it is necessary to have a complete device to read the metabolism of humans for the analysis and diagnosis of their problems.

- To improve the physiological warnings for the older persons living in the room a remote system is been developed and evaluated using the cognitive function.
- To minimize the gap between the individuals and the family members with the social network, so that they can experience the knowledge of the computer.
- To build up certain technologies that can make use of real-world circumstances for the metrics like psychometric measures, online day reconstruction, biological markers, and ecological assessments.
- Those patients using the wearable devices can provide feedback to the devices used to improve the rate of performance as if they use them daily like exercises.

The list of sensors is described structurally in Figure 11.4. The representation of the sensors in the human body is divided into two basic categories of applications:

i. Non-invasive (wearable)
ii. Invasive (implantable)

11.3.2 Non-Invasive Applications

The non-invasive applications are generally wearable sensors and electronic devices with artificial skin just like robots based, it should not penetrate and get into the human skin while it detects the dynamic signals during biometrics, and when dealing with a huge amount of biosensors devices for detecting environmental signals, and biophysical

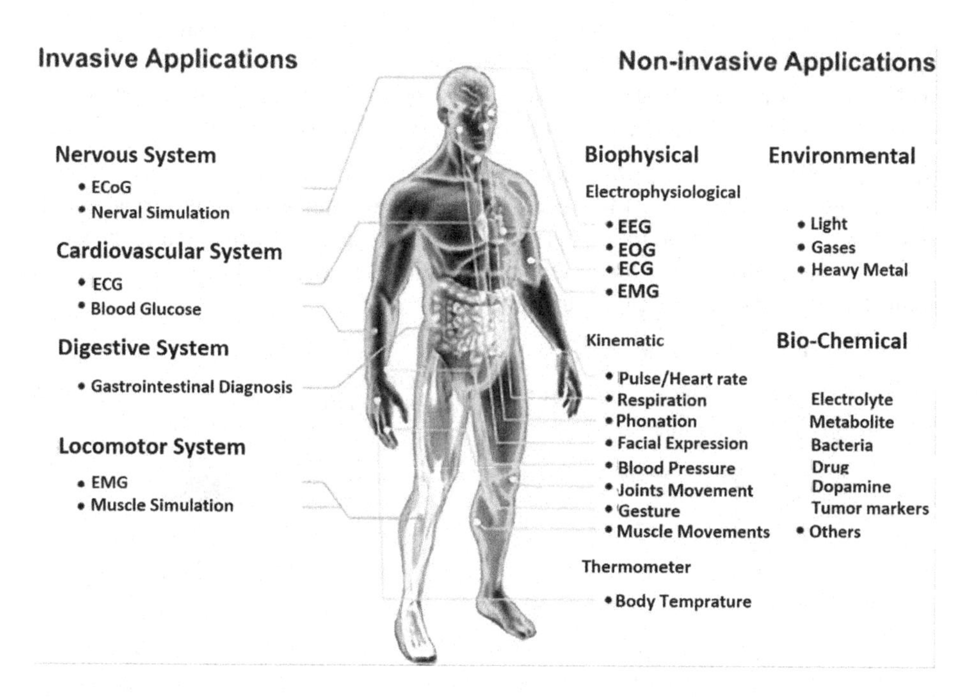

FIGURE 11.4 Non-invasive (wearable) and invasive (implantable) sensors.

signals and so on. This section gives the knowledge of these measurements and devices used in the body sensor, and how they work.

11.3.2.1 Electrophysiological Measurement

11.3.2.1.1 EEG – Electroencephalography

This is an electrophysiological process or measurement that helps in recording the changes in the electrical based activities happening in the brain. The function is studied due to the neurons in the brain, which perform the voltage level changes that happen due to iconic current within the cells or between the cells of the brain discussed in Figure 11.5.

The working of EEG sensor is based on recording the electrical activities in the brain. The EEG sensor can record up to thousands of snapshots of this electrical activity within a fraction of a second. These electrical activities are called to be brainwaves. The brainwaves are recorded, sent to amplifiers, and then the cloud platform or the computer devices are used to process the data, and also the graphical waves of the amplified signals are also stored.

The electrodes present in the EEG devices can capture the various EEG frequencies as raw signals of EEG which are unique in nature using the Fast Fourier Transform. Electrical oscillations per second for one cycle are the measurement of the frequency levels. With these frequency levels, the brainwaves are divided into different waves:

- Beta having a frequency limit from 14 to about 30 Hz.
- Alpha having a frequency limit from 7 to 13 Hz.
- Theta having a frequency limit from 4 to 7 Hz.
- Delta having a frequency limit of up to 4 Hz.

The performance of EEG in healthcare is that it shows the activity of the brain and obtained results give the information to diagnose the disorders in the brain. If the EEG data is abnormal, then the waves are displayed in an irregular pattern. This may lead to dysfunction, sleeping disorders, and head trauma, which are the various signs of this irregular EEG data.

11.3.2.1.2 ECG – Electrocardiogram

The performance of the ECG is to measure the representation of the electrical activities of the heartbeats in the form of electrical waves to the display function in the paper printed in the form of waves for easier understanding. The electrical activities are found in the basics of cardiac muscle contraction.

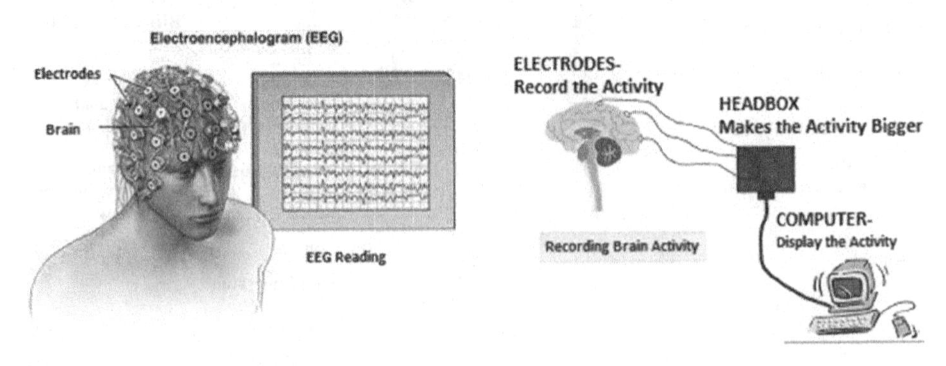

FIGURE 11.5 Electro encephalo graphy.

The "Surface ECG" is the one that is placed in the patient's skin area in form of electrodes. Electrical depolarization waves are spread by the passing of IVS to the ventricles in a downward fashion. It means the direction of the flow is measured with the exception of the substandard characteristic. This direction of depolarization in the heart is known as the electrical axis. If this axis is seen in the downward direction and faced to the left, it is possible to estimate accurately for the individual patients.

The basic working principle of ECG in Figure 11.6 is that the recording is done as positive or upward deflection if the depolarization travels toward a recording and it is noted to be negative or downward deflection if the polarization travels away from a recording.

For the results to obtain from the ECG, the depolarization of the ventricles always seems to be huge enough, which is why the mass of the muscle is greater specifically in the ventricles this is termed as QRS complex.

- The deflection in the form of negative or downward and the initial position wave called Q.
- The deflection in the form of upward means the cross of the isoelectric line that denotes the positivity is called R.
- The deflection in the form of downward that is the cross of isoelectric line and at the same time not returning to the baseline off isoelectric is called S.

11.3.2.1.3 EOG – Electrooculography

To measure the corneo-retinal placed in the front and back of the human electrooculogram signals are used. The basic functionality of this is to diagnose ophthalmology as well as the eye movements recording as in Figure 11.7.

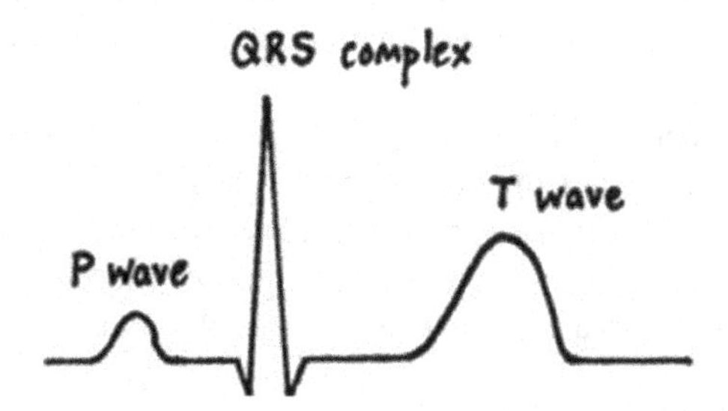

FIGURE 11.6 Electro cardiogram.

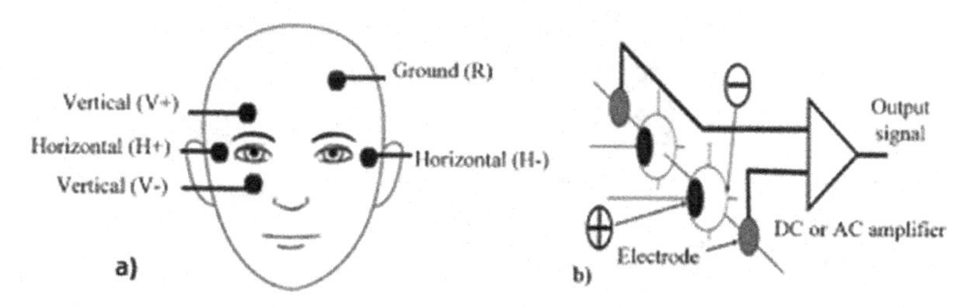

FIGURE 11.7 Electro oculography.

The movements of the eye are tracked by the principles of placing pair of electrodes in the right or left eye as above or below it. The moment the eye moves to any of the directions like left or right, the electrodes placed will track them as positive if it moves toward them or marks to be negative if it moves away from the retina. This provides a resulting in the differences between the electrodes while the assumption of the idle retina is constant and the movement is measured as per the ratio.

Epithelium pigment function is accessed and monitored using this EOG. When it is a dark resting area formally the potential decreases a little bit and when it reaches the minimum known as dark trough taking a long gap. The moment the lights are on the increase, potential resting occurs which is known as light peak; this breaks up within seconds as soon as the retina adjusts to the lightning effect. With the aspect of this light peak and dark trough, a ratio is obtained as the Arden ratio. This is almost to the similar function as mentioned above in the tracking of the eye by electrodes.

11.3.2.1.4 EMG – Electromyography

The outermost area of the body is also measured by electrical signals with the acknowledgment of the nerve's actions happening in the muscle. This makes finding the neuromuscular abnormalities possible. The functional part is carried out with the help of needles inserted into the muscle through the skin; these needles are also known as electrodes. The oscilloscope displays the content taken from the electrical activity is visualized in the form of waves and the audio is projected through an audio amplifier.

Generally, the EMG takes the measures when the movement of muscles is idle, with contraction of heavily loaded or very light movement. When the contraction of the muscle is idle, the tissues will not produce any signals. But once a signal electrode is inserted, then a complete activity will be taken into the consideration and displayed in the oscilloscope; later that signal will not be taken into consideration as devised in Figure 11.8.

The muscle contraction will be recognized once the electrode has been inserted into the skin like bending or lifting your legs, or hands. The potential verifications like shape and size of the wave are used in the basics of wave creations in the oscilloscope giving the necessary data regarding the responding of muscles when the nerves are enthused. Muscle fibers activate more in such a way that more muscle contractions are made.

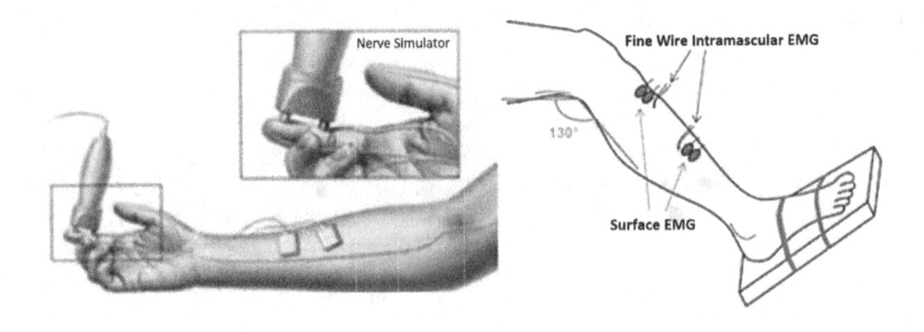

FIGURE 11.8 Electro myography.

11.3.2.2 Environmental, Biochemical and Biophysical Sensors

To know the health conditions of the human body biophysical measures provide an open space even though there seem to be many restrictions in the cases of assessments. Apart from the traditional measurements high developed measures are used in the biochemical analysis and in testing with various instruments, and well-experienced laboratory technicians in this field have used sampling like blood and biofluid as the main course for the diagnosis. Furthermore, conformist testing is classically expensive, invasive, time-consuming, and complicated in the Schematic diagram of the biosensor in Figure 11.9. Consequently, there are growing necessities in lucrative, unceasing, non-invasive, real-time, transferable, and wearable biochemical detection devices for fast, point-of-care detection to discourse these restrictions.

Biophysical sensors seem to be having three working components:

1. Bio-receptor
2. Transducer
3. Electronics circuit and display

The biosensor working methodology is summarized as given below:

- The physical input change is detected utilizing biological elements.
- The transformation to electrical signals from the physical range is done by the elements of the transducer.
- According to the desire of users' perspective, the output electrical signal is amplified and displayed.

Applying the methods of convolution, the biological materials are immobilized and is having contact with the transducer. The bind of analyte is formed using bound analyte by biological materials. This produces the response in an electronic pattern that can be measured uninterruptedly.

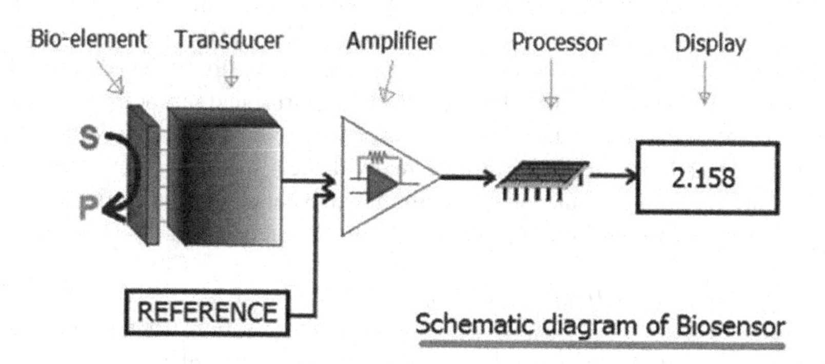

FIGURE 11.9 Schematic diagram of biosensor.

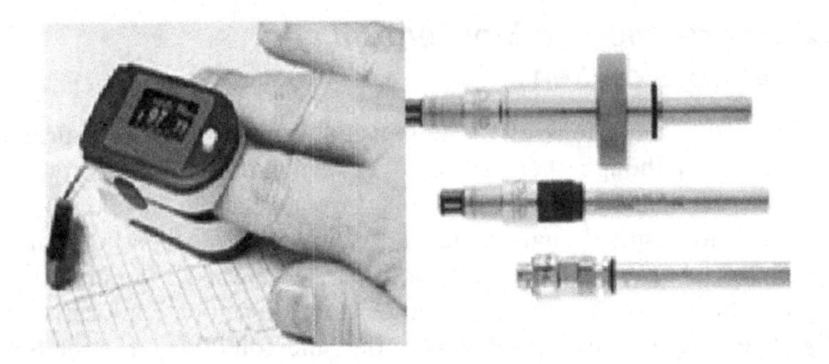

FIGURE 11.10 Oxygen sensor.

11.3.2.2.1 Oxygen Sensor

The technique for measuring oxygen in the blood is very necessary and vital in knowing and assessing the respiratory and circulatory conditions of a patient. In two different ways, the oxygen in the blood is carried out to various parts of the body from the lungs. The amount of oxygen that gets dissolved in the plasma is 2% which is propositional to the pO2 blood. This amount of oxygen is proved based on physiological conditions. The remaining 98% of the oxygen is carried out by the chemical combination of the blood components oxyhemoglobin and hemoglobin in this regard the level of oxygen is measured in two ways either by the polarographic sensor or using the oxygen saturation measurement oximeter in Figure 11.10.

11.3.2.2.2 Carbon Dioxide Sensor

It is also possible to measure the level of carbon dioxide in the blood to know the toxic range using the pH.

11.3.2.2.3 Heart Sound Sensor

The main core internal organ that makes the human alive is the heart. That main functionality of heart includes pumping out blood and this is major necessity to make a person alive. The beating of the heart takes place in different categories; according to the sound of the heart, few diseases can also be identified. The physical shrink and expansion of the heart artery produce some vibrations in the vein. This vibration goes under the surface of the thoracic cavity and sound of the heart is determined. This sound of the heart has a range from 20 to 200 Hz; the minimum range of the heart sound is 4 Hz and the maximum range of the sound reaches 1,000 Hz. Many heart sound sensors have come to the field that was divided into two basics direct conduction and air conduction heart sound sensor in Figure 11.11.

Two components in the heart sound sensor which is of air conduction called the normal sensor and air chamber can have many defects like less insensitivity, getting easily disturbances by the outer space environment. That is why the clinic and the most tightly surrounded area use these sensors as direct conduction to the heart sound sensor.

11.3.2.2.4 Piezoelectric Heart Sound Sensor

The sensing structure of a sensor for piezoelectric acceleration. Such a sensor is used for heart sound analysis. Its structure is very simple, consisting of a piezoelectric crystal and a vibratory mass block. To exert some stress on the vibration mass block between

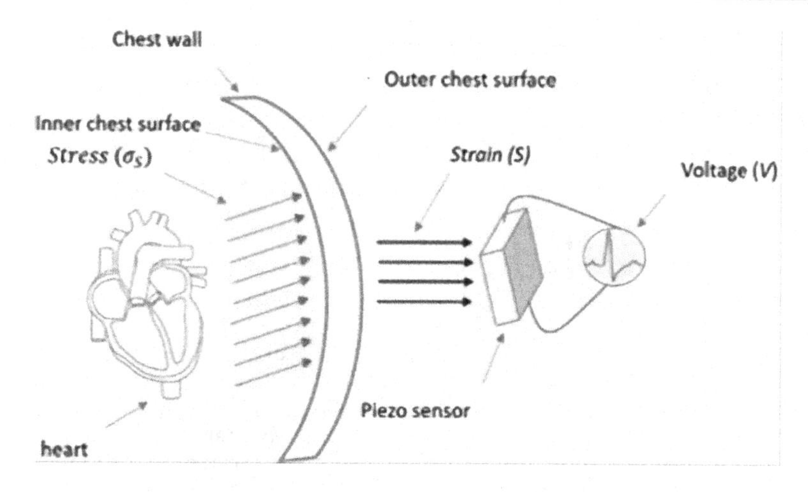

FIGURE 11.11 Heart sound sensor.

the top shell and mass block, a stress spring is used. Such a method could change the linear function of the sensing promptly.

11.3.2.2.5 Fetus Heart Sound Sensor

In clinical application, detecting fetal heart sound is very important for doctors who often need to understand the fetus' present body status. The piezoelectric thin-film PVDF sensor is used for the measurement of the heart sound of the fetus in Figure 11.12.

11.3.2.2.6 Blood Flow Sensor

If nutrients and oxygen have to influence tissues, the blood flow is mandatory to be maintained in the body. The output flow of cardiac is very often measured in the form of an index of cardiac performance, and the arterial graft helps to find blood flow and this ensures that the graft has been positively id during surgery, or the vein and peripheral arteries help in finding the blood flow that measures to evaluate vascular diseases.

This measurement happens in two different categories: first is a direct measurement, in that a transient blood flow is sensed by the blood pipe; second is an indirect measurement, placing a sensor outside the vein and sensing the blood flow by the structure linked to the blood flow like designed in Figure 11.13. These are the common pattern to find the measures of the blood flow that is taken into consideration clinically.

11.3.2.2.7 Respiration Sensor

The major source of human living is air. The functional organ that fits the inhaling of the air may also cause-effect due to unavoidable airflow this makes a sensing function to the respiratory organ. This respiration organ measurement has two classes namely: gas ingredient and physiological parameters.

A significant root of clinical or even non-clinical diagnosis is the quantity and consistency of the respiratory system, and it is important for the affected function in the fields of babies, surgical patients, sports medicine, monitoring of hospitalized patients,

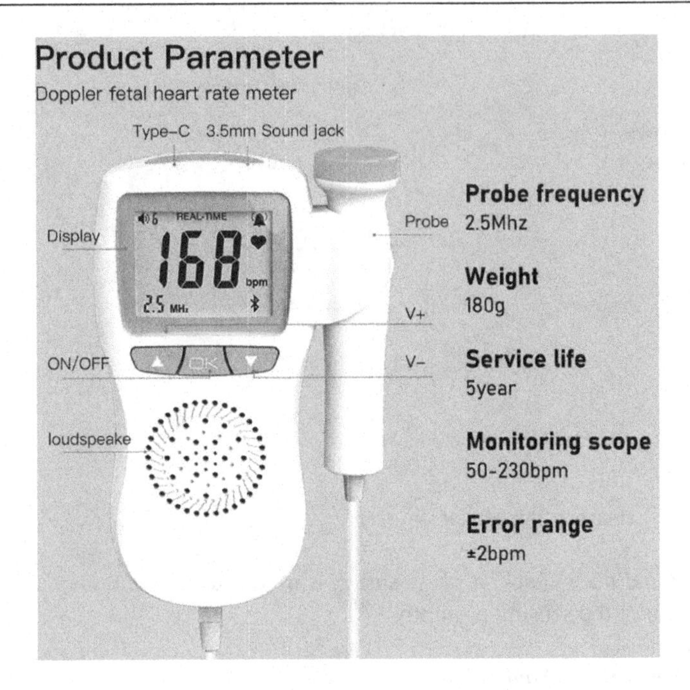

FIGURE 11.12 Fetus heart sound sensor.

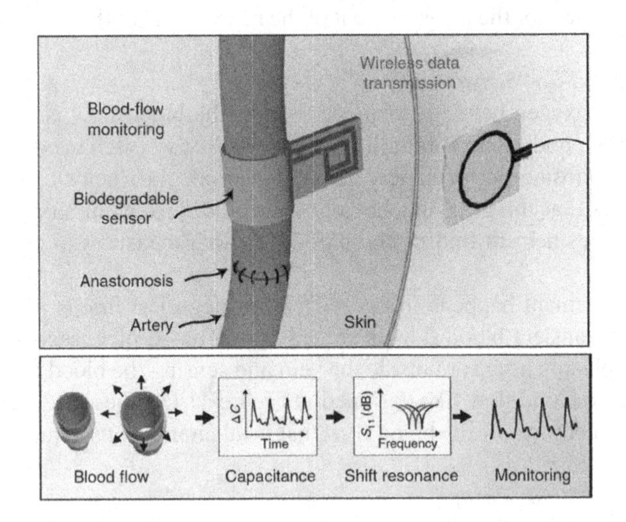

FIGURE 11.13 Blood flow S.

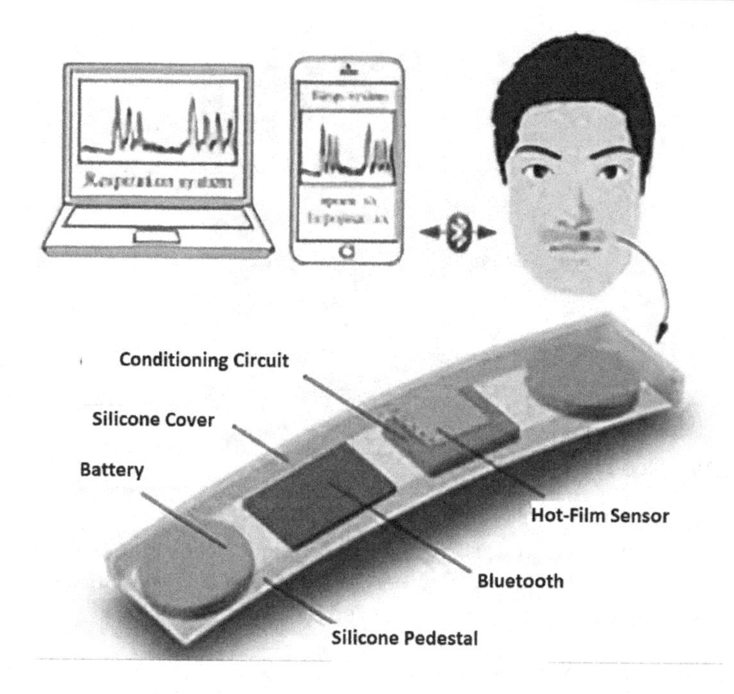

FIGURE 11.14 Respiration sensor.

and medical research. It was possible to classify the calculation of the respiratory system into three groups of parameters: respiratory flow, respiratory flow, and respiratory flow.

In biomedical research or clinic monitoring, the respiration frequency of a patient's requirements has to be occasionally detected to record the physiological status in Figure 11.14. A kind of sensor for respiration frequency is grounded on the thermistor sensing principle. The thermistor is straddling the front-end of a loose-leaf folder. When the loose-leaf folder clamps the nares, airstreams from the body flow through the superficial of the thermistor. Affording to the transformation of thermistor value, the breathing would be restrained.

11.3.2.2.8 BP Sensor

To have a good circulation of blood and to be maintained in the body, tissues play a better role in the form of perfusing the oxygen range. Evaluating the correct measurement of pressure seems to be applied as the basis in the vascular system. Very common means of sensing the BP are direct liquid coupling measurement, and pipe-end sensing measurement, occasionally in very critical cases we go for the indirect BP sensing measurement.

Direct fluid coupling estimation implies that the line loaded up with fluid is embedded into the deliberate part and that the pressing factor is estimated by fluid coupling of line end position in the body, which is the least difficult strategy. Line end detecting estimation utilizes a line end sensor to quantify pulse. A pipe-end sensor that can change over the pressing factor signal into an electronic sign is put on the deliberate part. And afterward, the electronic sign estimated is sent to the outside wire. Such a technique could keep away from the twisting the sign of circulatory strain. Line end detecting

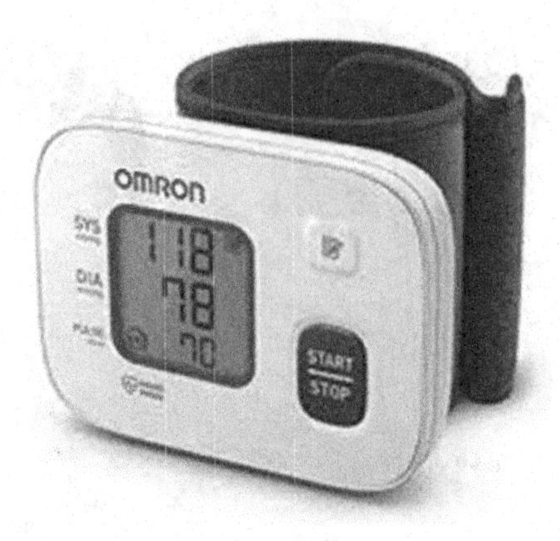

FIGURE 11.15 Blood pressure sensor.

estimation has a ton of preferences, yet such a technique needs to initiate the skin and relative sensors need to be put into the body. Henceforth aberrant pulse estimation is noted by individuals and constantly investigated like in Figure 11.15.

11.3.2.2.9 Temperature Sensors

Whatever the issues may be in the body, the first thing that has to be checked is the body temperature and whether the patient is a baby or an elder. The normal temperature in the body is 98°C. That is because the range in the temperature leads to the cause of the illness – if the temperature goes high, it leads to fever with high temperature that will overheat the body, and if the temperature is low enough, that leads to shivering and causes cold fever in the body. This variation in temperature itself tells the cause is due to viral or infections in the body. Nowadays, temperature checking has come to common people's hand itself. Utilizing a thermometer, we can to a certain extent confirm the cause and variations in temperature.

11.3.2.2.10 Blood Glucose Recording

World Health Organization (WHO) has given statistics that 9% of grownups worldwide presently agonize from diabetes. The cause of diabetics is decided by the range in glucose content in the body. So, there comes monitoring for this glucose content for assessing the health conditions of people with diabetes and must be performed frequently every day. Since diabetics is very commonly occurring disease and one device can be used for many. Glucose values for a healthy person and a person with diabetes are 70–100 and 80–130 mg/dL, respectively). In particular, biocompatible and biodegradable wearable glucose sensors have been widely used in human health monitoring and biomedical diagnostics because of their excellent biological functions. Besides, compared with traditional glucose sensors, biomultifunctional wearable glucose sensors can improve the safety of devices and reduce electronic waste.

11.3.2.2.11 Biomolecule Sensor

Lately, numerous adaptable biochemical sensors have been applied to real-time and in situ observing of human disease particles and protein markers. For example, a biocompatible silk/graphene-based spit sensor was accounted for by McAlpine and colleagues. The flexible salivation sensor perceives *Helicobacter pylori* cells in human spit, yet it additionally shows a low restriction of location (LOD) of roughly 100 cells when this gadget is moved to the outside of the tooth veneer. Albeit the adaptable salivation sensor shows high affectability to H. pylori cells, it is regularly influenced by dietary propensities, which restricts its capacity to precisely screen wellbeing. In this manner, the plan of another adaptable spit sensor with elite and protection from ecological impedance is vital.

11.3.2.2.12 pH Recording

As another significant property of bio-fluids, the arrangement of pH esteem is likewise firmly identified with human wellbeing. In particular, quick, real-time, and precise observing of the arrangement's pH worth can successfully guarantee an early finding of related diseases. A total of 123 as of now, the examination accentuation with respect to adaptable pH sensors is on creating high-performance and touchy detecting frameworks and viable perspiration inspecting strategies. Different wearable pH sensors have been accounted for, for example, adaptable FETs or sensors. A new amazing biodegradable pH sensor was accounted for by Rogers and associates. By consolidating single-crystalline silicon nanomembranes into a biodegradable elastomer (poly octane diol-co-citrate, POC), they got a biodegradable adaptable pH sensor that displayed a high strain limit (strain up to 30%) and high affectability (0.3 ± 0.02 and 0.1 ± 0.01 μS pH–1 for boron-doped Si nanoribbons (NRs) and phosphorus, separately). Furthermore, the pH sensor displayed incredible biodegradation, and all parts totally vanished following half a month.

11.3.2.2.13 Galvanic Skin Reaction (GSR) Sensor

Known as galvanic skin reaction (GSR), it alludes to changes in perspiration organ movement that are intelligent of the force of our passionate state, also called enthusiastic excitement. Skin conductance offers direct bits of knowledge into self-ruling enthusiastic guidelines as it isn't under cognizant control. For instance, in the event that you are terrified, upbeat, upset, or any enthusiastic related reaction, we will encounter an expansion in eccrine perspiration organ action which the sensor can get through the terminals and communicate to the expert gadget.

11.3.2.2.14 Light

Aside from the identification of natural compound signals, a graphene-based sensor is additionally applied in ecological actual signs estimation, among which light signals are of imperative significance. Thinking about the strength of human skin, UV sensors display a crucial job in identifying the level of UV power on the skin. Among the dynamic materials for estimating UV, two-dimensional graphene is one of the possible up-and-comers, showing remarkable adaptability and high versatility at room temperature. In view of the ZnO nanorods (NRs) and graphene, It was proposed that an adaptable UV field-impact semiconductor that demonstrated high photoconductive increase (8.3×106) under an entryway predisposition of 5 V, high adaptability of a curve of 12 mm without execution corruption, and high soundness with UV reaction after 10,000 twist cycles (Figures 11.7d and e). Concerning the expense and high reaction, It was proposed that an adaptable UV sensor using the rGO and hydrangea-like ZnO. This UV sensor showed a high photoresponse current (~1 μA), which was multiple times that of the ZnO UV sensor and a high on/off proportion (116/16), which expanded one request for a greatness contrasted and the first ZnO sensor.

11.3.2.2.15 Heavy Metals

Also, adhibitions of graphene are summed up by checking harmful hefty metal particles containing Cd^{2+}, Hg^{2+}, and Pb^{2+}. In view of electrospinning, It was already proposed that light-transmitting nanofibrous films including formed microporous polymers (CMPs)/polylactic corrosive (PLA). The imaginative synthetic sensor indicated high porosity, superb adaptability, and high surface-territory-to-volume proportion, which upgraded the capacity to identify nitroaromatic, oxidizing substantial metal particles, etc. A poisonous Hg^{2+} sensor with fluid door FET-type and straightforward graphene, which showed high adaptability and particularity of Hg^{2+} regardless of other compound substances.

11.3.2.2.16 X- Rays and Gamma Rays Sensors

Atomic medication tests are not quite the same as most other imaging procedures in that the principle capacity of the analytic test is to show the physiological capacity of the framework being explored, not normal for customary anatomical imaging, for example, MRI or CT. Atomic medication imaging examines are normally more organ or tissue-explicit (e.g.: lungs filter, heart check, bone sweep, and cerebrum examination) than those in conventional radiology imaging, which rather center around a particular segment of the body (e.g., chest X-beam, mid-region/pelvis CT filter, and head CT check). Additionally, there are atomic medication examines accessible that permit imaging of the entire body dependent on certain cell receptors or capacities. Models incorporate gallium checks, indium white platelet sweeps, MIBG and octreotide examines, and the more ordinarily realized entire body PET outputs or PET/CT filters.

An alteration called CT can be added to SPECT in the most developed tomographies, to accomplish a superior anatomical restriction of the needed articles and a superior picture quality. In PET/CT, for instance, this permits the area of tissues or tumors that could be seen on SPECT scintigraphy, yet are tricky to find concerning the other anatomical constructions.

11.3.2.2.17 Ultrasonic Sensors

Much clinical indicative imaging is done with X-beams. In light of the high photon energies of the X-beam, this sort of radiation is exceptionally ionizing, that is, X-beams are promptly fit for crushing atomic bonds in the body tissue through which they pass. This demolition can prompt changes in the capacity of the tissue in question or, in extraordinary cases, its destruction.

One of the significant points of interest in ultrasound is that it is a mechanical vibration and is thusly a nonionizing type of energy. In this way, it is usable in numerous touchy conditions where X-beams may be harming. Additionally, the goal of X-beams is restricted inferable from their extraordinary infiltrating capacity and the slight contrasts between delicate tissues. Ultrasound, then again, gives a decent differentiation between different sorts of delicate tissue.

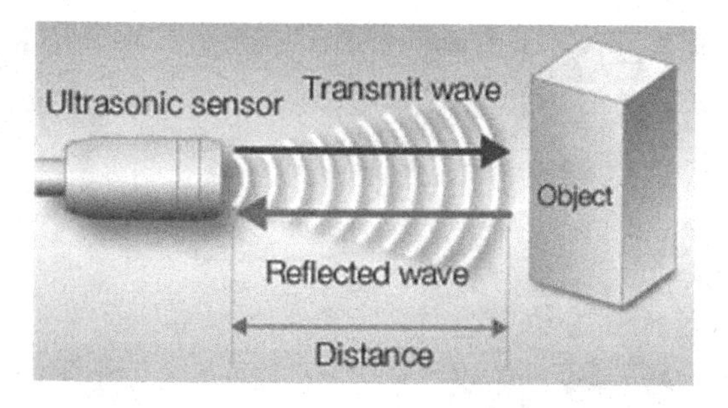

FIGURE 11.16 Ultrasonic sensor.

Ultrasonic filtering in Figure 11.16 in clinical analysis utilizes a similar rule as sonar. Beats of high-recurrence ultrasound, by and large over one megahertz, are made by a piezoelectric transducer and coordinated into the body. As the ultrasound navigates different interior organs, it experiences changes in acoustic impedance, which causes reflections. The sum and time deferral of the different reflections can be investigated to acquire data with respect to the inside organs. In the B-filter mode, a straight cluster of transducers is utilized to check a plane in the body, and the resultant information is shown on a TV screen as a two-dimensional plot. The A-filter strategy utilizes a solitary transducer to check along a line in the body, and the echoes are plotted as an element of time. This method is utilized for estimating the distances or sizes of interior organs. The M-filter mode is utilized to record the movement of inner organs, as in the investigation of heart brokenness. More prominent goal is acquired in ultrasonic imaging by utilizing higher frequencies, i.e., more limited frequencies. A limit of this property of waves is that higher frequencies will in general be significantly more unequivocally assimilated.

11.3.2.2.18 Body Fat Analyzer Machine

The BIA works by sending a little, innocuous electrical sign all through the body. (Very Star Trek, I know!) The gadget recognizes the measure of fat dependent on the speed at which the sign voyages. A slower voyaging signal as a rule shows a higher muscle-to-fat ratio.

Likewise, in Figure 11.17, a quicker speed shows less obstruction and a lower muscle-to-fat ratio percent. With respect to exactness, a handheld BIA gadget has about a±3.5%–5% wiggle room. To get the most precise perusing when utilizing a BIA, it is ideal to abstain from working out, showering, drinking liquor, drinking a lot of water, or eating a dinner 1–2 hours prior to perusing.

By and large, however, the BIA can get a decent gauge of a person's muscle versus fat percent while being speedy and effortless. Part of assuming responsibility for your wellbeing implies becoming more acquainted with your numbers, including muscle versus fat ratio and weight file.

11.3.2.2.19 Aerosol Therapy

Vaporized treatment like in Figure 11.18 has been utilized as a feature of the treatment for an assortment of respiratory infections. For sure, there is likewise huge interest in the

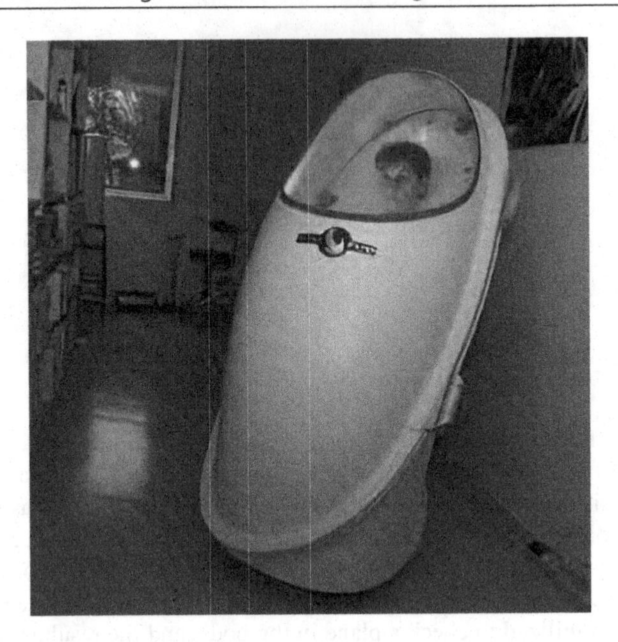

FIGURE 11.17 Body fat analyser.

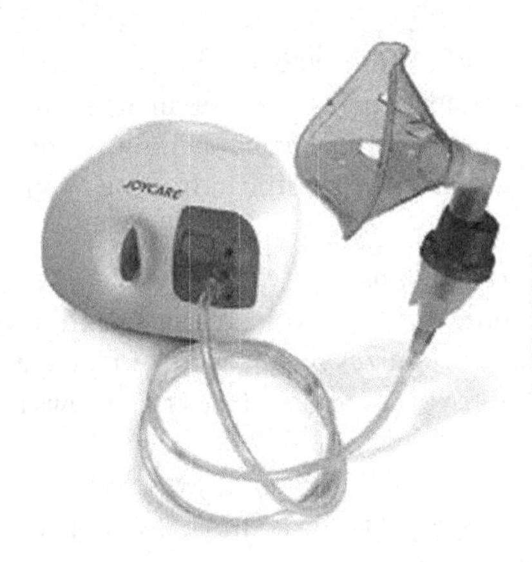

FIGURE 11.18 Aerosol therapy.

use of the respiratory framework as an entrance for foundational treatment of conditions that are not simply respiratory in nature.

Factors, for example, an enormous surface region, slight air–blood hindrance, and vascular epithelium combined with low first-pass digestion and enzymatic movement could accomplish high bioavailability for aerosolized medication treatment. The chance of accomplishing high neighborhood drug focuses at the restorative site for respiratory pathology, quick beginning of activity, and lower foundational results have accordingly prompted a reestablished interest in the field of aerosolized medication treatment in escalated care.

11.3.2.2.20 Image Sensors

The two fundamental sorts of computerized picture sensors are the charge-coupled gadget (CCD) and the dynamic pixel sensor (CMOS sensor), created in correlative MOS (CMOS), or N-type MOS (NMOS or Live MOS) advancements. Both CCD and CMOS sensors depend on MOS innovation, with MOS capacitors being the structure squares of a CCD, and MOSFET speakers being the structure squares of a CMOS sensor.

Cameras coordinated into little shopper items for the most part use CMOS sensors, which are typically less expensive and have lower power utilization in battery-fueled gadgets than CCDs. CCD sensors are utilized for top-of-the-line broadcast-quality camcorders, and CMOS sensors rule in still photography and customer merchandise where by and large expense is a significant concern. The two sorts of sensors achieve a similar undertaking of catching the light and changing over it to electrical signs.

Every cell of a CCD picture sensor is a simple gadget. At the point when light strikes the chip, it is held as a little electrical charge in each photosensor. The charges in the line of pixels closest to the (at least one) yield enhancers are intensified and yield, at that point each line of pixels moves its charges one line nearer to the speakers, filling the vacant line nearest to the speakers. This cycle is then rehashed until all the lines of pixels have had their charge enhanced and yield.

A CMOS picture sensor has a speaker for every pixel contrasted with the couple of enhancers of a CCD. This outcomes in less zone for the catch of photons than a CCD, yet this issue has been overwhelmed by utilizing microlenses before each photodiode, which shines light into the photodiode that would have in any case hit the speaker and not been detected (Bangash et al., 2014). Some CMOS imaging sensors additionally use back-side enlightenment to expand the number of photons that hit the photodiode (Mukhopadhyay et al., 2011). CMOS sensors can possibly be actualized with fewer parts, utilize less force, as well as give a quicker readout than CCD sensors (Kim et al., 2016). They are likewise less defenseless against friction-based electricity releases.

11.3.2.2.21 Ambient Sensors

Surrounding sensors can unpretentiously screen people in the home climate. Surrounding sensors can screen action designs, rest quality, washroom visits, and so forth and give alarms to parental figures when irregular examples are noticed. Such sensors are required to make the home of things come more intelligent and more secure for patients living with persistent conditions as in Figure 11.19.

This methodology has the quality of being absolutely subtle and of maintaining a strategic distance from the issue of losing or harming wearable gadgets. "Brilliant home" innovation that incorporates encompassing and ecological sensors has been fused into an assortment of recovery-related applications. One such application is surrounding helped living (AAL) which alludes to shrewd frameworks of wellbeing that help with the person's living climate. It covers ideas, items, and administrations that interlink and improve new advancements and the social climate. AAL advancements are inserted (disseminated all through the climate or straightforwardly coordinated into machines or furniture), customized (custom-made to the clients' requirements), versatile (receptive to the client and the client's current circumstance), and expectant (foreseeing clients' longings quite far without cognizant intercession). Stefanov et al. gave a rundown of

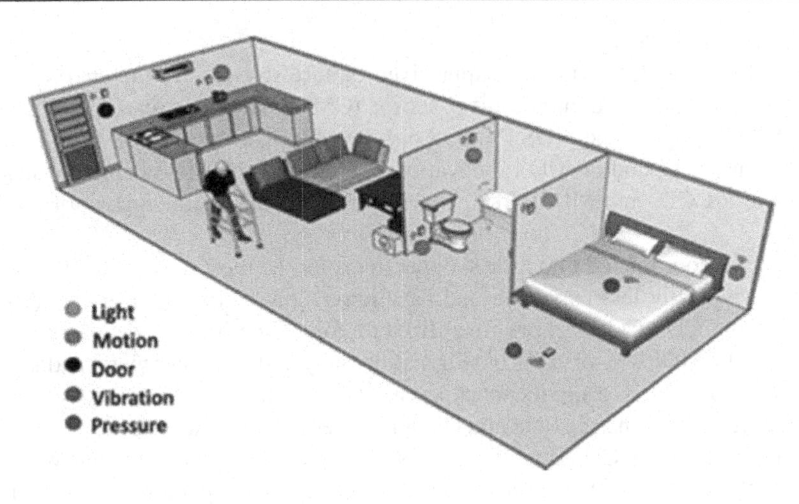

FIGURE 11.19 Ambient sensors.

the different kinds of gadgets that can be introduced in shrewd homes, and the related objective client populaces.

Distant observation of patient status and self-administration of persistent conditions address the frequently sought-after utilization of AAL advancements. The mix of wearable and encompassing sensors is being investigated and models are being created. A pertinent application in the field of recovery identifies with the distinguishing proof of a patient's examples of action and on giving recommendations concerning explicit practices and activities for self-administration of ailments. In this specific circumstance, data accumulated utilizing wearable sensors is enlarged by data assembled utilizing surrounding sensors. Information gathered utilizing, for example, body-worn accelerometers could be expanded by movement sensors circulated all through the home climate to decide the sort and force of the exercises performed by a person. As needs be, an individual going through checking who experiences, for example, the constant prominent pneumonic illness could get input about not overexerting himself/herself and the presentation of recovery practices that would be recommended to keep a palatable useful level.

Creative answers for perceiving crises in the home can be accomplished through a blend of checking indispensable boundaries of the individual living at home just as directing the states of homegrown apparatuses. Individual security can be improved if crucial information measures are joined with the observing and control of gadgets in the family unit. Far off observing expected wellsprings of risk expands the individual conviction that all is good and can make life a lot simpler and more agreeable (e.g., checking whether the oven or the espresso machine has been turned off and has the option to turn them off distantly if important). Sensors installed in electrical gadgets and inside and windows might be coordinated into a simple-to-utilize house-control framework that likewise gives improved individual wellbeing and security. An insightful framework may suggest to turn off gadgets as well as lights in a condo or not to fail to remember the pillbox or the portable terminal expected to educate companions or neighbors when fundamental.

11.4 FUTURE SCOPE IN HEALTHCARE

- Better cloud integration with existing technologies

  More than 90% of healthcare organizations are widely utilizing the cloud to host applications, according to a 2017 Healthcare Information and Management Systems Society (HIMSS) survey on cloud use. However, the industry is still using the cloud for separate functions, such as clinical apps, data hosting, and backup, and not in a holistic fashion. The HIMSS survey found that though there is a high level of cloud usage at healthcare organizations, the functionality is still limited.
- Deeper AI infusion

  Artificial intelligence has been a part of healthcare for years, but experts believe that in the next decade, it will be a regular part of the industry. A survey of 200 healthcare professionals by Intel Corporation, released in July 2018, found that 37% of respondents were using AI in limited ways, and 54% believe that there will be widespread AI adoption in the next 5 years.
- Infrastructure upgrades that make healthcare more accessible

  The ability for clinicians to meet with patients via web and mobile portals is essential for chronic care management, says Rhonda Collins, DNP, RN, a chief nursing officer at Vocera, and founder of the American Nurse Project.
- Smarter therapies

  Though smartphones, smartwatches, and other smart devices are being used by consumers, a focus on "smart" hasn't translated to healthcare solutions, says Kal Patel, MD, MBA, president of BrightInsight, a Flex company that provides biopharma and medical technology internet platforms.
- Enhanced personalized medical care

  The biggest thing is going to be our ability to use these advanced technology enablers to get much better at doing personalized medicine and personalized healthcare with our patients Jain says. "Because the back-end healthcare technology is crunching all of their clinical data and administrative data, looking at their genomic profile, looking at their social determines of health much faster than any human physician or clinician could, and combining that information in a trusted way."
- A workflow that mimics consumer technology

  When millennials go to work at a hospital, we are asking doctors, nurses, and care teams to step back 20 years and use landline phones, fax machines, pagers, and overhead calls-all which downgrade and add complexity to our millennial workforce. They carry a heavy burden every day working with patients in stressful hospital environments, and the very basic technology they're using only adds to the stress. Furthermore, we are adding to cognitive loads by forcing them to remember procedures and how to use outdated technologies they are not naturally accustomed to using. So, over time, antiquated technology that doesn't mirror what is used in our personal lives and is not secure will be eliminated. As younger people continue to enter the workforce, many hospitals will be forced to modernize.

11.5 FUTURE TRENDS

The material-based wearable body sensor organization will essentially propel understanding in the arising fields of biosensor plan, BSN, and biomedical registering, which will assume a vital part in the computerized age to meet the wide objectives of uses in both the medical clinic and home conditions. In particular, a complete framework system including correspondence and calculation should be created, including three significant segments: remote biosensors that can be appended to patients and additionally end-clients; proficient correspondence conventions that can send the physiological signs from the sensors to far-off workers; and effective sign preparing calculations that can extricate helpful data from the sensor information for clinical specialists to decide.

The new fast improvement of the IoT and remote correspondence network innovation and progressed cell phones and applications pretty much fulfill these requirements. WBSN gadgets screen essential signs, help the development of fake appendages, and capacity as smaller than usual "base stations" for the assortment and transmission of different physiological boundaries. In no time, smaller than expected transponders inserted in pills will empower specialists to track and screen drug use. As microchips become more modest and all the more remarkable, sometimes remote wearable innovations may have the option to screen or control essentially every real capacity and development. Numerous remote clinical gadgets speak with close-by beneficiaries that are associated with landline organizations, cell frameworks, or broadband offices that entrance the internet. Patients presently shouldn't be attached to one spot by tangled links, making a more secure working environment for clinical experts and a more agreeable climate for the patient, with a decreased danger of disease.

Remote observing licenses patients to remain outside clinical conditions, decreasing medical services costs and empowering doctors to get indispensable data on an ongoing premise without the requirement for office visits or clinic affirmations. Scrounging populaces, WBSN gadgets offer a significant answer for protection and oversaw care. Material-based wearable body sensor organizations will be an innovation foundation of future shrewd and associated wellbeing frameworks, particularly for universal wellbeing checking and registering. Novel calculations, hypothetical models, and rules for commonsense usage ought to be set up to empower lightweight and effective wellbeing observation. Future examination in the field will include signal handling and information preparation, sensor plan, remote medical care, displaying, recreation, and execution investigation. WBSNs will altogether affect nonintrusive walking wellbeing observing by incorporating lightweight sensor arrangements into detecting, correspondence, and figuring.

11.6 CONCLUSION

This chapter is having the content of the applications of the wireless BSN starting with a description of what is a sensor network, what is wireless BSN, and the architecture principles of the wireless sensor network. We have included the complete description of the sensors that play a major role in this BSN along with its applications. In the final part, we have stated how the growth is required and also the recent trends that were made and yet waiting for the results growth in this field.

BIBLIOGRAPHY

An, B., Ma, Y., Li, W., Su, M., Li, F., and Song, Y. (2016). Three-dimensional multi-recognition flexible wearable sensor via graphene aerogel printing. *Chemical Communications* Vol. 52, pp. 10948–10951. Doi: 10.1039/C6CC05910D.

Bangash, J.I., Abdullah, A.H., Razzaque, M.A., and Khan, A.W. (2014). Reliability aware routing for intra-wireless body sensor networks Hindawi publishing corporation. *International Journal of Distributed Sensor Networks*, Article ID 786537, 10 pages, Doi: 10.1155/2014/786537.

Bussooa, A., Neale, S., and Mercer, J. R. (2018). Future of smart cardiovascular implants. *Sensors* Vol. 18, pp. 1–11. Doi: 10.3390/s18072008.

Choi, J., Kim, S.-M., Ryu, R.-H., Kim, S.-P., and Sohn, J. (2018). Implantable neural probes for brain-machine interfaces – current developments and future prospects. *Experimental Neurobiology* Vol. 27, pp. 453–471. Doi: 10.5607/en.2018.27.6.453.

Gonzalez, E., Peña, R., Vargas-Rosales, C., Avila, A., and De Cerio, D.P.D. (2015). Survey of WBSNs for pre-hospital assistance: Trends to maximize the network lifetime and video transmission techniques. *Sensors*.

Huang, H., Su, S., Wu, N., Wan, H., Wan, S., Bi, H., and Sun, L. (2019). Graphene 26 based sensors for human health monitoring. *Frontiers in Chemistry*.

Kanagachidambaresan, G.R. (2015). Trustworthy architecture for wireless body sensor network. IGI Global, Chapter 2.

Kanagachidambaresan, G.R. (2018). Trustworthy architecture for wireless body sensor network. IGI Global, Chapter 17.

Kim, Y.K., Wang, H., and Mahmud, M.S. (2016). Wearable Body Sensor Network for *Health Care Applications*. Elsevier BV.

Mukhopadhyay, K., Sil, J., and Banerjea, N.R. (2011). A competency based management system for sustainable development by innovative organizations: A proposal of method and tool, Vision. *The Journal of Business Perspective*.

Sangwan, A., and Bhattacharya, P.P. (2015). Wireless body sensor networks: A review. *International Journal of Hybrid Information Technology* Vol. 8, No. 9, pp. 105–120 Doi: 10.14257/ijhit.2015.8.9.12.

Sentimental Analysis with Web Engineering and Web Mining

12

Rahul Malik and Sagar Pande
Lovely Professional University

Contents

12.1 Introduction 198
12.2 Constituents and Approaches 199
  12.2.1 Literature Aspects 199
12.3 Proposed Methodology 203
12.4 Outcomes and Explanations 208
  12.4.1 Movie Review Dataset 208
  12.4.2 OHSUMED Dataset 211
  12.4.3 Outcomes 213
12.5 Conclusion 215
Bibliography 216

Abstract

Across the globe, people are using mobile devices due to the swift evolution of Internet facilities that opened the doors for the utilization of many online platforms mainly for social connectivity among the people and for sharing their thoughts, activities, and ideas on worldwide issues. A lot of consumers got habituated to looking for reviews and ratings of a product on related well-known websites before going to buy that product. It turned out to be a common norm to have an analysis on buying a product from the perspective of believability and security. If we can get huge, data related to products and therefore by analyzing the obtained information, we need to recognize the necessary details of the products. In such a scenario, sentimental analysis acts handy for

DOI: 10.1201/9781003051022-12

recognizing or understanding the thoughts and reactions of consumers. Sometimes, sentiment analysis is also considered thought exploration or thought mining. SVD (singular value decomposition) and PCA (principal component analysis) are utilized in this chapter for the identification of sentiments or thoughts from text mining to improve accuracy and decrease the duration of implementation. The major contributions made in this analysis are as follows: the first one for the preprocessing of the information collected by using a proposed algorithm for creating a proper basis for classification of sentiments, the second is the supplementary attributes to improve the accuracy of the classification, the third is applying SVD and PCA on the obtained data, and the last one is to identify and associate the results of the performance by designing five segments based on various attributes with or without lemma. The test results demonstrate that the current approach is more reliable than other approaches and that its duration of implementation can be minimized by its current approach.

12.1 INTRODUCTION

Social media sites such as Facebook, Twitter, Instagram, etc.; e-commerce sites such as Amazon, Flipkart, etc., and many other online activities generate larger volumes of data every day that was categorized into big data as it will permit us the collection of the huge quantity of data both in the form of structured as well as unstructured format. Big data mainly deals with the extraction and analyzing the collected information to identify useful information. In this scenario, automated categorization of text methods of data analysis such as text mining, data mining, and web mining are suggested by many scholars. It is very difficult to analyze the case of the consumer datasets depending on their reviews and thoughts as it necessitates other methods for summarizing the reviews and opinions. A lot of platforms are already in existence for considering and recording the opinions and reviews of the customers.

That recorded information can be examined or analyzed for the identification of behavioral patterns and thoughts of customers for the policies and marketing promotions of the firms, goods preferred by customers, and for observing reputations (Saleh et al., 2011). The sites which are taking the reviews have become more popular and are valuable tools for views to be collected and analyzed. Moreover, a consumer who wishes to buy a good frequently finds information over the Internet for appropriate views. Consequently, examining the thoughts, perceptions, feelings, and attitudes has become a necessity globally to understand a customer; such kind of analysis is known as sentimental analysis (SA) (Medhat et al., 2014). In many sectors, SA has produced the best outcomes, notably in the areas of intelligent marketing (Li & Li, 2013), the satisfaction of consumers (Kang & Park, 2014), and retail prediction (Rui et al., 2013). Also, seeking appropriate text representation features is a concern.

Users generally can't read all feedback from the different review platforms. A summary is used by many scholars in analyzing aspects of the sentiments for larger durations. The study of sentiment analysis (SA) from various calibration approaches for mining the views, ideas, expressions, and attitudes articulated in the text formats (Medhat et al., 2014). The role of evaluating thoughts and emotions is also known as opinion mining. Besides, SA focuses on the various issues addressed in the summary or the text (Montoyo et al., 2012).

There are two key sections of text mining: (i) attributes to be collected and chosen and (ii) classification by using an algorithm. For film analyses, while removing and choosing items that use unigrams, i.e., individual words, bigrams, i.e., two words in order leaving one after the other word from the starting of the text and parts of speech (POS). Furthermore, (Rahate & Emmanuel, 2013) displays their results with POS labeling by utilizing n-gram sequences (Moraes et al., 2013), training the classifier model by entering the data that can be transformed by the matrix with the help of the term frequency-inverse document frequency (TF-IDF) approach. It is very difficult to define the appropriate functionality to process a file. Because reviews usually include less than 300 characters, the features reflecting the individual are difficult to identify. Furthermore, it reveals that certain initiatives have no standardized laboratory environments. This chapter proposes additional functionality and a singular value decomposition (SVD) and after that primary component analysis (PCA) methodology to boost accuracy and minimize the time of deployment for text mining and design five-segment tests with specific characteristics to evaluate results and explore variables that affect the classification accuracy. In brief, this analysis has the following aims:

1. Present a sentimental of text mining, focused on a supplementary method of enhancing the accuracy of classification in big data analyzes propose an abstraction algorithm to facilitate the exact description of feelings;
2. To the data dimensionality and deployment time, using an effective SVD and after that primary component analysis (PCA) methodology.

The chapter is arranged as follows: Section 12.2 presents related works such as quality tests, sentiment mining, extraction and selection tools, SVD, and primary component analysis methods. Section 12.3 introduces the analysis principle and the suggested process. The experimental findings are discussed in Section 12.4. Lastly, Section 12.5 points out the inference.

12.2 CONSTITUENTS AND APPROACHES

12.2.1 Literature Aspects

Such related literature and definitions are discussed briefly in the following sections including user reviews, opinion processing, abstraction, and collection of features and graders. Such related literature and definitions are discussed briefly in the following sections including user reviews, opinion processing, abstraction, and collection of features and graders.

Reviews or feedbacks of products – A platform that reports customer feedback on goods, resources, and businesses offers an electronic analysis of items. Owing to Internet 2.0, a huge amount of individual usage of the word online through the mouth to share the corresponding impressions and desires for several items. To order to fully grasp customer opinions and desires, online product reviews provide marketers with more open knowledge. Many previous opinion mines have been evaluating their commodity properties by examining the relevant details for customers to decide if they can or should not purchase the product and by which the amount of product knowledge customers may allow decision-making. Indeed, comments are viewed as a screening method to minimize quality confusion. Suggested an approximation econometric model of preferences derives the desires of the users from online product analysis. Besides, Archak, Ghose, and Ipeirotis also reported that customers' rating reviews are beneficial for company strategies.

Extraction of sentiment – Analyzing the ideas or thoughts or sentiments analysis is a common technique related to the processing of texts, which operates text data analysis to interpret the opinions expressed. The text is subjective and is typically carried through individuals with normal thoughts, moods, views, and perceptions. For social network studies, SA is commonly employed for various methods for natural language processing (NLP), data recovery (DR), and organized/unorganized information extraction. The greatest obstacle is the unorganized real-time information. In recent years there have been several attempts to collect valuable and usable knowledge from such unorganized datasets. The effort of (Ravi & Ravi, 2015) can be narrowly separated into six different aspects: The study of emotions/thoughts/views/sentiments should be mentioned as follows:

1. Definition of subjectivity (Pang et al., 2002);
2. Description of emotions (Ravi & Ravi, 2015; Li & Tsai, 2013; Tan et al., 2009; Bollegala et al., 2013);
3. Calculation of value analysis (Tsytsarau & Palpanas, 2012);
4. Development of word lists i.e., Lexicons (Niles & Pease, 2003);
5. Spam identification thoughts (Ravi & Ravi, 2015); and
6. Extraction of the word of opinion (Medhat et al., 2014; Ravi & Ravi, 2015.

The first phase in sentiment mining from literature, data collection, and preprocessing involves this whole process. The second stage is to derive characteristics from the raw data and to use a master classification system.

Feature mining and collection – The extraction and collection of features have been explored and studied for a very long time in text mining. Records are to be interpreted as multidimensional vectors for feature extraction (Liu et al., 2005). Type collection or object extraction methods are used to reduce corpus dimensionality and increase the classifier training period. Function selection is used to remove new functionality from all sets of apps with dynamic mapping (Whitelaw et al., 2005). The main problem with role extraction is that it is difficult to understand its output while extracted features are useless (Abbasi et al., 2011). The selection of functions increases the efficiency of the classification by reducing corpus dimensionality without reducing its precise nature. In the literature, several unattended methods of role selection were suggested. The most popular approaches are frequency document (DF), variance term (TV), the benefit of information (BI), shared data (SD), and so on. This has been shown that the benefit of intelligence is more efficient than other strategies (Rui et al., 2013; Abbasi et al., 2011).

Singular value decomposition (SVD) – It is a lower computational matrix within the region of linear algebra, on which the decomposition of its value may be used only on the existing matrix. The SVD approach is utilized to derive the corresponding eigenvalues of the existing matrix and the corresponding eigenvectors of the existing matrix (Kang et al., 2016). In other terms, matrix, let us say A, can be factorized into the three matrices, and of that one is a diagonal matrix and other two are orthonormal matrices and these can be taken as the multiplication of an orthonormal matrix with diagonal matrix and multiplied by the transport of another orthonormal matric that can be represented as $A = U*D*VT$, which is orthonormal in U and V and diagonal in matrix D with a positive real number. For many instances, matrix A is similar to an establishable and reasonable solution for the data matrix; we may get another matrix, let us say B, of the rank j, and it is the best-approximated matrix for A; besides, everyone may seek and j for various claims. An SVD can be applied in several fields. Besides, defining a lot of spectral decomposition based on the methodologies of linear algebra, an SVD is defined on all matrices. In SVD, eigenvalues may be used to evaluate the matrix size for data reduction as a decision criterion.

Principle component analysis (PCA) – It is one of the major methodologies used for the decrease in the dimensions (Yu et al., 2014) of the existing dataset that is used to divide broad classes of variables into smaller sets in such a way that much of the knowledge from the original data is contained in the chosen key components. "PCA" is a mathematical equation that modifies the relevant attributes into the less uncorrelated key (PC) elements. The first primary variable accounts for most variation, and the remaining volatility is the greatest amount for each following portion. In multivariate statistics, the "PCA" methodology is almost analogous to the factor analysis methodology. The number of attributes in the results is usually less than the number of earlier existing attributes. "PCA" can be considered an ellipsoidal data in the space of n-dimensions that represent the main components on each ellipsoid axis. It measures and determines the covariance matrix to modified information and the corresponding eigenvalues matrix, as well as the corresponding eigenvectors of the matrix of the modified data. Finally, the collection of orthogonalizing and standardization of the unit vectors will be completed. The global algorithms that are capable of removing the most significant dataset characteristics are SVD and the principal component analysis (PCA). The covariance matrix is a matter of concern for the PCA methodology, whereas the dataset itself is a matter of concern for SVD methodology (Liu et al., 2016).

Machine learning classifier models – The work mentioned in this chapter mainly deals with four rising classifier models, several of these classifier models are used in the categorization of sentiments. All four categorizers are Naïve Bayes (NB), support vector machines (SVMs), maximal entropy (ME), and Random Forest (RF). Now, we are going to discuss these methods in the following way.

Naïve Bayes (NB) – Based on the theorem of Bayes, the naïve Bays classification (Lewis, 1998) is especially appropriate when the inputs' dimensionality is strong since they are simpler probabilistic classification. From the theorem of Bayes, consider the possibility that a given paper, *d*, is allocated to a class ai, and bi, that is a single word in that particular article. The training data are used to measure *P(aj)* and *P(bi)*, and the *P(bi)* also reflects a conditional likelihood that *xi* will appear in the class *aj* paper. Although this approach has a conditional presumption of freedom that does not encompass real-time scenarios, the advantages of these conditions are very clear and remarkably very strong [28].

Maximum entropy (ME) – It is a valuable methodology to estimate any probability distribution in many NLP fields (Nigam et al., 1999). It can be used in the text classification was shown to be a feasible and efficient algorithm. The ME principle is that the distribution should be considered uniform when nothing is known. This research is concerned with classification ME which, for document classification, is often stronger than NB (Juan et al., 2007). ME model aims to identify the parameters that render all the training data as probable as possible. The ME model approximation of *P*(c|*d*) was started as an exponential form (Nigam et al., 1999). If suppose the considered word to be "happy", wherever that particular word, then the view of the document considered to be positive if that word appears elsewhere, for example, the function is triggered. The ME model of the classifier is an exponential form of probabilistic classifier. The ME classification doesn't say, unlike the NB classification, that the features are unconditionally distinct. The ME model of the classifier can be solved irregular text classification issues including identification of language, classification of texts, analyzing the opinions/ideas/sentiments, etc.

Support vector machine (SVM) – These (Vapnik, 1995; Joachims, 1998) are located in the space of domain with dimensions of n, where the points of the data are precisely marked. There are several potential hyperplanes to be chosen to differentiate between various types of data points, and the aim is to achieve the maximum possible margin hyperplane. Vectors generated by this methodology are information points that will enter into the hyperplanes which influence hyperplane locations and orientations. The vectors generated by this methodology are used to optimize the margin classifications. The release of the vectors which are generated will change the locations of the hyperplane' since this helps us to develop the SVM. Texts must be converted into vectors to enable the function of SVM text classification. Let us consider the class, *Dj*, the methodology's Lagrange multiplier is utilized for the parameters of primordial derivatives, w in a text document {1, –1} (responding positive and negative) to find the solution (Vapnik, 1995). If you can't split the data points into a high dimension, instead of using the kernel alternative the information points can be modified into a separable hyperplane.

Random forest (RF) – It is an adaptable and simple learning methodology (Breiman, 2001). Owing to its simplicity and its utilization for classification, i.e., categorization as well as for the methodology of regressions, RF is perhaps one of the most popular algorithms. Normal RF classification is used for random samples in the database of different decision trees. One of these will divide the information details and at the same time generates the list with nodes that deals with decisions and terminal nodes. All trees are combined in the RF for creating a less fluctuating layout. The RF can effectively operate on broad datasets and manage several variables without missing variables. Therefore the advantages of these classifier models are (i) to decrease the overfitting by considering the mean of the several trees and (ii) numerous trees can be considered to decrease the disparity or the instability in the performance of the classification where the classification data is different from that between training and test data. These classifier models use essential techniques such as bagging and bootstrapping (Cheng, 2016).

12.3 PROPOSED METHODOLOGY

The purpose of the classification of sentiments is to categorize the available information in the form of text and transform it into already identified categories such as positive, negative, pleased, and sad. The difficult task for the categorization of thoughts/sentiments is to identify a way of improving outcome accuracy. Several factors, including various preprocessing measures, the classification of sentiments (documents or sentences), different attributes, lexicons, and various methodologies of machine learning, can influence the study. In earlier studies, the outcomes of the attribute selection methodologies such as unigrams, i.e., individual words, bigrams, i.e., the pair of words considered in the order from starting of the text one after the other, and part-of-speech (POS) tag (Rahate & Emmanuel, 2013), POS tagging n-graph sequences (Pang et al., 2002), and TF-IDF (Moraes et al., 2013) have mentioned variations in numbers.

Ravi et al. (Ravi & Ravi, 2015) have demonstrated that several studies don't have an identical investigational framework. So, this article was centered (Cheng, 2016) to enhance the tests on other accuracy characteristics to apply the "first SVD then PCA" approach to decrease the dimensions and reduce text classification runtime. In addition, this research uses five segments of the experiment to compare the output of each segment with other segments with different characteristics and to reveal the factors that influence the classification accuracy.

Figure 12.1 shows the proposed methodology procedure of implementation. Initially, in the primary step, the gathered data is utilized for the sentiment classification. In the secondary step, the necessary and primary steps of preprocessing will be considered by R statistics are acquired. In the tertiary step, the defined attributes are mined inclusive of TF-IDF, the positive frequencies, the negative frequencies, the number of adjectives, and the number of adverbs will be considered by the sentiment score of each document. In the quarternary step, training and predicting the data by utilizing the classifier is made. Finally, the evaluation of outcomes or results will be made.

The algorithm proposed – For easy understanding of the suggested methodology, implementation of methodology is classified into five major steps for mentioning the algorithm proposed procedure.

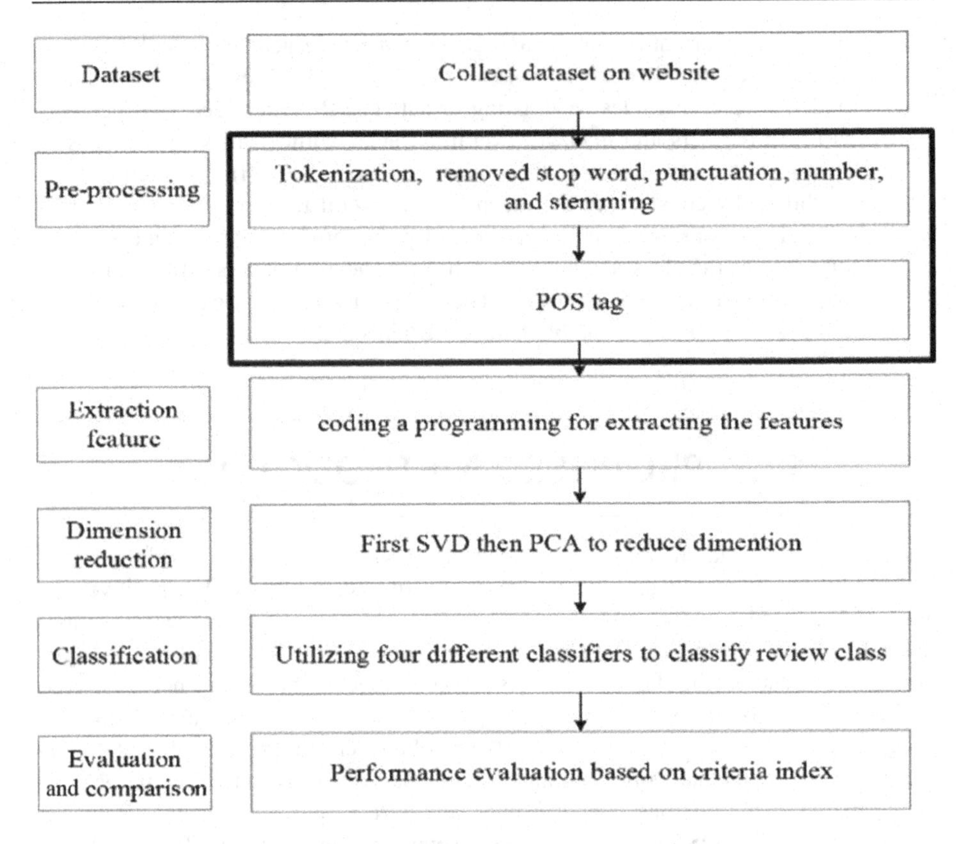

FIGURE 12.1 The proposed methodology.

Step 1 Gathering of the dataset – Initially, we gathered the most frequently used dataset, the movie dataset comprises documents related to sentimental scenarios (Parmar et al., 2014; Movie reviews dataset). The movie reviews are very difficult to classify than the other reviews. This data comprises positive and negative reviews each of 1,000. Transforming the text file into excel format for the labeled data was made possible through an Excel VBA (Microsoft) program that was specially coded by us.

Simultaneously, another dataset gathered from OHSUMED source was produced (Hersh et al., 1994; Ohsumed Dataset). This data comprises 23 various categorical cardiovascular diseases. The five classes are selected from the source that are related to the peripheral nervous system blood vessels such as C03, 11, 12, 15, and C21. The related description of the data is considered as mentioned in Table 12.1.

Step 2 Preprocessing – In practice, the noise will always be attached to the data gathered from the source. So, there is a need for processing the data beforehand and going for further implementations of learning methodologies. The various preprocess techniques to be done in this scenario such as tokenization, removing stop words, lemming, POS tagging, and feature mining (Ravi & Ravi, 2015). The first technique we applied to raw data is

TABLE 12.1 The Descriptions of Various Categories of the OHSUMED Dataset

VARIOUS DISEASES CATEGORIES	*CLASS*	*FEATURE*
Virus	C03	402
Nervous systems	C11	1,673
Eye	C12	375
Cardiovascular	C15	2,661
Immunologic	C21	1,330

TABLE 12.2 Descriptions of Various Attributes

ATTRIBUTES	*DESCRIPTIONS*
TF-IDF	Text converted into term frequency-inverse document frequency (TF-IDE), from 0 to 1 range
Frequency of negative	Negative terms count in a document
Frequency of positive	Positive terms count in a document
Frequency of an adverb	POS terms count is an adjective
Frequency of an adjective	POS term count is an adverb

the removal of punctuations which is referred to as tokenization that is not impacting the classifier's accuracy. Next to this step, the necessary thing is to remove the stop word such as articles, conjunction, etc., which is called the removal of stop words as these words will disintegrate the outcomes. Next to this step, the necessary thing is to identify the root word of the words in the document by ignoring the POS of the word is called the lemming. Later in this step, the necessary thing is to recognize words' parts of speech in the document that is known as POS tagging. This step is essential as the noise always exists along with the data. Finally, from the data, we need to withdraw some useful information in the form of attributes to achieve the appropriate information. The packages are used in the R language to implement the step of POS tagging with the help of openNLP and RTexTools. The subsequent section will discuss feature mining. Besides feature mining, feature selection is also very important, which can impact the result significantly while analyzing the scenario.

Step 3 Feature mining and combining the other features – An attribute set inclusive of TF-IDF, negative terms frequency, positive terms frequency, adverbs frequency, and adjectives frequency as mentioned in Table 12.2 defined by the study. Conversion of all documents into weights is the major scenario of this step. So, we transform into weights of TF-IDF, simultaneously, positive frequencies and the negative frequencies will be formed into another set of attributes. Subsequently, we use POS tag to count the number of adverbs, adjectives, and a combination of other attributes that will be added.

The descriptions of the parameters of TF-IDF were mentioned in Table 12.3, and the suggested feature mining process is shown in Algorithm 12.1.

TABLE 12.3 Term Frequency-Inverse Document Frequency (TF-IDF) Parametric Descriptions

PARAMETRIC	*DESCRIPTION*
removeSparseTerms	Matrix terms deleted; % of sparse empty elements
minDocfreq	Words seems under set amount in files dumped with document matrix
lemmaWords	Logical parameter, make use of dialect specified to specify for lemma words

Algorithm 12.1: Feature Mining

Representation: ***M***: labeled text or sentiment containing matrix, T_i: it represents the ith text in the matrix M, ***PL***: it represents the word list with positive words. ***NL***: it represents the word list with negative words, W_i: it represents the *i*th text containing word list, ***PM***: it represents the common terms that occur in T_i as well as PL, ***NM***: it represents the common terms that occur in T_i as well as NL. ***Tag***: it represents the POS tag containing the form of a matrix.

Output: It represents the frequency or count of different words such as PM, TF-IDF, frequency of adjectives, NM, and adverbs

1. For each text (ith text)T_i in (labeled matrix)M:
   a. Discriminate each $T_{i(\text{ith text})}$ into unigram, i.e., one word at a time will be considered and saved into W_i.
   b. Match the word list W_i with already existing word lists such as PL and NL, and then transform it into the word lists PM and NM.
   c. Return the word lists PM as well as NM.
2. For each text (*i*th text) T_i in (labeled matrix) M:
   a. Annotate the *T*th with word_token_annotator by using the package of the R language such as openNLP.
   b. Annotate the $T_{i(\text{ith text})}$ with POS after $T_{i(\text{ith text})}$ is annotated by using word_token_annotations.
   c. Count the number of adjectives as well as adverbs in tag.
   d. Return the outcome.

Step 4: The decrease in dimensions – It is important as the TF-IDF is in the form of a matrix that is very large and sparse with a lot of elements of the matrix are zero which will be made difficult to analyze this matrix (Table 12.4). So, in this step, first, we will apply SVD and then PCA methodologies for the decrease in dimensions of the matrix. Once feature mining is done, we will obtain the preprocessed data in the form of a matrix which will be used as the input for SVD methodology. It decomposes the matrix in such a way that the values of elements in the matrix will be transformed to zero or close to zero. After this methodology application, the other technique, i.e., PCA methodology will be applied to decrease the dimensions of the matrix. The final resultant of PCA was mentioned in Table 12.5. Finally, after decreasing the dimensions, the size of the movie dataset was decreased from 2,000 rows and 46,467 columns to 2,000 rows and 2,000 columns.

TABLE 12.4 Explanation of the Outputs of the Principal Component Analysis (PCA)

PARAMETERS	*DESCRIPTION*
x	Individuals principal components coordinate
Scale	Feature standard deviations
Center	Feature means
Rotation	Feature loadings matrix
Sdev	PC standard deviations

TABLE 12.5 The Classifier Settings of the Parameters

Naïve Bayes	na.action	A feature to establish the excitement to be considered in case NAs are found. The default activity isn't to count them because of the computation of the probability elements.
	laplace	Good two-fold controlling Laplace smoothing. The default (zero) disables Laplace smoothing.
ME	verbose	Logical specifying whether to offer descriptive result. about the training system. Defaults to False, and no output.
	use_sgd	A rational indicating that SGD parameter evaluation must be used. Defaults to False.
SVM	type	C-classification.
	shrinking	The choice if you should work with the shrinking heuristics (default: TRUE).
	epsilon	Epsilon within the insensitive loss feature (default: 0.1).
	tolerance	Tolerance of termination criterion (default: 0.001).
	cost	Cost of constraints violation (default: 1).
	gamma	The parameter required for most kernels except linear (default: 1/(data dimension)).
	kernel	Linear for two labels. The radial time frame for several labels.
	scale	A rational vector indicating the variables being scaled. Per default, information are scaled internally (both x and y variables) to 0 mean plus unit variance.
RF	cutoff	A vector of length equal to the number of courses. The "winning" category for an observation is the person with the optimum ratio of the proportion of votes to cutoff. Default is $1/k$ wherein k will be the number of classes.

Step 5: Classification – The methodologies such as ME, NB, RF, and SVM are used or utilized for training the treated data for the classification of text in classes. The comprehensive description and settings of the parameters are mentioned in Table 12.6. The parameters of this learning will be set to the default values for ten-fold cross-validation and ten times the random sampling for verifying the accuracy.

TABLE 12.6 Sentiment Classification Used Confusion Matrix

	PREDICTED	
	NEGATIVE	*POSITIVE*
Actual negative	Negative true	Positive false
Actual positive	Negative false	Positive true

TABLE 12.7 Characteristics of the Used Datasets OHSUMED and Movie

	OHSUMED DATASET	*MOVIE DATASET*
Number of review	6,198	4,000
Class	C03(402), C11(1,673), C12(375), C15(2,661), C21(1,331)	Negative (2,000) Positive (2,000)
Feature	Unigram	Unigram
Number of attributes	30,496, five attributes	47,578, five attributes
Vector space	6,198 3,0496	4,000 4,7578

Step 6: Evaluation – For evaluation of the performance of the classification uses accuracy and it can be calculated with the help of the confusion matrix in this step as mentioned in Table 12.7. For accuracy calculation, positive and negative labels will be considered from the confusion matrix (Moraes et al., 2013). Equation 12.1 represents the equation for the calculation of accuracy and it utilizes in this experimentation by using the confusion matrix with the help of positive and negative labels.

$$\text{accuracy} = \text{True Positive} + \text{True Negative} \tag{12.1}$$

12.4 OUTCOMES AND EXPLANATIONS

This learning gathers two datasets that are open and uses the various segments of the experiment to implement the segments and relate the outputs with the existing methodologies. The gathered datasets are considered from the widely popular sites used in text mining scenarios. For this experimentation, we used two datasets, one of them is the movie review and the other is the cardiovascular disease abstracts, i.e., OHSUMED. The details of the characteristics of both datasets are mentioned in Table 12.8.

12.4.1 Movie Review Dataset

The various settings of the parameters of the TF-IDF employ the study a lemming to achieve various attributes frames into the design of five segments of the experiment

TABLE 12.8 Segments of the Experiment (Movie Dataset)

ID	*PARAMETER*	*NUMBER OF ATTRIBUTES*	
		NO LEMMING	*LEMMING*
Segment 1 (M1)	All attributes	47,578	31,699
Segment 2 (M2)	MinDocfreq=3	13,457	9,858
Segment 3 (M3)	MinDocfreq=4	6,077	5,553
Segment 4 (M4)	RemoveSparseTerm=0.98	4,477	3,778
Segment 5 (M5)	RemoveSparseTerm=0.96	974	1,159

to relate with the existing methodologies and deliberate about the factors which will impact the accuracy of classifiers. Five segments of the experiment have various attributes and settings, as mentioned in Table 12.8. The attribute set has about features of 47,578 are achieved in the suggested algorithm, later to steps 2 and 3. For testing the consequences of the various settings, the SVD and then the utilization of PCA methodologies is related to the existing methodologies. The various settings are ten folded cross-validation and ten times random sampling for testing the performance.

As mentioned in Table 12.9, the suggested methodology with the combination of other attributes is performing better than the methodology without the combination of other attributes from the perspective of the average accuracy of the five classifiers. The ME and the linear SVM methodologies are performing better than the other methodologies from the perspective of accuracy. As mentioned in Table 12.10, indicates the decrease in dimensions, demonstrates the comparison of the outcomes between with and without the decrease in dimensions for segments 1 and 4 with no lemming. The suggested methodologies with the combination of other features are performing better than without the combination of other attributes for the cases of having dimensions and without the decrease in dimensions. The SVM and ME classifiers are having better accuracy in most of the settings. The outcomes of the experimentations can be explained in the following three ways:

1. Feature mining – Segment 1 is the best suited for the proposed methodology in the performance aspect as demonstrated in Table 12.9. Segment 4 achieves the best accuracy for all of the experimentations, and the number of attributes decreases to 9.4% as demonstrated in Table 12.8. The impact of lemmatization is not very obvious in this experimentation, as demonstrated in Table 12.9.
2. Combining the other features – the mean of the five classifiers for segments 1–5 are used to identify the impact of combining the other features as demonstrated in Table 12.9 and Figure 12.2. Later, combining the other features into the attribute set, the outcomes demonstrates the other features can achieve better outcomes with a methodology such as SVM-RBF.
3. The decrease in dimensions – The outcomes demonstrate that the accuracy with the decrease in dimensions technique and the accuracy without the decrease in dimensions technique is mentioned in Table 12.10. Better accuracy is achieved with the usage of the movies dataset with the suggested

TABLE 12.9 Relating the Outcomes Without the Decrease in Dimensions (Dataset Movie)

	LEMMA	*ADD ATTR.*	*SVM-RBF*	*SVM-LINEAR*	*ME*	*RF*	*NB*	*AVE.*
M1	Yes	Yes	0.5407	0.8407	0.8523	0.8298	0.6269	0.7421
		No	0.5491	0.8480	0.8513	0.8240	0.6337	0.7416
	No	Yes	0.5545	0.8550	0.8551	0.8329	0.6254	0.7448
		No	0.5099	0.8548	0.8536	0.8244	0.6262	0.7338
M2	Yes	Yes	0.5520	0.7648	0.7820	0.7668	0.5127	0.6859
		No	0.5520	0.7477	0.7502	0.7463	0.5118	0.6638
	No	Yes	0.5625	0.7681	0.7712	0.7596	0.5053	0.6733
		No	0.5126	0.7512	0.7591	0.7472	0.5219	0.6566
M3	Yes	Yes	0.7457	0.7266	0.7272	0.7354	0.5001	0.6880
		No	0.6537	0.6759	0.6796	0.6932	0.4994	0.6444
	No	Yes	0.7472	0.7191	0.7277	0.7332	0.4998	0.6878
		No	0.6989	0.6847	0.6936	0.6757	0.4992	0.6524
M4	Yes	Yes	0.6814	0.8353	0.8527	0.8375	0.7293	0.7875
		No	0.5681	0.8289	0.8496	0.8291	0.7359	0.7627
	No	Yes	0.6738	0.8484	0.8624	0.8296	0.8362	0.7893
		No	0.5390	0.8512	0.8691	0.8322	0.8353	0.7658
M5	Yes	Yes	0.7175	0.8345	0.8327	0.8218	0.7685	0.7940
		No	0.5476	0.8143	0.8131	0.8226	0.7580	0.7513
	No	Yes	0.7181	0.8293	0.8257	0.8217	0.7548	0.7899
		No	0.5490	0.8295	0.7918	0.8253	0.7551	0.7525

TABLE 12.10 The Decrease in Dimensions Outcomes in the Dataset Movie

			SVM-RBF	*SVM-LINEAR*	*ME*	*RF*	*NB*	*AVE.*
M1	RD Yes	Attr. Yes	0.4901	0.8572	0.8408	0.7494	0.5412	0.6997
		Attr. No	0.5121	0.8563	0.8465	0.7201	0.5419	0.6974
	RD No	Attr. Yes	0.5545	0.8561	0.8465	0.8329	0.6254	0.7448
		Attr. No	0.4999	0.8548	0.8536	0.8244	0.6262	0.7338
M4	RD Yes	Attr. Yes	0.8431	0.8494	0.8490	0.7235	0.5345	0.7501
		Attr. No	0.8384	0.8380	0.8385	0.7188	0.5459	0.7561
	RD No	Attr. Yes	0.6738	0.8484	0.8624	0.8207	0.7362	0.7803
		Attr. No	0.5390	0.8512	0.8601	0.8322	0.7353	0.7658

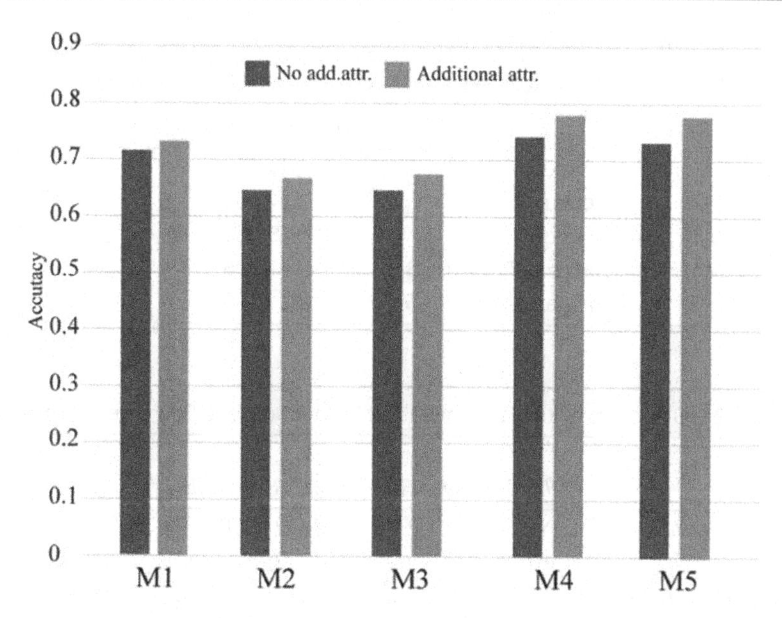

FIGURE 12.2 The impact of combining the other features for different segments in the movie dataset.

combining the other features. Hence, the performance in sentimental classification has improved by combining the other features and the SVD then PCA techniques.

12.4.2 OHSUMED Dataset

The source for the second dataset is the OHSUMED corpus, generated in (Movie reviews dataset; Hersh et al., 1994). The used dataset has documents of 51,327 divided into 24 categories. The documents are classified into classes CO3, 11, 12, 15, and 21 based on the multiple class classification. Total features of about 30,496 were there in the full feature set. Table 12.8 demonstrates the characteristics of the OHSUMED dataset, Table 12.11 demonstrates the five segments of experimentations having various settings and attributes for testing the impact of various feature mining settings of the dataset. Later PCA is applied for measuring the impact of the decrease in dimensions; SVD is applied as well. Lastly, the various settings are categorized into ten groups randomly for cross-validation.

Table 12.12 demonstrates the dataset without the decrease in dimensions, Table 12.13 demonstrates the dataset with the decrease in dimensions, and Figure 12.3 demonstrates the outcomes of OHSUMED dataset. Segments 1 and 4 can be considered the best experimentation segments which can be observed in Table 12.12. On the whole, the impact of lemmatization is not even evident. Table 12.13 demonstrates the outcomes of segments 1 and 4 for the decrease in dimensions for comparison for the same purpose of the depiction of the Movie dataset. On the whole, segment 1 obtained the best accuracy of 78.79% without the decrease in dimensions and obtained an accuracy of 71.26%

TABLE 12.11 The Decrease in Dimensions Outcomes in the Dataset OHSEMED

SEGMENT ID	*PARAMETER*	*NUMBER OF ATTRIBUTES*	
		NO LEMMATIZATION	*LEMMATIZATION*
Segment 1 (M1)	All attributes	30,496	22,022
Segment 2 (M2)	MinDocfreq=3	14,705	12,242
Segment 3 (M3)	MinDocfreq=4	8,482	7,318
Segment 4 (M4)	RemoveSparseTerm=0.98	1,405	1,337
Segment 5 (M5)	RemoveSparseTerm=0.96	237	269

TABLE 12.12 Outcomes of the OHSUMED Dataset without the Decrease in Dimensions

		ADD ATT.	*SVM-RBF*	*SVM-LINEAR*	*ME*	*NB*	*AVE*
M1	Lemma	Yes	0.7959	0.7492	0.7594	0.0814	0.6251
		No	0.7980	0.7512	0.7523	0.0798	0.6234
	No Lemma	Yes	0.7948	0.7504	0.7483	0.0781	0.6263
		No	0.7985	0.7519	0.7551	0.0792	0.6244
M2	Lemma	Yes	0.7857	0.7240	0.6961	0.0667	0.5914
		No	0.7703	0.7239	0.6965	0.0672	0.5990
	No Lemma	Yes	0.7607	0.7328	0.6985	0.0617	0.5983
		No	0.7736	0.7227	0.6997	0.0617	0.5995
M3	Lemma	Yes	0.7363	0.6296	0.6222	0.0590	0.5446
		No	0.7323	0.6470	0.6350	0.0590	0.5534
	No Lemma	Yes	0.7269	0.6397	0.5991	0.0570	0.5445
		No	0.7196	0.6384	0.6204	0.0570	0.5453
M4	Lemma	Yes	0.7899	0.6883	0.6218	0.4568	0.6559
		No	0.7813	0.6868	0.6284	0.4571	0.6556
	No Lemma	Yes	0.7643	0.6847	0.6225	0.4440	0.6460
		No	0.7772	0.6837	0.6279	0.4451	0.6405
M5	Lemma	Yes	0.6754	0.6264	0.6535	0.4948	0.6158
		No	0.6720	0.6243	0.6528	0.4932	0.6144
	No Lemma	Yes	0.6519	0.5973	0.6412	0.4843	0.5921
		No	0.6107	0.4442	0.6300	0.4821	0.5635

TABLE 12.13 Outcomes of the OHSUMED Dataset with the Decrease in Dimensions

		ADD ATT.	*SVM-RBF*	*SVM-LINEAR*	*RF*	*ME*	*NB*	*AVE*
M1	Yes RD	Yes	0.2927	0.7238	0.6315	0.5489	0.3239	0.4942
		No	0.2730	0.7193	0.6174	0.5376	0.3266	0.4940
	No RD	Yes	0.7988	0.7504	0.7297	0.7483	0.0881	0.6251
		No	0.7985	0.7519	0.7261	0.7551	0.0792	0.6244
M4	Yes RD	Yes	0.6994	0.6613	0.4934	0.5963	0.4214	0.5764
		No	0.6946	0.6490	0.4958	0.5826	0.4252	0.5716
	No RD	Yes	0.7843	0.6847	0.7184	0.6225	0.4440	0.6489
		No	0.7732	0.6837	0.7197	0.6279	0.4451	0.6499

with a decrease in the dimensions. Whenever the feature set is large enough, usage of SVM with the usage of kernel methodology radial basis function (RBF) will result in the best accuracy.

12.4.3 Outcomes

Some outcomes have been identified from the implemented work which is summarized in the following manner:

1. Feature mining – The first major outcome of the experimentation demonstrates that segments 4 and 5 perform better than the other segments with the usage of the movie dataset which can be identified in Figure 12.2 and Table 12.9. In this experimentation, the higher accuracy are obtained in segment 4, and the number of features are reduced to 9.4% which can be identified in Table 12.8. Besides this observation, we can also identify that the impact of lemmatization is not so evident in the experimentation which can be identified in Table 12.9. Segments 1 and 4 are performing better than the remaining segments with the usage of the OHSUMED dataset which can be identified in Figure 12.3 and Table 12.12. In this experimentation, the higher accuracy is obtained in segment 4 and the number of features is decreased to 9.4% which can be identified in Table 12.11. From the above two indications, we are able to deduce that segment 4 showing an impact due to lemmatization, i.e., it results in decreasing features, and the speed of the computation has improved.
2. Combining the other features – The suggested methodology mainly deals with the combination of other features with the existing ones for the improvement of the accuracy of the text classification model which implies increasing the concentration or the amount of negative and positive adverbs and adjectives. For testing the effect of combining the other features in this experimentation, the accuracy mean of all the classifiers from segments 1 to 5 is computed with the help of the movie dataset which can be identified in Table 12.10. In general, we can identify the improvement in the accuracy due to combining

other features and it is more significant in SVM with the kernel as RBF, i.e., SVM-RBF classifier that can be identified in Figure 12.3 and Table 12.10. In this experimentation with the usage of the OHSUMED dataset, we can identify that the best segments are achieved with segment 1 and 4, and it improved the mean accuracy marginally due to the impact of lemmatization that can be identified in Figure 12.3 and Table 12.13. Besides this, we are also able to identify that segments 1–5 are performing better in the case of combining other features rather than without combining the other features along with without lemmatization, which can be identified from Figure 12.3.

3. The decrease in dimensions: In this experimentation, the performance in terms of accuracy with the decrease in dimensions is almost the same as that without the decrease in dimensions with both scenarios of usage of the movie and the OHSUMED datasets which can be identified in Tables 12.11 and 12.14. In the case of the movie dataset, combining the other features after the decrease in dimensions can achieve better performance. The duration of implementation will be decreased in sentimental text mining to test the SVD and then the PCA. This experimentation was made with the help of two datasets such as the movie dataset and the OHSUMED dataset, implemented in the environment of the *R* platform, Intel i7–3770k, 3.5GHz CPU, and Windows 10 system. The complete duration of implementation of five segments, i.e., classifiers are shown in Table 12.14. Four segments of all five segments can decrease the duration of implementation, which is not the same in segment 5. So, for the suggested methodology, combining the other features and the decrease in dimensions are sustainable.

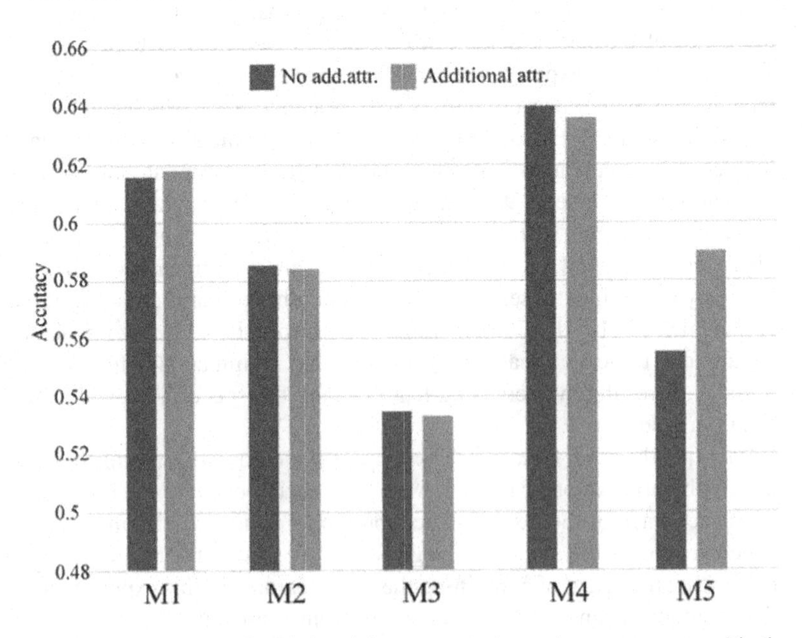

FIGURE 12.3 The impact of additional features on the various segments with the usage of the OHSUMED dataset.

TABLE 12.14 The Complete Duration of the Implementation of Five Classifiers (Time Units: Seconds)

			OHSUMED DATASET		*MOVIE DATASET*	
			WITH PCA	*WITH SVD+PCA*	*WITH PCA*	*WITH SVD+PCA*
M1	No Lemma	YesAtt	266,002.14	21,596.71	70,005.35	4,403.22
		NoAtt	266,813.91	23,490.81	70,581.12	4,659.92
	Lemma	YesAtt	221,885.62	22,968.26	38,010.79	4,965.30
		NoAtt	229,822.73	23,902.31	47,126.49	4,812.58
M2	No Lemma	YesAtt	282,563.75	19,055.63	23,672.66	4,847.24
		NoAtt	301,525.22	21,297.45	26,594.84	4,799.63
	Lemma	YesAtt	214,123.26	20,113.75	16,098.47	4,861.77
		NoAtt	213,700.29	19,710.19	18,718.4	4,763.38
M3	No Lemma	YesAtt	193,607.56	14,390.69	13,492.03	4,530.97
		NoAtt	202,189.82	14,821.74	23,483.63	4,681.32
	Lemma	YesAtt	204,986.23	15,550.37	11,960.84	4,927.53
		NoAtt	517,883.10	15,647.37	13,838.9	4,956.53
M4	No Lemma	YesAtt	61,632.98	14,676.93	7,334.4	4,632.05
		NoAtt	58,965.67	14,359.62	7,326.07	4,527.28
	Lemma	YesAtt	14,569.32	15,280.09	4,679.0	4,837.77
		NoAtt	15,346.36	14,897.09	4,714.28	4,606.77
M5	No Lemma	YesAtt	7,734.47	4,236.95	1,496.55	2,016.95
		NoAtt	4,096.38	7,364.83	1,523.34	1,924.54
	Lemma	YesAtt	1,532.65	9,473.87	1,614.8	2,451.70
		NoAtt	1,467.96	8,822.11	1,675.97	2,288.86

12.5 CONCLUSION

The main focus of this chapter is to improve the accuracy and decrease the duration of implementation in SA by using methods such as SVD and PCA. The concentration or the amount of negative and positive adverbs and adjectives are also considered the additional attributes in this scenario. The outcomes of these experimentations demonstrate that the method that was proposed performed better in terms of accuracy when compared with the existing techniques. Besides, due to its additional features, it also impacted the accuracy in the direction of improvement, particularly in the case of SVM with kernel RBF classifier. When we compare all the models across, the best-preferred model can be an SVM classifier and ME classifier in the scenario of SA. There are more issues still in existence that can be analyzed in the future to improve the concept of SA. Some of them are as follows:

1. In the attribute choosing technique, the major concerns are as follows: (i) usage of a word list that specifies a domain for identifying or filtering the attributes, (ii) improving the accuracy with the help of assigning various

weights to attributes, and (iii) noticing the association between the document and the words is also prominent.
2. Applying the suggested approach to various domains of applications, such as management of reputation and identification of social emotions.

BIBLIOGRAPHY

Abbasi, A., France, S., Zhang, Z., & Chen, H. (2011). Selecting attributes for sentiment classification using feature relation networks. *IEEE Transactions on Knowledge and Data Engineering*, 23(3), 447–462.

Archak, N., Ghose, A., & Ipeirotis, P. G. (2011). Deriving the pricing power of product features by mining consumer reviews. *Management Science*, 57(8), 1485–1509.

Bollegala, D., Weir, D., & Carroll, J. (2013). Cross-domain sentiment classification using a sentiment sensitive thesaurus. *IEEE Transactions on Knowledge and Data Engineering*, 25(8), 1719–1731.

Breiman L. (2001, 10). Random forests. *Machine Learning*, 45(1): 5–32.

Cheng C.H. (2016). A text mining based on refined feature selection to predict sentimental review. In *Proceedings of the Fifth International Conference on Network, Communication and Computing*, 150–154, Dec 17–21, 2016 Kyoto, Japan.

Hersh, W., Buckley, C., Leone, T.J., Hickam, D. (1994). OHSUMED: An interactive retrieval evaluation and new large test collection for research. In *Proceedings of the 17th Annual International ACM SIGIR conference on Research and Development in Information Retrieval*, 192–201, July 03–06, 1994, Dublin, Ireland.

Joachims, T. (1998, April). Text categorization with support vector machines: Learning with many relevant features. In *European Conference on Machine Learning*, 137–142. Springer, Berlin, Heidelberg.

Juan, A., Vilar, D., & Ney, H. (2007). Bridging the gap between Naive Bayes and maximum entropy text classification. In *PRIS*, 59–65.

Kang, B., Lee, K., & Choe, J. (2016). Improvement of ensemble smoother with SVD- assisted sampling scheme. *Journal of Petroleum Science and Engineering*, 141, 114–124.

Kang, D., & Park, Y. (2014). Based measurement of customer satisfaction in mobile service: Sentiment analysis and VIKOR approach. *Expert Systems with Applications*, 41(4), 1041–1050.

Li, S. T., & Tsai, F. C. (2013). A fuzzy conceptualization model for text mining with application in opinion polarity classification. *Knowledge-Based Systems*, 39, 23–33.

Li, Y. M., & Li, T. Y. (2013). Deriving market intelligence from microblogs. *Decision Support Systems*, 55(1), 206–217.

Liu, L., Kang, J., Yu, J., Wang, Z. (2005) A comparative study on unsupervised feature selection methods for text Clustering. In *Proceeding of NLP-KE*, Vol. 9, 597–601.

Liu, Y. Y., Wang, Y., Walsh, T. R., Yi, L. X., Zhang, R., Spencer, J., et al. (2016). Emergence of plasmid-mediated colistin resistance mechanism MCR-1 in animals and human beings in China: A microbiological and molecular biological study. *The Lancet Infectious Diseases*, 16(2): 161–168. Doi: 10. 1016/S1473-3099(15)00424-7 PMID: 26603172.

Lewis, D.D. (1998). Naive (Bayes) at forty: The independence assumption in information retrieval. In *Proceedings of the European Conference on Machine Learning (ECML)*.

Medhat, W., Hassan, A., & Korashy, H. (2014). Sentiment analysis algorithms and applications: A survey. *Ain Shams Engineering Journal*, 5(4), 1093–1113.

Montoyo, A., Martı´Nez-Barco, P., & Balahur, A. (2012). Subjectivity and sentiment analysis: An over-view of the current state of the area and envisaged developments. *Decision Support Systems*, 53(4), 675–679.

Moraes, R., Valiati, J. F., & Neto, W. P. G. (2013). Document-level sentiment classification: An empirical comparison between SVM and ANN. *Expert Systems with Applications*, 40(2), 621–633.

Movie reviews dataset, http://www.cs.cornell.edu/people/pabo/movie-review-data/.

Nigam, K., Lafferty, J., & McCallum, A. (1999). Using maximum entropy for text classification. In *Proceedings of the IJCAI-99 Workshop on Machine Learning for Information Filtering*, 61–67.

Niles, I., & Pease, A. (2003, June). Linking Lixicons and ontologies: Mapping wordnet to the suggested upper merged ontology. In *Ike*, 412–416.

Ohsumed Dataset, https://www.mat.unical.it/OlexSuite/Datasets/SampleDataSets- download.htm.

Pang, B., & Lee, L. (2004). A sentimental education: Sentiment analysis using subjectivity summarization based on minimum cuts. In *Proceedings of the 42nd Annual*, 271.

Pang, B., Lee, L., & Vaithyanathan, S. (2002, July). Thumbs up ?: sentiment classification using machine learning techniques. In *Proceedings of the ACL-02 Conference on Empirical Methods in Natural Language Processing. Association for Computational Linguistics*, Vol. 10, 79–86.

Parmar, H., Bhanderi, S., & Shah, G. (2014). Sentiment mining of movie reviews using random forest with tuned hyperparameters. In *International Conference on Information Science*, Kerala.

Rahate, R. S., & Emmanuel, M. (2013). Feature selection for sentiment analysis by using SVM. *International Journal of Computer Applications*, 84(5), 24–32.

Ravi, K., & Ravi, V. (2015). A survey on opinion mining and sentiment analysis: Tasks, approaches and applications. *Knowledge-Based Systems*, 89, 14–46.

Rui, H., Liu, Y., & Whinston, A. (2013, 11). Whose and what chatter matters? The effect of tweets on movie sales. *Decision Support Systems*, 55(4), 863–870.

Saleh, M. R., Martín-Valdivia, M. T., Montejo-Ráez, A., & Ureña-López, L. A. (2011). Experiments with SVM to classify opinions in different domains. *Expert Systems with Applications*, 38(12), 14799–14804.

Tan, S., Cheng, X., Wang, Y., & Xu, H. (2009, April). Adapting naive bayes to domain adaptation for sentiment analysis. In *European Conference on Information Retrieval* (pp. 337–349). Springer, Berlin, Heidelberg.

Tsytsarau, M., & Palpanas, T. (2012). Survey on mining subjective data on the web. *Data Mining and Knowledge Discovery*, 24(3), 478–514.

Turney, P.D. (2002). Thumbs up or thumbs down? Semantic orientation applied to unsupervised classification of reviews. In *ACL*.

Vapnik, V. (1995). *The Nature of Statistical Learning Theory*. Springer: New York.

Weathers, D., Swain, S. D., & Grover, V. (2015). Can online product reviews be more helpful? Examining characteristics of information content by product type. *Decision Support Systems*, 79, 12–23.

Whitelaw, C., Garg, N., & Argamon, S. (2005, October). Using appraisal groups for sentiment analysis. In *Proceedings of the 14th ACM International Conference on Information and Knowledge Management*, 625–631, ACM.

Xiao, S., Wei, C. P., & Dong, M. (2016). Crowd intelligence: Analyzing online product reviews for preference measurement. *Information & Management*, 53(2), 169–182.

Yu, X., Chum, P., & Sim, K.B. (2014). Analysis the effect of PCA for feature reduction in non-stationary EEG based motor imagery of BCI system. *Optik-International Journal for Light and Electron Optics*, 125(3): 1498–1502.

Big Data in Cloud Computing - A Defense Mechanism

13

N. Ramachandran
Indian Institute of Management

Salini Suresh, Sunitha and Suneetha V
Department of Computer Applications
Dayananda Sagar College

Neha Tiwari
IIS University

Contents

13.1 Introduction 221
13.2 Overview of Cloud 221
13.2.1 Important Characteristics 222
13.2.1.1 On-Demand Self-Service 222
13.2.1.2 Broad Network Access 222
13.2.1.3 Resource Pooling 222
13.2.1.4 Rapid Elasticity 222
13.2.1.5 Measured Service 222

DOI: 10.1201/9781003051022-13

13.2.2 Deployment Models 223
13.2.2.1 Private Cloud 223
13.2.2.2 Community Cloud 223
13.2.2.3 Public Cloud 223
13.2.2.4 Hybrid Cloud 223
13.2.3 Service Models 223
13.2.3.1 Software as a Service (SaaS) 223
13.2.3.2 Platform as a Service (PaaS) 223
13.2.3.3 Infrastructure as a Service (IaaS) 223
13.3 Big Data: Overview 223
13.3.1 Characteristics of Big Data 224
13.3.1.1 Volume 224
13.3.1.2 Veracity 224
13.3.1.3 Value 224
13.3.1.4 Variety 224
13.3.1.5 Velocity 224
13.3.2 Significance of Big Data 225
13.3.3 Big Data in Cloud 225
13.4 Security Issues Faced by the Big Data in Cloud 226
13.4.1 Confidentiality 227
13.4.2 Integrity 227
13.4.3 Authenticity 227
13.4.4 Availability 228
13.4.5 DoS and DDoS Attacks 228
13.4.6 MitM Attack 228
13.4.7 Sniffer Attacks 229
13.4.8 Spoofing 230
13.4.9 SQL Injection Attack 230
13.4.10 Cross-Site Scripting (XSS) 230
13.4.11 Vulnerability in Data Security 231
13.4.12 Data Breach 231
13.5 Security Measures for Big Data in Cloud 232
13.5.1 Encryption 232
13.5.2 Hashing 232
13.5.3 Digital Signature 233
13.5.4 DDoS Prevention 233
13.5.5 Secure Sockets Layer (SSL)/Transport Layer Security (TLS) 233
13.5.6 Prevention of SQL Injection 234
13.5.7 Prevention of Cross-Site Scripting (XSS) Attacks 234
13.5.8 Physical Server Security 234
13.5.9 Virtual Machine (VM) Security 234
13.6 Conclusion 234
Bibliography 235

13.1 INTRODUCTION

Data is crucial in all spheres of the present world, and its uninterrupted generation eventually leads to creating an enormous amount of information. Data is accessible in various formats such as audio, video, and doc and different types such as structured, unstructured and semi-structured. Processing and analyzing massive data in the traditional way of storing it is challenging, thus giving rise to the Big Data concept. Big Data can be stored, processed and analyzed in cloud computing.

Big Data in cloud computing is an emerging technology in many fields such as IoT, healthcare, multimedia, and social networks. But the stakeholder's worry is about data security when vast data is uploaded to the cloud. Security issues may arise for various reasons such as unauthorized access, impersonation, hackers, malware, virus, and Denial of Service (DoS) attacks. The cloud providers' responsibility is to outfit proper security mechanisms and draw stakeholders' confidence to develop their business.

The study makes available various defense mechanisms for securing Big Data in cloud computing. This systematic investigation will help pave the way for future research to promote Big Data in cloud computing technology.

To achieve this objective, in Section 13.2, we describe the overview of cloud; Section 13.3 describes Big Data and its significance; Section 13.4 describes the security issues faced by the Big Data in cloud; Section 13.5 describes the defense mechanism for overcoming the security issues. Finally, in Section 13.6, we point out the conclusion of the study we have carried out.

13.2 OVERVIEW OF CLOUD

National Institute of Standards and Technologies (NIST), USA, defined CC as "a model for enabling ubiquitous, convenient, on-demand network access to a shared pool of configurable computing resources (e.g., networks, servers, storage, applications and services) that can be rapidly provisioned and released with minimal management effort or service provider interaction" (Mell & Grance, 2011). It possesses five important characteristics – on-demand self-service, broad network access, resource pooling, rapid elasticity and measure service; four types of deployment model – private cloud, community cloud, public cloud and hybrid cloud; and three service models – Platform as a Service (PaaS), Software as a Service (SaaS) and Infrastructure as a Service (IaaS). Figure 13.1 illustrates an overview of cloud computing.

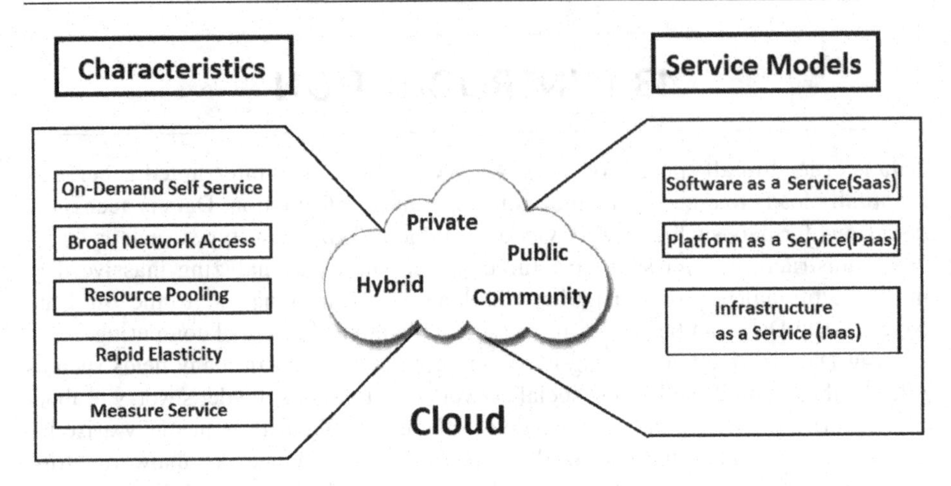

FIGURE 13.1 Overview of cloud.

13.2.1 Important Characteristics

13.2.1.1 On-Demand Self-Service

Customers can automatically utilize computing resources such as network, server, storage, and applications as and when required by a cloud provider.

13.2.1.2 Broad Network Access

It refers to the ability to access cloud resources over the network through any gadgets such as workstations, laptops, tablets and mobile phones.

13.2.1.3 Resource Pooling

The cloud provider's computing resources serve several customers by multi-tenancy. The resources are allocated and reallocated based on the customer's needs.

13.2.1.4 Rapid Elasticity

Resources are elastically scaled inward or outward and automatically provide and release based on the customer's need. Resources are available unlimited as per his demand at any time to the customers.

13.2.1.5 Measured Service

Customer utilized resources can be controlled, monitored and automatically generate access reports by the provider. Both customer and cloud provider can access reports, and it provides transparency.

13.2.2 Deployment Models

13.2.2.1 Private Cloud

An organization that installed a cloud for its exclusive use is called a private cloud. A third party or a company may own it, and it may reside on-campus or off-campus.

13.2.2.2 Community Cloud

This cloud installed for a community and exclusively shared within their community is called a community cloud. It may be owned by a single or many organizations in the community; even a third party may hold and maintain it for the community, and it may exist on-campus or off-campus.

13.2.2.3 Public Cloud

The cloud resources that are open to use by the general public are called public clouds.

13.2.2.4 Hybrid Cloud

Two or more different cloud deployment models combined are called a hybrid cloud.

13.2.3 Service Models

13.2.3.1 Software as a Service (SaaS)

Any application running on cloud is accessible by the customer using any gadgets anywhere, anytime. Customers need not install anything for accessing those applications.

13.2.3.2 Platform as a Service (PaaS)

Customers can create and use virtual platforms for their requirement using the cloud without installing the equipment.

13.2.3.3 Infrastructure as a Service (IaaS)

Customers can use the network, storage or any other computing resources for their requirements using the cloud without installing the equipment.

13.3 BIG DATA: OVERVIEW

Big Data refers to the enormous amount of data in Peta Bytes or Zeta Bytes generated from different sources such as mobile communications, Internet of Things (IoT), scientific applications, healthcare, multimedia and social networks. As the number of devices connected to IoT continues to rise to unpredicted levels, Big Data is experiencing exponential growth, generating large amounts of data that are required to convert into valuable information. Big Data is a mix of structured, unstructured and semi-structured data.

- Structured data – This is organized data stored in the form of tables or databases.
- Unstructured data – Represent in the form of text, messages, videos, images, etc.
- Semi-structured data – The data is structured but not in traditional relational databases or data tables (e.g., XML, JSON).

13.3.1 Characteristics of Big Data

Big Data's essential characteristics are represented in 5Vs: Volume, Veracity, Value, Variety and Velocity (Hadi et al., 2014) as discussed below.

13.3.1.1 Volume

Volume represents the amount of data produced from multiple sources and continuing to expand. The massive amounts of data in petabytes and zeta bytes must be further processed and transformed into useful information.

13.3.1.2 Veracity

In a broad sense, veracity is how precise or authentic a data set can be. Authenticity, accountability, availability and reliability of data contribute to data veracity, whereas ambiguous and uncertain input data and imprecise modeling cause a veracity challenge.

13.3.1.3 Value

Value refers to the methods used to discover hidden information that is useful from massive data sets.

13.3.1.4 Variety

The extremely diverse types of data gathered through IoT devices, mobile devices, social media communications may include text, image, audio, video, etc., which could be in a structured, semi-structured and unstructured format.

13.3.1.5 Velocity

Velocity refers to the speed at which data accumulates. Big Data inflow should be evaluated and used as they flow into organizations to optimize the value of information for time-sensitive operations. Faster processing of enormous data and real-time responses maximizes the practical usage of information. Figure 13.2 illustrates the 5Vs of Big Data.

FIGURE 13.2 5Vs of Big Data.

13.3.2 Significance of Big Data

The exhaustive use of the Internet, the IoT, cloud computing and other evolving IT advancements have significantly contributed to the increase of data of diverse types and structures, generated from various sources at exponential. Big Data can play an essential part in supporting nations' macroeconomic stability and improving enterprises' competitive edge. With the emergence of Big Data, researchers have easy access to information and knowledge directly applied to tackling societal problems such as public health and economic growth. Smarter forecasts of future trends can be accomplished through efficient integration and concise analysis of multi-source heterogeneous Big Data. Big Data analytics enables us to examine and analyze the tremendous amount of data collected and offers better insights for the business. On a large scale, Big Data helps companies make appropriate decisions simultaneously, leading to better customer service and successful business outcomes.

13.3.3 Big Data in Cloud

With an enormous number of IoT devices interconnected, the flow, storage, access, and processing of the tremendous amount of data they will produce must be given specific attention. The proliferation of smart devices and the prevalence of sensors, in essence, call for scalable compute environments. Transforming Big Data in cloud gives a significant advantage to the company's business. The ability to hold vast amounts of data in various forms, and processing it at substantial speeds guides businesses to develop fast. Cloud computing is a perfect platform for storing a massive quantity of workloads (Hadi et al., 2014). While traditional storage mechanisms cannot meet Big Data storage requirements (Tian & Zhao, 2014), cloud technology addresses the problems due to its enormousness. Cloud technology can store massive data, and it keeps the data in more than one location. Cloud computing also offers elasticity and high computational power for the increased volume of data generated from various sources. It provides efficiency in cost, speed, scalability, increased productivity, reliability and mostly security. Cloud platforms serve as the most powerful framework for organizations in mobile communications, IoT, healthcare, multimedia, social networks to process a massive amount of business data (Australia Government, 2013,2011; Cai et al., 2016). Also, to tackle the enormous amount of data in scientific applications, cloud computing technology is realized as the most compelling platform (Heath). Big Data offers the opportunity to leverage commodity computing for processing distributed queries through multiple datasets and produce resultant data sets resourcefully. Cloud computing delivers the underlying engine through multiple nodes that run in a cluster for processing data.

13.4 SECURITY ISSUES FACED BY THE BIG DATA IN CLOUD

> IoT without security=internet of threats.
>
> *Stephane Nappo*

Security and data privacy are the most challenging task in Big Data technology, besides data size. The security issues are often intensified due to the enormous volume and variety of Big Data that needs to be processed, different data sources and acquisition techniques. A significant problem hindering Big Data usage in cloud environments is security (Shau et al., 2015). Security issues that may arise for various reasons should be addressed appropriately for Big Data applications within cloud computing. The massive cloud infrastructures and inter-cloud migration also add to the security threats posed to the data (Andhare & Sonone, 2016). Thus efficient and reliable security mechanisms regarding information, data, application, and network are in dire need to help establish the credibility for Big Data applications hosted in the cloud (Schmidt; Yao et al., 2010).

Most cloud service providers and cloud users consider Confidentiality, Integrity and Availability as the essential security requirements for the services they deliver. In the cloud computing era, data owners have minimal control of virtualized storage, ensuring confidentiality, integrity and data availability prevail as a critical issue. Cloud clients need to be aware that the implementation of faulty cloud-based strategies results in security risks. A detailed understanding of the security issues, followed by a comprehensive assessment of the service provider's security policies, is imperative to choose a cloud service.

13.4.1 Confidentiality

Confidentiality refers to the characteristics of data by which it is accessible only by authorized parties. In cloud computing platforms that host Big Data, confidentiality pertains to controlling the access to data in-store and transit (Eri et al., 2014). A cloud framework that offers confidentiality is when unauthorized persons or applications do not access sensitive data (Hu & Vasilakos, 2016).

The enterprise data flow between the SaaS application and storage poses security threats like packet sniffing, IP spoofing and man-in-the-middle attack. Sensitive information is contained in most information systems. It could be confidential business data that rivals may use to benefit from personal details about the employee or customer. Confidential data always has significance as hackers search for flaws to exploit, and systems are under prevalent threat. Physical attacks such as password theft and capturing network traffic or social engineering and phishing are also potential threats. Not all violations of confidentiality are deliberate. There are many unintentional violations, including sending sensitive data to a faulty recipient by email, disclosing private information to public web servers or displaying confidential data on an unsecured computer. An excellent deal of separation must exist between non-sensitive and sensitive data to protect it from unauthorized users. The isolation of data can be constituted through access control and encryption schemes. Fine-grained access control mechanisms that are attribute, identity, and time-based can be implemented to provide data isolation. Lack of such access control mechanisms deployed by the IaaS providers leads to VM attacks and forfeits confidentiality. The cloud service delivery model must implement security policies to safeguard the data in transit and storage and to ensure confidentiality by protecting the information from unauthorized access and misuse.

13.4.2 Integrity

Integrity refers to the absence of any malicious activity that alters data. In addition to access control mechanisms, the maintenance of integrity also needs to ensure that users can only modify information that they are genuinely allowed to alter. Unauthorized alteration of the data affects the accuracy and completeness of data and thereby affects its integrity. Also, accidental alteration, such as user input errors or loss of data due to a system malfunction, causes a threat to data integrity. Effectual integrity countermeasures guarantee a cloud consumer that data transmitted, stored and processed remain unaltered.

13.4.3 Authenticity

Authenticity refers to having been offered by an authorized source. Authenticity includes non-repudiation, which proves that the processes involved are connected to authentic sources.

Big Data from various sources poses the risk of theft and tampering when stored in a cloud model. Lack of a proper authentication mechanism can cause attackers to access data through cloud platforms. The issues that can be classified at the user authentication level pertain to encryption/decryption methods, user authentication like administrative rights and authentication of applications and nodes, and the log-in process.

13.4.4 Availability

The SaaS model focuses on the cloud provider to achieve security features. In contrast, the IaaS model intends the security functions to be deployed from the user's end and availability to be taken care of by the service provider. PaaS providers should ensure the platform's security and the client application deployed on it while the data integrity is the service provider's responsibility.

The primary security issues faced by Big Data in the cloud are listed as follows: DoS, distributed denial of service (DDoS) attacks, man-in-the-middle (MitM) attack, sniffer attacks, spoofing, SQL injection attack, cross-site scripting (XSS), the vulnerability in data security and data breach.

13.4.5 DoS and DDoS Attacks

In a DoS attack, the services will be made inaccessible to authentic users as the malicious attackers and deliver large quantities of requests to the target computer at a time. Flooding the target machine results in DoS (Raghavendra et al., 2015). A DDoS is a method of attack launched by multiple attackers to deplete the available resources, which causes server inaccessibility and service unavailability, and genuine users are not able to access the resource. Attackers generally target public clouds to initiate and execute DDoS attacks (Vlajic et al., 2019). Thus, DoS and DDoS attacks are designed to target cloud services, and infrastructure is identified as the most significant cloud computing vulnerability (Figure 13.3).

13.4.6 MitM Attack

An insecure network causes attackers to use flawed services running on the device to attack other IoT network devices (Raghavendra et al., 2015). MitM attack is a prevalent attack in the SaaS cloud framework (Kavvadia et al., 2015). In MitM attacker intercepts, communications between two sides steal the data secretly and modify the traffic traveling between the two parties. Attackers will launch a communication between two computer nodes, observe the traffic, listen silently and monitor the network to steal the user's secret credentials from the web. Even though the attack can be protected by encryption, intruders reroute traffic to phishing sites that look genuine. Figure 13.4 shows MitM attack in the cloud environment.

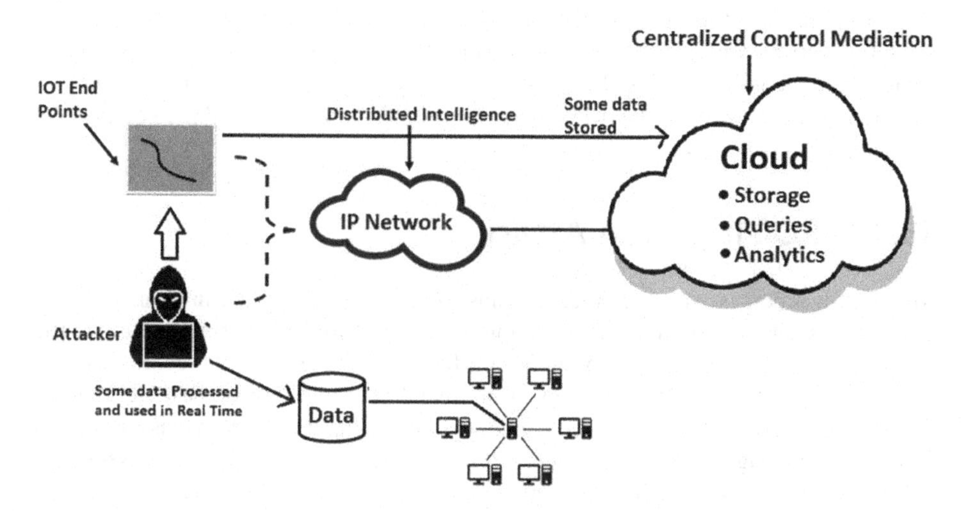

FIGURE 13.3 DDoS attack.

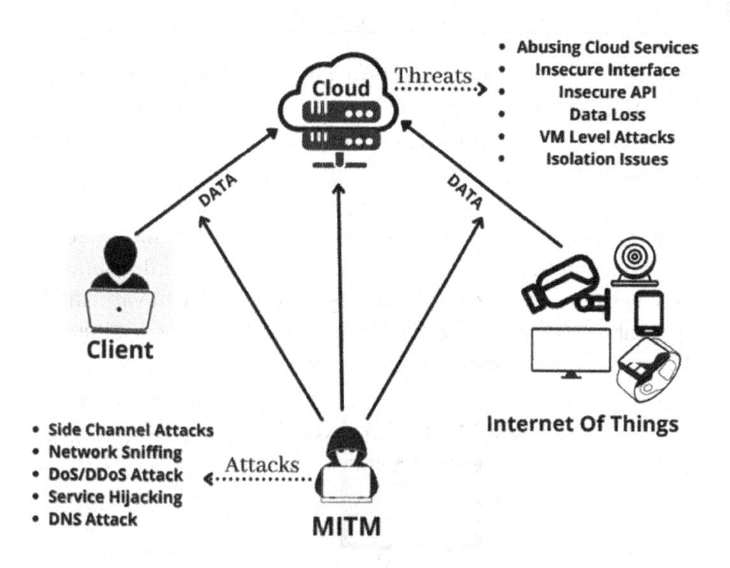

FIGURE 13.4 Man-in-the-middle attack in cloud environment.

13.4.7 Sniffer Attacks

Sniffer attacks are launched on data packets with non-decrypted data in transit in a network. The user's sensitive data flow and confidential information is monitored by the attacker (Gupta & Badve, 2017).

13.4.8 Spoofing

Spoofed IP packets transmitted in and out from a public cloud provider's physical and virtual servers to misuse these machines. Hackers easily target the service providers' public cloud for spoofing IP packets or as intermediate packet reflectors in the network. DoS and DDoS attacks are launched on public cloud servers using IP packet spoofing (Somani et al., 2017; Osanaiye et al., 2016).

13.4.9 SQL Injection Attack

An SQL injection is when an attacker runs SQL statements that can control a web application's database server. SQL injection attacks (SQLIA) add harmful data into the database layer through various codes. It modifies, deletes the database records without the user's awareness and can even insert unofficial data into the database. It compromises data integrity. The sensitive and confidential data in the data storage is exposed to SQL attacks through malicious queries over the web application (Niharika & Ashutosh, 2019). Figure 13.5 shows a SQL injection attack.

13.4.10 Cross-Site Scripting (XSS)

In a cross-site scripting attack, a malicious script is injected into the target's cloud storage server. The attacker imports the malicious code to the directed cloud server, and it will propagate to the whole organization that has access to that server. The user assumes that the remote storage server is entirely secure and accesses it through the browser. Thus, the malicious code will automatically run until it has been removed from the browser and gather the user's data (Nithya et al., 2015). Since most data analytics and business intelligence applications may be interactive web-based solutions that receive and execute inputs from users, and if the inputs are used without validation, there exists a significant probability of cross-site scripting attacks. Figure 13.6 illustrates cross-site scripting in Big Data cloud storage server.

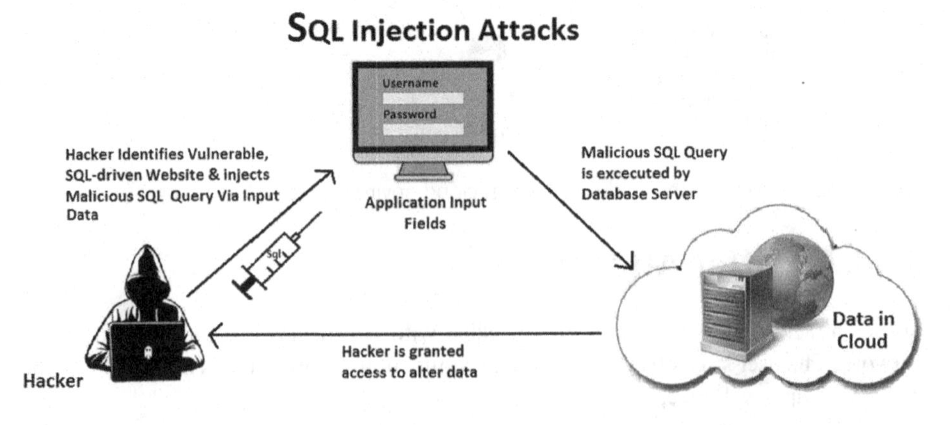

FIGURE 13.5 SQL injection attack.

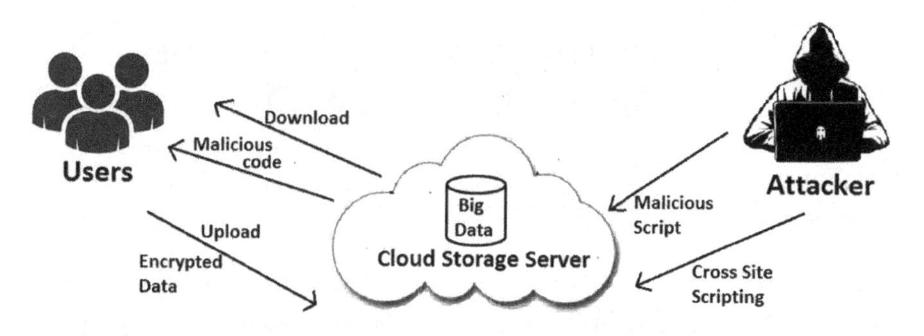

FIGURE 13.6 Cross-site scripting in Big Data cloud storage server.

13.4.11 Vulnerability in Data Security

Vulnerability refers to weakness caused due to inadequate security controls or deficiencies in security architecture, which may be exploited by adversaries. The potential vulnerabilities which concern significant data users are privacy, security and lack of proper security standards (Fernando, 2018). Lack of authentication mechanisms and insecure channels to cloud storage are substantial reasons for causing vulnerability in Big Data security (Moreno et al., 2016).

13.4.12 Data Breach

A data breach is an occurrence that can uncover private data from an unofficial party. Data breaches may occur because of many reasons such as data theft and leak. Data theft is when sensitive data is stolen accidentally or explicitly. Compromised data security in the cloud may lead to attacks such as data leak.

Cybercriminals can access and transport sensitive data when the cloud service or a connected device is breached. This way of transporting cloud data electronically or physically is data leak. Data breaches in the cloud are occurring due to various aspects, for example, malicious attacks, insider threats, malicious insiders, compromised or stolen credentials, and misconfigurations.

Personal Health Information (PHI), Personally Identifiable Information (PII), and Proprietary Innovations and Intellectual Properties (IP) are the frequently targeted data breach areas, and these require the security measures of the highest levels of security in cloud computing. Cloud computing infrastructure delivers on-demand services relating to frameworks, software programming and platforms, and it is defenseless for many types of data breaches.

Data owners cannot rely upon the cloud infrastructure unless there is an assured and optimized, secure mechanism in place. The data owners can undoubtedly have many security concerns. Even though cloud service organizations are endeavoring to guarantee the assurance of data security of their clients, there were adequate episodes identified with data breaches that raised a problem for a most reputed organization.

13.5 SECURITY MEASURES FOR BIG DATA IN CLOUD

Privacy is not an option, and it shouldn't be the price we accept for just getting on the Internet.

Gary Kovacs

Security attack and defense always remain a challenging problem between spear and shield. Even though the current defense mechanisms seem to be very arduous and powerful, new threats are ever-evolving day by day, especially with various data formats and multi-tenancy user groups; identifying the security holes and resolving them is a herculean task.

However, when the tree's root is adequately secured in the soil, it yields sweet fruits. Likewise, suppose significant security issues are appropriately addressed at the root level, and the necessary security requirements are met appropriately. In that case, we can get the real benefits of Big Data in the cloud.

This section addresses the defense mechanisms of significant security threats described in the previous section.

13.5.1 Encryption

In general, the data is readable and is in plaintext format; when transmitting these data through a network, unauthorized persons can easily access the data for malicious activity. The encryption mechanism converts the plaintext into an unreadable format. The encryption uses an algorithm called cipher for converting plain text into an encrypted format. While encrypting, the string of characters called encryption key is added with the plaintext and the encryption key is shared with the authorized user. The user can change the encrypted data into its plaintext format by using the encryption key. Therefore, even though unauthorized users access the data, the data will be in an encrypted form. It cannot be decrypted without an encryption key easily. It preserves the cloud user's authenticity, confidentiality, and integrity. Also, it mitigates eavesdropping, malicious attack and unauthorized access (Eri et al., 2014).

13.5.2 Hashing

The hashing mechanism is used for data protection of one-way and non-reversible formats. Once hashing technology is applied to data, it will be locked, and it cannot be unlocked using any key; this mechanism is used to store passwords.

In general, the hash is shorter than the original input message with a fixed length. The sender appends the message digest with the message. Once the recipient receives it, the same hash function can be applied to verify whether the message digest is identical to the one attached to the message. If any intruder has altered the original message in transit, it results in an entirely different message digest and clearly shows tampering. It mitigates malicious attacks and unauthorized access.

13.5.3 Digital Signature

The digital signature mechanism uses both encryption and hash technology, and it includes a message digest that is encrypted with a private key and attached to the original message. The receiver decrypts the digital signature by using the public key, which produces a message digest, and also extracts the message digest from the original message using the hash mechanism. If both processes give identical results, then it shows that the data is maintaining its integrity.

The digital signature must be attached before sending the message; afterward, any unauthorized user changes the message, which will solidify invalid. A digital signature proves that the message received is sent by the authentic sender. The digital signature mechanism provides integrity and authenticity through authentication and non-refutation. It helps to mitigate malicious intruders, unauthorized users and crossing trust boundaries security threats.

13.5.4 DDoS Prevention

Prevention of DDoS attack is a proactive measure, and prevention techniques are applied to all genuine and illegal users (Somani et al., 2017). A legitimate user's system may be affected by malware or virus and start sending packets without user knowledge. Therefore, all the requests need to be filtered in advance; otherwise, requests may cause a threat to the server by applying standard mitigation methods such as synchronized cookies, limiting the bandwidth, number of connections and connection timings (Gupta & Badve, 2017).

Another most common DDoS attack prevention technique is the graphical Turing test. This test displays an image and a question related to that, and the image can contain image, distortion, an arc, audio and text. This graphical CAPTCHA may have a set of multiple images or moving images.

Yet another technique is Crypto puzzles, which assess the computational capability of a client. It raises the question of seeking an answer to given inputs, and the user has to answer back in a stipulated time. The server should compute and verify the answer given by the user.

All the above techniques used for mitigating the DoS and DDoS attacks in the cloud environment will ensure data access anytime with ease.

13.5.5 Secure Sockets Layer (SSL)/ Transport Layer Security (TLS)

Secure Sockets Layer (SSL) and Transport Layer Security (TLS) are called cryptographic protocols widely used to create a secure connection between the client and the server over the public network. It prevents many network security attacks such as sniffer, spoofing and MitM (Ray et al., 2020).

13.5.6 Prevention of SQL Injection

Harmful SQLIA can happen in vulnerable applications by using different mechanisms. For avoiding such a dangerous threat to database-driven applications, good security scripts can be brought into the application. It protects the attacker to inject additional SQL code and allows only with required parameters (Niharika & Ashutosh, 2019).

13.5.7 Prevention of Cross-Site Scripting (XSS) Attacks

Using several methods such as content-based data leakage prevention and active content filtering aids to detect and fix the issues.

13.5.8 Physical Server Security

In the cloud environment, the physical server is a vital component as data is stored on it. Since the data has to save and retrieve fast in the cloud, the server has to be secure appropriately; if any untoward attack happened, the entire service would be affected. The server has to update with the latest released patches and antivirus software properly. It is recommended that the regular backup has to take proper data management. A different set of access control systems should be in place based on user roles. Only authorized users should be allowed to access data with a robust authentication mechanism. The premise's essential safety measures where the server is located also must be considered (Somani et al., 2017).

13.5.9 Virtual Machine (VM) Security

Since VM plays a vital role in cloud computing, VM should be secured appropriately. In VM, unwanted software should be removed to limit the potential vulnerabilities that intruders can exploit. To protect from the attackers, except for the critical ports, all other ports should be closed, and all unnecessary services should be disabled. VM cloning should be taken at regular intervals as a backup to restore the data if anything goes wrong. VM should isolate guest OS and host operating systems. This process is employed at the hypervisor level. Thus, virtual machines operating on a hypervisor would not be at any stake (Somani et al., 2017).

13.6 CONCLUSION

Cloud computing is considered the best option for Big Data as it provides all IT resources such as software, hardware and network, in a massive way to users anywhere, anytime. On the other hand, other cloud computing features such as multi-tenancy and heterogeneity may create a security vulnerability. This chapter discussed elaborately major security issues and the defense mechanism of Big Data in the cloud. Moreover, we would like to emphasize that security measures to be taken by the cloud provider alone are not enough. Users should also be more vigilant about their gadgets' security measures, like updating software regularly, installing proper antivirus, using a strong password, etc. To reap the full benefit of Big Data, both cloud providers and users together have to put adequate security measures.

BIBLIOGRAPHY

C. Andhare and S. Sonone. 2016. Survey on security primitives in big data on cloud computing, *International Journal of Innovative Research in Computer and Communication Engineering (IJIRCCE)* 4(3).

Australia Government. Department of finance and deregulation, big data strategy—issues paper 2013. Available: http://agimo.gov.au/files/2013/03/.

Australia Government. Department of finance and deregulation, cloud computing strategic direction paper: Opportunities and applicability for use by the Australian government 2011. Available: http://agimo.gov.au/files/2012/.

H. Cai, B. Xu, L. Jiang and A.V. Vasilakos. 2016. IoT-based big data storage systems in cloud computing: Perspectives and challenges. *IEEE Internet of Things Journal* 4(1):75–87.

T. Eri, Z. Mahmood and R. Puttini. 2014. Cloud computing concepts, technology and architecture. pearson. In text: (Eri, Mahmood and Puttini 2014).

A. Fernando. 2018. Big Data: Concept, potentialities and vulnerabilities. *Emerging Science Journal* 2(1):1–10.

B.B. Gupta and O.P. Badve. 2017. Taxonomy of DoS and DDoS attacks and desirable defense mechanism in a Cloud computing environment. *Neural Computing and Applications* 28:3655–3682.

H.J. Hadi, A.H. Shanin, S. Hadishaheed & A.H. Ahmad. 2014. Big data and five V'S characteristics, 29–36.

N. Heath. CERN: Cloud computing joins hunt for origins of the universe. Available: http://www.techrepublic.com/blog/european-technology/cern-cloud-computing-joins-hunt-for-origins-of-the-universe/262Journal.

J. Hu and A.V. Vasilakos. 2016. Energy big data analytics and security: Challenges and opportunities. *IEEE Transactions on Smart Grid* 7(5):2423–2436.

E. Kavvadia, S. Sagiadinos, K. Oikonomou, G. Tsioutsiouliklis and S. Aïssa. 2015. Elastic virtual machine placement in cloud computing network environments. *Computer Networks* 93(3):435–447.

P. Mell & T. Grance. 2011. The NIST definition of cloud computing - Recommendations of the national institute of standards and technology. Special Publication 800-145, available at: http://csrc.nist.gov/publications/nistpubs/800-145/SP800-145.pdf (accessed 12 April 2014).

J. Moreno, M. Serrano and E. Fernández-Medina, 2016. Main issues in big data security. *Future Internet* 8(44):1–16.

S. Niharika and S. Ashutosh. 2019. SQL-Injection vulnerabilities resolving using valid security tool in cloud. *Pertanika Journal of Science and Technology* 27:159–174.

V. Nithya, S. LakshmanaPandian, V. Nithya, S. LakshmanaPandian and C. Malarvizhi. 2015. A survey on detection and prevention of cross-site scripting attack. *International Journal of Security and Its Applications* 9:139–152.

O. Osanaiye, K-K Raymond Choo and M. Dlodlo. 2016. Distributed denial of service (DoS) resilience in cloud: Review and conceptual cloud DDoS mitigation framework. *Journal of Network and Computer Applications* 67:147–165.

K. Raghavendra, K.P. Konugurthi, A. Arun, R.R. Chillarige and R. Buyya. 2015. The anatomy of big data computing. *Software: Practice and Experience*:79–105.

S. Ray, K. N. Mishra and S. Dutta. 2020. Big data security issues from the perspective of IoT and cloud computing: A review. *Recent Advances in Computer Science and Communications* 12(1):1–22.

S.E. Schmidt, Security and privacy in the AWS Cloud. Available: http://aws.amazon.com/apac/awssummit-au/.

S.K. Shau, L. Jena and S Satapathy, 2015. Big Data Security issues and challenges in cloud computing environment. *International Journal of Innovation in Engineering and Technology* (IJIET) 6(2):297–306.

G. Somani, M.S. Gaur, D. Sanghi, M. Conti and R. Buyya. 2017. DDoS attacks in cloud computing: Issues taxonomy, and future directions. *Computer Communications* 107:30–48.

W. Tian & Y. Zhao. 2014. Optimized cloud resource management and scheduling: Theories and practices. *Optimized Cloud Resource Management and Scheduling: Theories and Practices*: 1–266.

N. Vlajic, M. Chowdhury and M. Litoiu, 2019. IP Spoofing in and out of the public cloud: From policy to practice. *Computers* 8(4):81.

J. Yao, S. Chen, S. Nepal, D. Levy and J. Zic. 2010. Truststore: Making Amazon S3 trustworthy with services composition. *Proceedings of the 10th IEEE/ACM International Conference on Cluster, Cloud and Grid Computing, CCGRID'10*, Melbourne, Australia, 600–605.

J. Zhao, L. Wang, J. Tao, et al. 2014. A security framework in G-Hadoop for big data computing across distributed Cloud data centre. *Journal of Computer and System Sciences* 80(5):994–1007.

14 Sound and Precise Analysis of Web Applications for Injection Vulnerabilities

Chitsutha Soomlek
Khon Kaen University

Krit Kamtuo
Buzzebees, Co., Ltd

Ekkarat Boonchieng
Chiang Mai University

Contents

14.1 Introduction 238
14.2 Related Work 239
  14.2.1 Injection Vulnerabilities 239
  14.2.2 SQL Injection 239
  14.2.3 Roslyn: Microsoft.NET Compiler Platform 241
  14.2.4 Microsoft Azure Machine Learning (Azure ML) 242
14.3 Proposed Architecture 243
14.4 Data Collection and Preparation 245

DOI: 10.1201/9781003051022-14

14.4.1 Independent Variable 245
14.4.2 Dependent Variable 245
14.4.3 Feature Selection 246
14.5 The Implementation of the Framework 246
14.6 Experimental Results 251
14.6.1 Evaluation of the Models 251
14.6.2 Verification and Validation of the Compiler Platform 252
14.7 Conclusions and Future Work 265
Bibliography 266

14.1 INTRODUCTION

Injection vulnerabilities are the major flaws that allow attackers to pass an untrusted input, a piece of malicious code, or a command modifier through an application to exploit a computer system; for example, SQL queries, system calls, and shell commands. Injection vulnerabilities are commonly found in poorly designed applications, especially web applications. A web application usually employs features and functions provided by operating systems and external programs to conduct its functionalities. When passing a request or an information from a web application to another system, an extra care is required; otherwise, malicious code might be able to bypass to other trusted systems. Lacking of defensive design and programming in software development leaves the application suffering from injection attacks.

Since each programing language has various ways to respond to a call to external commands and external resources, the best approach to determine if an application has injection vulnerabilities is to analyze its source code to search for all requests to run external commands and calls to external resources. Both static and dynamic code analysis can be employed to identify the security flaws. In addition, code reviews is one of good practices for a software development team to find potential flaws and improve the software quality, then, further protection steps can be included to prevent undesirable consequences. However, finding injection vulnerabilities through code reviews is not a simple task even for experienced developers. Static code analysis tools play an important role during code reviews to help developers identify bugs, security flaws, code smells, etc. The effectiveness of those static code analysis tools does affect the accuracy of bugs and vulnerabilities detection. Therefore, a sound and precise code analysis approach for detecting injection vulnerabilities is required.

SQL injection is a frequently found injection attack against web application vulnerability. The security flaw could be unintentionally included in the source code of a web application during the software development phase. To guarantee the adoption of secure coding practices and defensive programming for vulnerability prevention before they are inadvertently added to the production code, the framework of SQL injection prevention using compiler platform and machine learning is proposed. The framework is comprised of machine learning models and a compiler platform. The trained machine learning models on Azure ML are responsible for SQL injection identification. The compiler platform is an enhanced version of Roslyn, which is an open-source compiler for Microsoft.NET. The compiler platform can be integrated with Microsoft Visual Studio IDE in order to support a software developer to identify certain types of SQL injection vulnerabilities and perform code refactoring at the early stage of software development.

This chapter is organized as follows: the next section gives some background information and related work. Section 14.3 explains the architecture and workflow of the proposed framework. Section 14.4 presents how the data were collected and prepared. The implementation of the framework is described in Section 14.5, followed by experimental results. Conclusions and future work are discussed in the last section.

14.2 RELATED WORK

14.2.1 Injection Vulnerabilities

Injection attacks occur when an attacker can send a hostile input to a target system to interpret. Generally, an application is subjected to injection attacks when

1. Carelessly handling inputs – user-supplied/untrusted data is not validated, filtered, or sanitized before the input is interpreted or used.
2. Not using context-aware escaping – dynamic queries or non-parameterized calls are used directly in the interpreter. However, using an unsafe structure template may result in a risk of injection vulnerabilities as well. Therefore, well understanding of the precise context in which the variables are being inserted is very important during an implementation phase.
3. Hostile data/parameters are directly used or concatenated – malicious commands can be added to commands, dynamic queries, and stored procedures.

In order to prevent injection attacks, data should be examined or be separated from queries and commands, for example, input validation, enforcing prepared statements with parameterized queries, escaping all user-supplied input, code-level defenses, platform-level defense, etc.

14.2.2 SQL Injection

SQL injection vulnerabilities usually happen when the input values received from the front-end through a web form, cookie, input parameter, etc., are not validated or encoded before passing them to the SQL queries to be executed on a database server. There are various types of SQL injection vulnerabilities. The followings are the most commonly found SQL injection attacks and examples of vulnerable SQL queries:

1. Illegal/logically incorrect queries

   By inserting a malicious SQL query, as shown in Example 14.1, the query causes the CGI tier replies an error message.

Example 14.1

```
SELECT * FROM user WHERE id='1111' AND password= '1234'
AND CONVERT (char, no) --';
```

2. Union queries

   An attacker uses the "*UNION*" operator in between two or more SQL queries. The union queries attack performs unions of malicious queries and a regular SQL query with the "*UNION*" operator. The examples are shown in Example 14.2.

Example 14.2

```
SELECT * FROM user WHERE id='1111' UNION
SELECT * FROM member WHERE id='admin' --' AND
password='1234';
```

   In Example 14.2, all of the subsequent strings after the "*UNION*" operator are also interpreted. The results of the query reveal the DBMS administrator's information.

3. Piggybacked queries

   Malicious SQL queries can take a piggyback ride on a normal SQL query. An attacker can append malicious SQL queries as many as possible to be processed by adding the operator ";" after each query, as shown in Example 14.3. Then, the operator ";" is added at the end of the query. The result of Example 14.3 is to delete the user table.

Example 14.3

```
SELECT * FROM user WHERE id='admin' AND
password='1234'; DROP TABLE user; --';
```

4. Stored procedures

   DBMS provides a method called stored procedures, which allows users to store their own functions and could use the functions as needed. Example 14.4 depicts a collection of SQL queries that included a function resulting in a vulnerability in a stored procedure. Note that injection vulnerabilities in stored procedures are not tested in this research; this is left for future work.

Example 14.4

```
CREATE PROCEDURE DBO @userName varchar2, @pass
varchar2,
AS
```

```
EXEC("SELECT * FROM user WHERE id='" + @userName + "'
and password='" + @password + "');
GO
```

14.2.3 Roslyn: Microsoft.NET Compiler Platform

Roslyn is an open-source C# and Visual Basic compiler with code analysis APIs. Microsoft Visual Studio IDE can integrate Roslyn to leverage the features of the IDE, e.g., code refactoring and code analysis.

Roslyn allocates API layers in the C# and Visual Basic compiler's code analysis. The APIs consist of two main layers, which are the Compiler APIs and Workspaces APIs.

1. Compiler APIs

   The compiler layer contains object models and information relative to both syntactic and semantic of the compiler pipeline in each phase. The complier layer contains source code files, compiler options, and assembly references. In addition, this particular layer does not have any dependency on Visual Studio IDE components.

   The mechanism of the compiler APIs is in the following sequence: first, a source code is tokenized and parsed to a syntax, which relies on the grammar of the programming language. After that, the declaration phase analyzes the declaration from the source code to named symbols. Then, the binding phase matches the identifiers in the code with the symbols. Finally, the emit phase sends out all information from the complier as an assembly.
2. Workspaces API

   This is the operational session in code analysis and refactoring for the entire solution. Workspace APIs assist in gathering all project information in a solution into a single object model and access to the compiler layer without parsing files, configuring options, or managing project-to-project dependencies. Moreover, implementing code analysis and refactoring tools within the Visual Studio IDE can be done by using the Workspace APIs, e.g., the Find All References, Formatting, and Code Generation APIs.

As shown in Figure 14.1, the Workspaces APIs diagram shows the relations to the host environment and tools. The number in each block stands for the number of instance. The solutions could be adapted by constructing new instances based on the existing solutions and specific changes such as with syntax trees and compilations. Once the workspace explicitly applied the changed solution back to the workspace, this would reflect changes.

A project as a part of the unchangeable solution model represents all source code documents, parse, and compilation options of both assembly and project-to-project references. This allows access to the corresponding compilation without determining parse or project dependencies from any other source files. Moreover, the model contains a

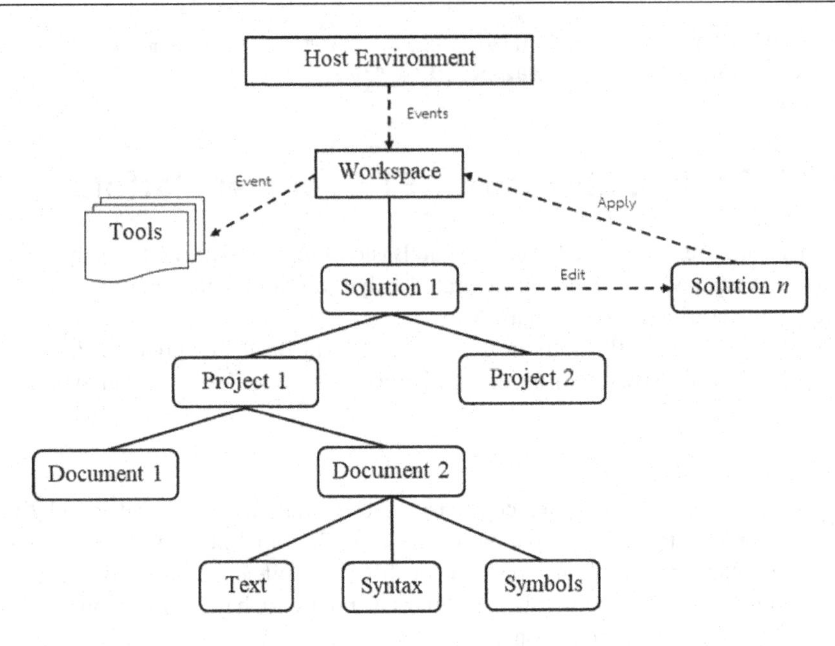

FIGURE 14.1 The host environment and tools of Workspaces APIs.

document showing a single source file, which could be able to access the text of file, syntax tree, and semantic model.

14.2.4 Microsoft Azure Machine Learning (Azure ML)

Microsoft Azure Machine Learning provides a large library of algorithms ranging from classification, recommender systems, clustering, anomaly detection, and regression, to text analytics families on the cloud. Each of these is designed for different groups of machine learning problems. The service grants supervised learning, unsupervised learning, reinforcement learning, and web services as a fully managed model. Additionally, Azure ML provides tools for launching predictive analytics solutions, e.g., Net#, neural network specification language. The basic workflow of Azure ML starts with Azure ML retrieves data from different data sources to process the analytics and create predictive models. Then, the service would process the web services by using the validated predictive analytics models. Thus, they could be connected to other websites or applications via an API, which is the ended point of Azure ML. It could be used together with business intelligence (BI) tools to provide an insight.

In order for this research to classify vulnerable SQL queries and to evaluate the models, Azure ML provides six two-class classification modules to predict between two categories from a non-linear dataset as follows:

1. Two-class support vector machine (SVM) – This module is based on the SVM algorithm. The supervised learning algorithm can predict two possible outcomes from both categorical and continuous variables. During the training process, SVM analyzes the training data and create a hyperplane which is a multi-dimensional feature space. All inputs are represented as points in the space and are mapped to certain output categories. The categories are divided by a wide and clear space. New examples are linked to the predicted category based on which side of the indicated space.
2. Two-class locally deep SVM – This is an optimized version of SVM. The supervised learning model is scaled to support larger training datasets while still can work with both linear and non-linear classification tasks. The kernel function used for mapping data points to feature space is re-designed to reduce the time needed for training while maintaining the accuracy of classification.
3. Two-class boosted decision tree – This module is based on the boosted decision tree algorithm. The algorithm is an ensemble learning method in which the second tree corrects the errors of the first tree, the third tree corrects the errors of the first and second trees, and so on. Prediction results are based on the entire ensemble of the trees together. The limitation of the module is the implementation holds everything in the memory. As a result, the module uses intensive memory and can handle limited size of datasets.
4. Two-class decision forest – Decision forest algorithm is an ensemble learning method. Instead of having one decision tree, decision forest module has multiple decision trees. Each classification tree is created by using the entire dataset but with different starting point. Each tree produces a non-normalized frequency histogram of labels. Then, at the aggregation process, the histograms are summed and the results are normalized to get the probabilities of each label. The trees having high prediction confidences will have a greater weight than others will. The results of the decision forest algorithm are from voting.
5. Two-class decision jungle – This module is an extension of two-class decision forest. It allows tree branches to merge and make a decision out of directed acyclic graphs (DAGs). However, decision jungle has lower memory utilization and a better performance than decision tree.
6. Two-class neural network (NN) – This module creates an artificial neural network to predict a target with only two values.

14.3 PROPOSED ARCHITECTURE

This section proposes a framework designed for SQL injection prevention using compiler platform and machine learning. The framework is designed to validate SQL statements

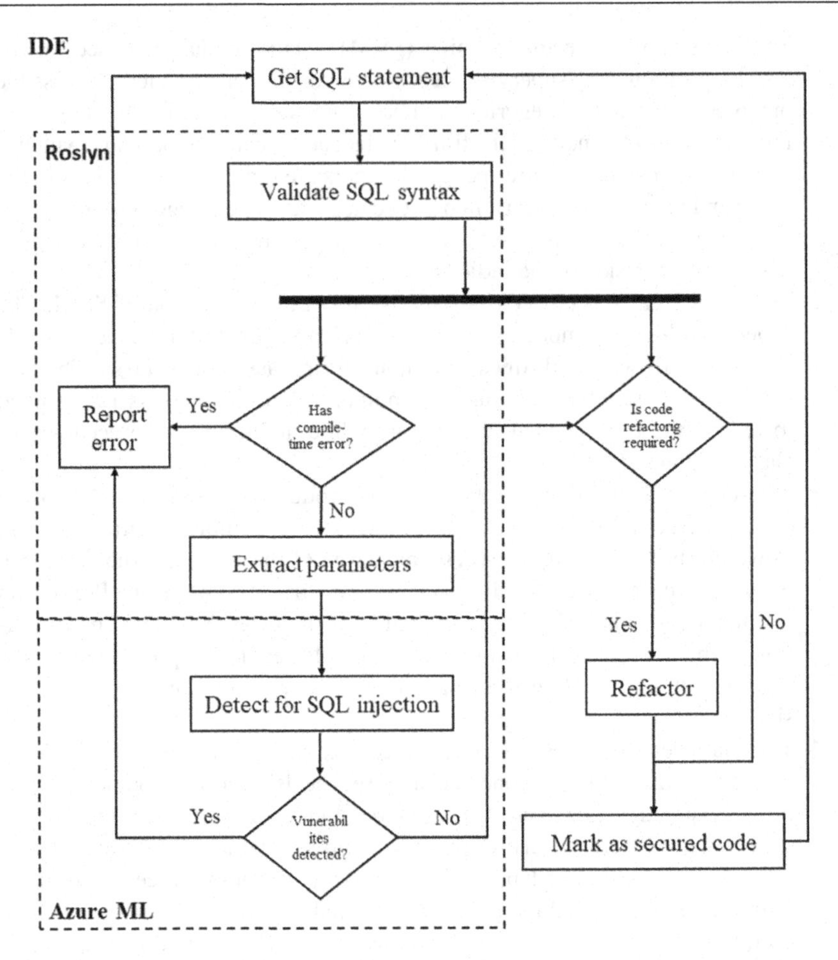

FIGURE 14.2 The basic workflow of the framework.

and send parameters to predict the possibility to have SQL injection using our enhanced version of Roslyn and Azure ML. The framework is integrated into Microsoft Visual Studio IDE to support a developer during the software development phase. Figure 14.2 depicts the basic workflow of the framework.

Once a SQL statement is detected in a source code, the syntax of the statement is validated. Next, the parameters in the validated SQL statement are extracted. The extracted parameters will be sent to the analytics web services, which already include the trained and evaluated machine learning models. In case that a SQL injection flaw is detected, the prediction result will be reported to the IDE in order to suggest the possibility of having an injection vulnerability to a developer. Meanwhile, the modified Roslyn will ask the IDE to use the results of SQL syntax validation and injection vulnerability detection to check for the need of code refactoring. Shortly thereafter, the IDE will inform a developer about the flaw and the need of code refactoring. Code refactoring would help a developer to fix the flaw found in their source code at the early stage of software development and, therefore, support SQL injection prevention and code quality improvement.

14.4 DATA COLLECTION AND PREPARATION

Since SQL injection is an injection vulnerability that is commonly found in web applications, this research employed open-source content management systems (CMS) to create both training and testing datasets. The data extraction method is an inspired and modified from. SQL statements in the CMSs were extracted from Wordpress 3.9.1, Drupal 7, Joomla 3.6.0, and Simple Machine Forum (SMF) 2.0.5. An experienced senior software developer, then, marked the vulnerable SQL statements manually. Both independent and dependent variables in the SQL statements were identified. The marked vulnerable SQL statements are manually classified into three types of SQL injection: illegal/logically incorrect queries, union queries, and piggybacked queries.

14.4.1 Independent Variable

An independent variable is a variable that obtains value internally within the server-side scripts. This type of variable is not directly involved with SQL injection.

14.4.2 Dependent Variable

A dependent variable is a variable that obtains a value from the user-supplied input; thus, the SQL injection could be possible. The example below shows the dependent variables in an SQL statement.

```
SELECT *FROM db_user WHERE username ='$input_username'AND
password = '$input_password'
```

This statement has two variables inputted by a user: *$input_username* and *$input_password.* If the user injected malicious inputs as follows: *$input_username* =*"krit"* and *$input_password*= " *'; drop table users --* ", then, the executed SQL statement will be:

```
SELECT * FROM db_user WHERE username='krit' AND password = ' ';
drop table users -- '
```

Suppose that *"krit"* exists in the targeted database. The injected SQL statement will allow the user to bypass the system and remove the user's table from the targeted database.

This study collected the attributes and marks vulnerable SQL statements in the server-side scripts of the well-known CMS applications line-by-line according to vulnerable SQL statements presented while considering independent and dependent variables.

Since there are certain patterns and characteristics in SQL statements of the server-side scripts that could introduce injection vulnerabilities to a system, those patterns and characteristics could be used as attributes for classification analysis. The attributes of vulnerable SQL commands are presented in Table 14.1.

After that, the dataset was formulated from the extracted SQL statements. 0 and 1 indicate the availability of each attribute in the extracted SQL statements, where 0 means unavailable attribute and 1 means available attribute. The samples of the dataset are illustrated in Figure 14.3. However, the number of non-vulnerable SQL statements is four times larger than the number of vulnerable SQL statements, which indicated that the dataset was imbalanced. To generate a new balanced dataset, synthetic minority over-sampling technique (SMOTE) was applied to increase the number of cases in the dataset in a balanced way with SMOTE percentage value of 252.

14.4.3 Feature Selection

In order to testify that the 19 input attributes are correlated to SQL injection vulnerabilities, namely, illegal/logically incorrect queries, union queries, and piggybacked queries, linear correlation was performed by using three datasets. Each of the datasets has 1,000 vulnerable SQL statements corresponding to each type of SQL injection vulnerability.

Figures 14.4–14.6 present Pearson correlation coefficients of the 19 input attributes presented in Table 14.1 in relation to each type of SQL injection vulnerability. For the illegal/logically incorrect queries, the 19 input attributes correlate with SQL injection vulnerability in the positive direction. In the case of the union queries, the *UNION* command in a query is the only attribute that has a strong positive linear relationship with this particular type of SQL injection, which is a very sensible result. A semicolon is the only attribute that has a strong positive linear relationship with piggybacked queries as well. Other attributes have weak to neutral levels of either positive or negative relationship with SQL injection vulnerabilities; therefore, they were not cut off. All of the 19 attributes were employed in this framework to identify SQL injection vulnerabilities and to classify the detected vulnerabilities into illegal/logically incorrect queries, union queries, or piggybacked queries.

14.5 THE IMPLEMENTATION OF THE FRAMEWORK

The framework is comprised of both Microsoft.NET compiler platform and Azure ML. Once the machine learning experiments on Azure ML have finished, the trained models can be deployed as a web service API. The web service API allows a software developer

TABLE 14.1 Attributes Used for Classifying SQL Injection

NO.	*ATTRIBUTE NAMES*	*DESCRIPTION*	*EXAMPLES*
1	Single-line comment	Single line comment. Ignore the remainder of the statement.	--Select all:
2	Semicolon	A query termination.	SELECT * FROM students WHERE username=' ';
3	Three single quotes	Three single quote (''') in a query.	SELECT * FROM test.members WHERE user_name='''
4	Two single quotes	Two single quotes ('') in a query.	SELECT * FROM students WHERE username=''; TRUNCATE TABLE Username; --' AND password=''
5	Separated two single quotes	Separated two single quotes (' ') in a query.	SELECT user_name, password FROM test. members WHERE user_name= ' '%31%27%20%4F%52%20 %27%31%27%3D%27%31'
6	Number equals to the same number, e.g., 0=0	Condition that always returns true.	SELECT * FROM table WHERE username=' ' or 0=0 #' and password=''
7	Number equals to the same number, e.g., 1=1	Condition that always returns true.	SELECT * FROM test.members WHERE user_name=''or 1=1;
8	Character equals to the same character, e.g., 'x'= 'x'	Condition that always returns true.	SELECT * FROM newsletter WHERE email=' ' or 'x'='x'
9	Variable equals to the same variable, e.g., a=a	Condition that always returns true.	SELECT * FROM custTable WHERE User=' ' or a=a--' OR 1=1'-- AND Pass=?
10	Character equals to the same character, e.g., 'a'= 'a'	Condition that always returns true.	SELECT * FROM product WHERE PCategory=' a' or 'a'='a'
11	Double quote	Double quote (") in SQL statement.	SELECT * FROM user WHERE id =''ddd"
12	Comment delimiter	Comment delimiter (/*) in a query. Text within comment delimiters is ignored.	"*/' and password=''

(Continued)

TABLE 14.1 (*Continued*) Attributes Used for Classifying SQL Injection

NO.	*ATTRIBUTE NAMES*	*DESCRIPTION*	*EXAMPLES*
13	Semicolon and SET IDENTITY_INSERT commands	; SET IDENTITY INSERT commands in a query	SELECT 1,2, 'or 1=1; SET IDENTITY_INSERT Students ON INSERT INTO Students (StudentID, Username, Password) VALUES (attacker studentid, 'attacker username','attacker password');-- FROM some_table
14	Semicolon and TRUNCATE TABLE commands	; TRUNCATE TABLE commands in a query	SELECT * FROM test.members WHERE user_name=''; TRUNCATE TABLE Username; --'
15	Semicolon and DROP TABLE command	; DROP TABLE commands in a query	SELECT 1,2, '; DROP table Username -- FROM some_table WHERE ex=ample
16	Semicolon and UPDATE command	; UPDATE commands in a query	SELECT email FROM users WHERE email=''or 1=1; UPDATE Grades SET grade='attacker grade' WHERE studentId=attacker studentid;--'
17	Semicolon and INSERT INTO command	; INSERT INTO commands in a query	SELECT id FROM Users WHERE username=''; INSERT INTO Username (username, password, user_type) value('admin2', 'admin2', '1'); --'
18	Semicolon and DELETE command	; DELETE command in a query	SELECT id FROM Users WHERE username=''or 1=1; DELETE FROM Grades WHERE course='attacker class' AND studentId=attacker studentid;--'
19	UNION command	UNION commands in a query	SELECT * FROM 'news' WHERE 'id'='' UNION SELECT password FROM users --' ORDER BY 'id' DESC LIMIT 0,3

SQL Query	Single Line Comment	Semi colon	Three Single Quote	Two Single Quote	Two Singapore Quote	True Case Zero	True Case One	True Case CharX	True Case VarA	True Case charA	Double Quote	Multiple Line Comment	SET IDENTITY _INSERT	TRUNCATE TABLE	DROP table	UPDATE	INSERT into	DELETE	union	Union Query
SELECT * from test.members where user_name=' ' or 0=0 #	0	0	0	0	1	1	0	0	0	0	0	0	0	0	0	0	0	0	0	0
SELECT * FROM customers WHERE username = ''; INSERT into Username (username, password, user_type) value('admin2', 'admin2', '1'); --' and password = ''	1	1	0	1	0	0	0	0	0	0	0	0	0	0	0	0	1	0	0	0
SELECT * FROM product WHERE PCategory=''; DROP table Username --'	1	1	0	1	0	0	0	0	0	0	0	0	0	0	1	0	0	0	0	0
SELECT * FROM newsletter WHERE email = ' 'sqlvuln'	0	0	0	0	1	0	0	0	0	0	0	0	0	0	0	0	0	0	0	0
SELECT * from test.members where user_name=' 1' AND non_existant_table = '1'	0	0	0	0	0	0	0	0	0	0	0	0	0	0	0	0	0	0	0	0

FIGURE 14.3 The samples of a dataset.

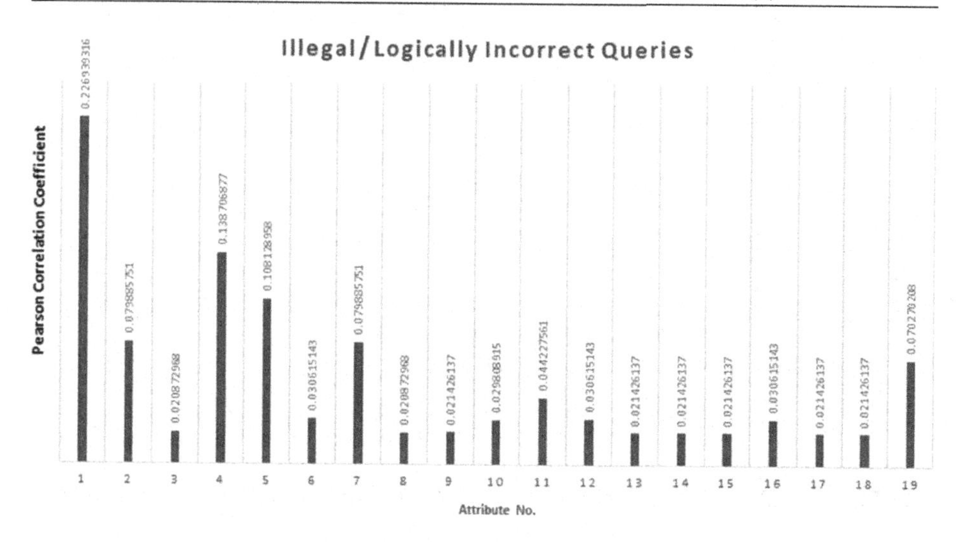

FIGURE 14.4 Pearson correlation coefficients of each attribute and illegal/logically incorrect queries.

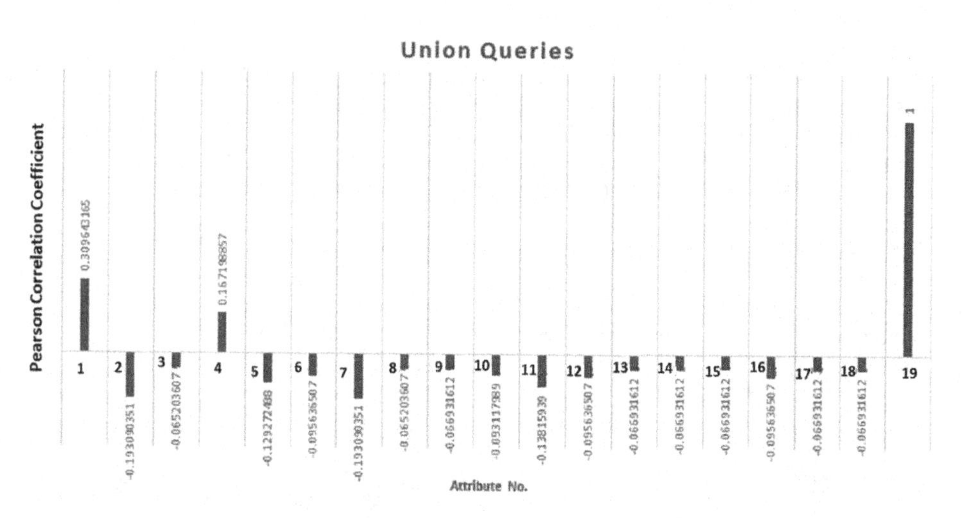

FIGURE 14.5 Pearson correlation coefficients of each attribute and union queries.

to develop an application to connect and send input parameters to the web service API by using JSON request and the prediction result will be sent back in JSON response body as well. In addition, the request headers must be defined for authentication and authorization purposes before sending the parameters. The samples of JSON request and response body are as follows:

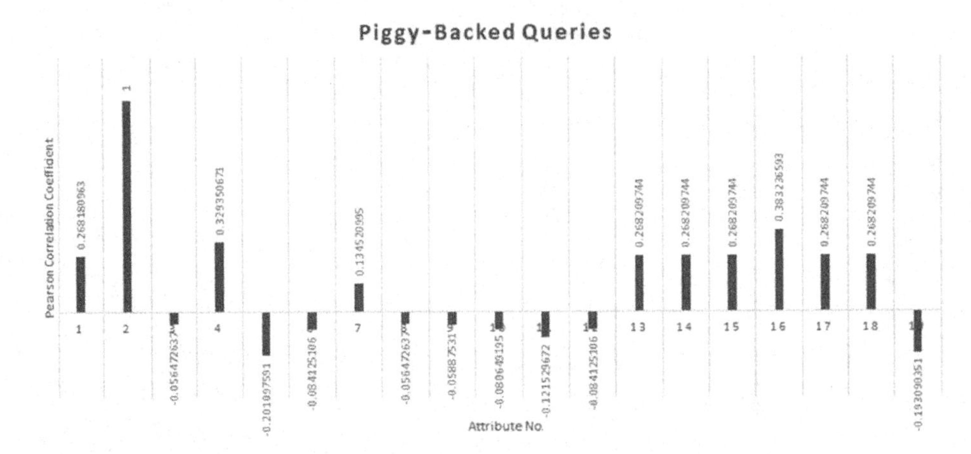

FIGURE 14.6. Pearson correlation coefficients of each attribute and piggybacked queries.

SAMPLE: JSON REQUEST BODY


```
{
"Inputs": {
"input1": {
  "ColumnNames": [ "SingleLine Comment","Semicolon", "Three
Single Quote", "Two Single Quote", "Two Singapore Quote",
"True Case Zero", "True Case One", "True Case CharX", "True
Case VarA", "True Case CharA", "Double Quote", "Multiple
Line Comment", "SET IDENTITY_INSERT", "TRUNCATE TABLE",
"DROP table", "UPDATE",
"INSERT into", "DELETE", "union" ],
  "Values": [
  [
"0", "0", "0", "0", "0", "0", "0","0", "0", "0", "0", "0",
"0", "0", "0",
  "0", "0","0", "0"
  ],
  [
"0", "0", "0", "0", "0", "0", "0","0", "0", "0", "0", "0",
"0", "0", "0",
"0", "0","0", "0"
  ]
  ]
  }
  },
  "GlobalParameters": {}
}
```

The proposed framework was not only designed for identifying SQL injection vulnerabilities but also for validating SQL statements to provide a recommendation for code refactoring as well. The process starts with parameter extraction. An SQL parser was implemented to detect SQL statements. The IDE extracts parameters from SQL statements. Then, the extracted parameters are sent as the input parameters to the authorized web service API. At the same time, the Nuget package manager, which is the component for the middle-tier, is used to support communication between the external component and the authorized web service. The finished product is considered an improved version of Roslyn. The product is wrapped and published as a package using Visual Studio Extension Template. The package can be embedded into Microsoft Visual Studio IDE.

14.6 EXPERIMENTAL RESULTS

SAMPLE: JSON RESPONSE BODY


```
{"Results": {
  "output1": {
  "type": "DataTable",
  "value": { "ColumnNames": [ "Scored Labels", "Scored
Probabilities"
  ],"ColumnTypes": ["Numeric", "Numeric"],
  "Values": [ ["0", "0"], [ "0", "0" ] ]
  }
  }
  }
}
```


In order to confirm the validity and performance of our framework, the experiments are divided into two parts: the first part explains the evaluation of the machine learning models and the second part presents the verification and validation of the compiler platform, which is the modified version of Roslyn integrated into Microsoft Visual Studio IDE to support code refactoring and SQL injection prevention during a software development phase.

14.6.1 Evaluation of the Models

This study evaluated six machine learning models, namely, SVM, locally deep SVM, boosted decision tree, decision forest, decision jungle, and artificial neural network, by using 70% of the formulated datasets as the training datasets and 30% as the testing datasets to find high-performance models to be used in the framework. The study initiates the initial settings in each of the machine learning models as recommended by Azure ML. Then, hyperparameter tuning with ten-fold cross-validation was performed on each model to find the best combination of settings while using accuracy as the metric. In order to have effective machine learning models that can identify SQL injection vulnerabilities, the resulted models were evaluated and compared in terms of high accuracy with a low false alarm. The processing time was also taken into consideration. The following details show the metrics used for performance measurement of the prediction models:

1. Probability of detection (*Pd*) or recall = $tp / (tp + fn)$
2. Probability of false alarm (*Pf*) or false positive rate = $fp / (fp + tn)$
3. Precision = $tp / (tp + fp)$
4. Accuracy = $(tp + tn) / (tp + fp + fn + tn)$
5. F1-Score = 2*(Recall * Precision)/(Recall + Precision)
6. Area under the receiver operating characteristics curve (AUC)
7. Processing time

Pd represents the efficiency of the prediction model to find the actual vulnerabilities. *Pf* represents the possibility of having a false alarm. Precision represents the ratio of correctly predicted actual vulnerabilities to the total predicted positive vulnerabilities. Accuracy indicates the correctness of the model prediction. F1-Score represents the balance between precision and *Pd*. AUC presents the degree of separability. Processing time describes the computation time for each model in seconds. Table 14.2 presents the confusion matrix used in this research.

Tables 14.3–14.5 present the experimental results. In case of illegal/logically incorrect queries, SVM and locally deep SVM outperform the other machine learning models with the perfect scores. F1-score and AUC values also confirm the performance of both classification models. However, the processing time of both SVM and locally deep SVM are much longer, which will affect the processing time of the proposed framework. In the case of union queries and piggybacked queries, there is no significant difference in terms of efficiency, correctness, and sensitivity. However, in terms of processing time, decision forest spent less time than the other classification models for all the three datasets. As a result, the decision forest is a promising candidate for the classification model to be integrated into the proposed framework.

14.6.2 Verification and Validation of the Compiler Platform

The compiler platform is designed to work with the trained machine learning models and to be integrated with an IDE in order to support a software developer to identify the

TABLE 14.2 Confusion Matrix (Shar and Tan 2013)

	ACTUAL	
PREDICTED BY CLASSIFIER	*VULNERABLE*	*NON-VULNERABLE*
Vulnerable	True positive (*tp*)	False positive (*fp*)
Non-vulnerable	False negative (*fn*)	True negative (*tn*)

TABLE 14.3 Trained Models Using Initial Settings and Hyperparameter Tuning with Cross-Validation – Illegal/Logically Incorrect Queries

	MODEL NAME	*TUNED PARAMETERS*	*PD*	*PF*	*ACCURACY*	*PRECISION*	*F1-SCORE*	*AUC*	*PROCESSING TIME (SECONDS)*
Initial Settings	SVM	Number of iterations=1,000 Lambda=0.001	0.996587	0	0.996667	1	0.998291	0.998294	2.972
	Locally deep SVM	Depth of the tree=3 Lamda W=0.1 Lambda Theta=0.01 Lambda Theta Prime=0.01 Sigmoid sharpness=1 Number of iterations=15000	0.996587	0	0.996667	1	0.998291	0.998294	2.641
	Boosted decision tree	Maximum number of leaves per tree=20 Minimum number of samples per leaf node=10 Learning rate=0.2 Number of trees constructed=100	0.996587	0	0.996667	1	0.998291	0.998294	3.702

(Continued)

TABLE 14.3 (*Continued*) Trained Models Using Initial Settings and Hyperparameter Tuning with Cross-Validation – Illegal/Logically Incorrect Queries

MODEL NAME	*TUNED PARAMETERS*	*PD*	*PF*	*ACCURACY*	*PRECISION*	*F1-SCORE*	*AUC*	*PROCESSING TIME (SECONDS)*
Decision forest	Resampling method=Bagging Number of decision trees=8 Maximum depth of the decision trees=32 Number of random splits per node=128 Minimum number of samples per leaf node=1	0.996587	0	0.996667	1	0.998291	0.998294	2.575
Decision jungle	Resampling method=Bagging Number of decision DAGs=8 Maximum depth of the decision DAGs=32 Maximum width of the decision DAGs=128 Number of optimization steps per decision DAG layer=2,048	0.996587	0	0.996667	1	0.998291	0.998294	2.6819
ANN	Number of hidden nodes=100 Learning rate=0.1 Number of learning iterations=100 The initial learning weight=0.1	0.996587	0	0.996667	1	0.998291	0.998294	5.7633

(*Continued*)

TABLE 14.3 (*Continued*) Trained Models Using Initial Settings and Hyperparameter Tuning with Cross-Validation – Illegal/Logically Incorrect Queries

	MODEL NAME	*TUNED PARAMETERS*	*PD*	*PF*	*ACCURACY*	*PRECISION*	*F1-SCORE*	*AUC*	*PROCESSING TIME (SECONDS)*
Tuned Parameters	SVM	Number of iterations=1 Lambda=0.01	1	0	1	1	1	1	8.688
	Locally deep SVM	1. Depth of the tree=1 2. Lamda W=0.001 3. Lambda Theta=0.1 4. Lambda Theta Prime=0.01 5. Sigmoid sharpness=1 6. Number of iterations=10000	1	0	1	1	1	1	9.859
	Boosted decision tree	1. Maximum number of leaves per tree=128 2. Minimum number of samples per leaf node=1 3. Learning rate=0.4 4. Number of trees constructed=20	0.998540	0	0.998571	1	0.999270	0.999319	5.632
	Decision forest	Number of decision trees=1 Maximum depth of the decision trees=16 Number of random splits per node=128 Minimum number of samples per leaf node=4	0.998540	0	0.998571	1	0.999270	0.998881	2.579

(*Continued*)

TABLE 14.3 (*Continued*) Trained Models Using Initial Settings and Hyperparameter Tuning with Cross-Validation – Illegal/Logically Incorrect Queries

MODEL NAME	*TUNED PARAMETERS*	*PD*	*PF*	*ACCURACY*	*PRECISION*	*F1-SCORE*	*AUC*	*PROCESSING TIME (SECONDS)*
Decision jungle	Number of decision DAGs=8 Maximum depth of the decision DAGs=32 Maximum width of the decision DAGs=8 Number of optimization steps per decision DAG layer=16,384	0.998540	0	0.998571	1	0.999270	0.999465	5.755
ANN	Learning rate=0.04 Number of learning iterations=40 The initial learning weight=0.1 Loss function=CrossEntropy	0.998540	0	0.998571	1	0.999270	1	5.912

TABLE 14.4 Trained Models Using Initial Settings and Hyperparameter Tuning with Cross-Validation – Union Queries

	MODEL NAME	*TUNED PARAMETERS*	*PD*	*PF*	*ACCURACY*	*PRECISION*	*F1-SCORE*	*AUC*	*PROCESSING TIME (SECONDS)*
Initial Settings	SVM	Number of iterations=1,000 Lambda=0.001	1	0	1	1	1	1	2.495
	Locally deep SVM	Depth of the tree=3 Lamda W=0.1 Lambda Theta=0.01 Lambda Theta Prime=0.01 Sigmoid sharpness=1 Number of iterations=15000	1	0	1	1	1	1	2.540
	Boosted decision tree	Maximum number of leaves per tree=20 Minimum number of samples per leaf node=10 Learning rate=0.2 Number of trees constructed=100	1	0	1	1	1	1	2.855
	Decision forest	Resampling method=Bagging Number of decision trees=8 Maximum depth of the decision trees=32 Number of random splits per node=128 Minimum number of samples per leaf node=1	1	0	1	1	1	1	2.875

(*Continued*)

TABLE 14.4 (*Continued*) Trained Models Using Initial Settings and Hyperparameter Tuning with Cross-Validation – Union Queries

	MODEL NAME	*TUNED PARAMETERS*	*PD*	*PF*	*ACCURACY*	*PRECISION*	*F1-SCORE*	*AUC*	*PROCESSING TIME (SECONDS)*
	Decision jungle	Resampling method=Bagging Number of decision DAGs=8 Maximum depth of the decision DAGs=32 Maximum width of the decision DAGs=128 Number of optimization steps per decision DAG layer=2,048	1	0	1	1	1	1	2.719
	ANN	Number of hidden nodes=100 Learning rate=0.1 Number of learning iterations=100 The initial learning weight=0.1	1	0	1	1	1	1	2.665
Tuned Parameters	SVM	Number of iterations=100 Lambda=0.00001	1	0	1	1	1	1	6.195
	Locally deep SVM	1. Depth of the tree=5 2. Lamda W=0.001 3. Lambda Theta=0.001 4. Lambda Theta Prime=0.1 5. Sigmoid sharpness=0.1 6. Number of iterations=10000	1	0	1	1	1	1	6.491

(*Continued*)

TABLE 14.4 (*Continued*) Trained Models Using Initial Settings and Hyperparameter Tuning with Cross-Validation – Union Queries

MODEL NAME	*TUNED PARAMETERS*	*PD*	*PF*	*ACCURACY*	*PRECISION*	*F1-SCORE*	*AUC*	*PROCESSING TIME (SECONDS)*
Boosted decision tree	1. Maximum number of leaves per tree=32 2. Minimum number of samples per leaf node=50 3. Learning rate=0.2 4. Number of trees constructed=100	1	0	1	1	1	1	6.235
Decision forest	Number of decision trees=1 Maximum depth of the decision trees=16 Number of random splits per node=128 Minimum number of samples per leaf node=16	1	0	1	1	1	1	5.757
Decision jungle	Number of decision DAGs=8 Maximum depth of the decision DAGs=32 Maximum width of the decision DAGs=8 Number of optimization steps per decision DAG layer=16,384	1	0	1	1	1	1	6.585
ANN	Learning rate=0.02 Number of learning iterations=20 The initial learning weight=0.1 Loss function=CrossEntropy	1	0	1	1	1	1	6.513

TABLE 14.5 Trained Models Using Initial Settings and Hyperparameter Tuning with Cross-Validation – Piggybacked Queries

	MODEL NAME	*TUNED PARAMETERS*	*PD*	*PF*	*ACCURACY*	*PRECISION*	*F1-SCORE*	*AUC*	*PROCESSING TIME (SECONDS)*
Initial Settings	SVM	Number of iterations=1,000 Lambda=0.001	1	0	1	1	1	1	2.522
	Locally deep SVM	Depth of the tree=3 Lamda W=0.1 Lambda Theta=0.01 Lambda Theta Prime=0.01 Sigmoid sharpness=1 Number of iterations=15000	1	0	1	1	1	1	2.630
	Boosted decision tree	Maximum number of leaves per tree=20 Minimum number of samples per leaf node=10 Learning rate=0.2 Number of trees constructed=100	1	0	1	1	1	1	2.433
	Decision forest	Resampling method=Bagging Number of decision trees=8 Maximum depth of the decision trees=32 Number of random splits per node=128 Minimum number of samples per leaf node=1	1	0	1	1	1	1	2.724

(Continued)

TABLE 14.5 (*Continued*) Trained Models Using Initial Settings and Hyperparameter Tuning with Cross-Validation – Piggybacked Queries

	MODEL NAME	*TUNED PARAMETERS*	*PD*	*PF*	*ACCURACY*	*PRECISION*	*F1-SCORE*	*AUC*	*PROCESSING TIME (SECONDS)*
	Decision jungle	Resampling method=Bagging Number of decision DAGs=8 Maximum depth of the decision DAGs=32 Maximum width of the decision DAGs=128 Number of optimization steps per decision DAG layer=2,048	1	0	1	1	1	1	2.541
	ANN	Number of hidden nodes=100 Learning rate=0.1 Number of learning iterations=100 The initial learning weight=0.1	1	0	1	1	1	1	2.799
Tuned Parameters	SVM	Number of iterations=100 Lambda=0.00001	1	0	1	1	1	1	6.447
	Locally deep SVM	1. Depth of the tree=5 2. Lamda W=0.001 3. Lambda Theta=0.001 4. Lambda Theta Prime=0.1 5. Sigmoid sharpness=0.1 6. Number of iterations=10000	1	0	1	1	1	1	6.298
	Boosted decision tree	1. Maximum number of leaves per tree=32 2. Minimum number of samples per leaf node=50 3. Learning rate=0.2 4. Number of trees constructed=100	1	0	1	1	1	1	5.632

(Continued)

TABLE 14.5 (*Continued*) Trained Models Using Initial Settings and Hyperparameter Tuning with Cross-Validation – Piggybacked Queries

MODEL NAME	*TUNED PARAMETERS*	*PD*	*PF*	*ACCURACY*	*PRECISION*	*F1-SCORE*	*AUC*	*PROCESSING TIME (SECONDS)*
Decision forest	Number of decision trees=1 Maximum depth of the decision trees=16 Number of random splits per node=128 Minimum number of samples per leaf node=16	1	0	1	1	1	1	5.603
Decision jungle	Number of decision DAGs=8 Maximum depth of the decision DAGs=32 Maximum width of the decision DAGs=8 Number of optimization steps per decision DAG layer=16,384	1	0	1	1	1	1	5.678
ANN	Learning rate=0.02 Number of learning iterations=20 The initial learning weight=0.1 Loss function=CrossEntropy	1	0	1	1	1	1	6.237

```
namespace SecurityLibrary
{
    2 references
    public class SqlQueries
    {
        0 references
        public object TestingQuery(
            string connection, string name, string password)
        {
            SqlConnection someConnection = new SqlConnection(connection);
            SqlCommand someCommand = new SqlCommand();
            someCommand.Connection = someConnection;

            someCommand.CommandText = "SELECT AccountNumber FROM Users " +"WHERE Username='" + name +
"' AND Password='" + password + "'";

            someConnection.Open();
            object accountNumber = someCommand.ExecuteScalar();
            someConnection.Close();
            return accountNumber; } }
```

FIGURE 14.7 Vulnerable SQL statements detected by the proposed framework.

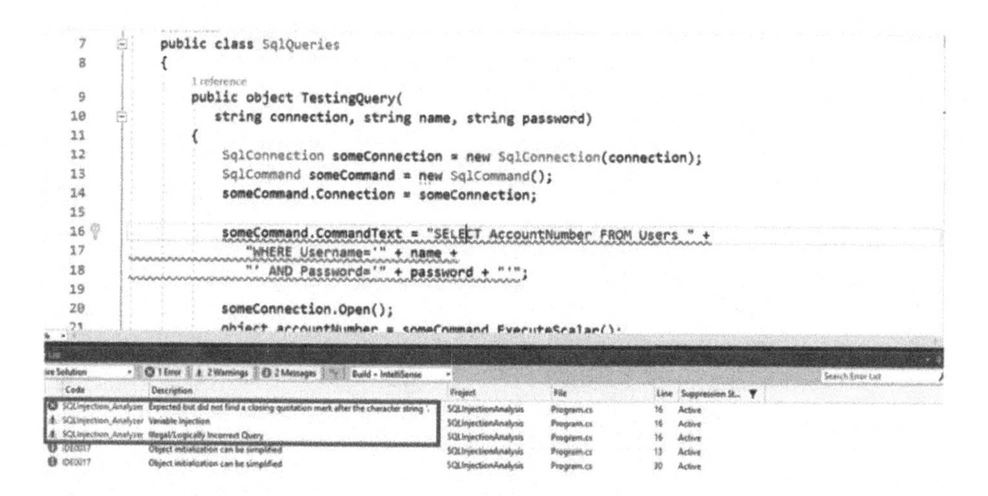

FIGURE 14.8 Framework specifies SQL syntax errors, vulnerable SQL statement, and classified types of SQL injection vulnerabilities.

possibility of having SQL injection vulnerabilities so that they can refactor their code at the early stage of software development. The framework scans a source code for SQL statements automatically to extract input attributes for vulnerabilities prediction, and therefore, it can suggest where to refactor the code. Figure 14.7 shows a vulnerable SQL statement detected by the proposed framework.

Figure 14.8 confirms that the framework not only can detect the vulnerable SQL statement but also can identify SQL syntax errors simultaneously. In addition, the type of SQL injection vulnerabilities is also classified.

Moreover, the commercial version of Microsoft Visual Studio IDE provides code analysis feature, which can be used to testify the performance of SQL injection detection. The feature can also detect vulnerable SQL statement as presented in

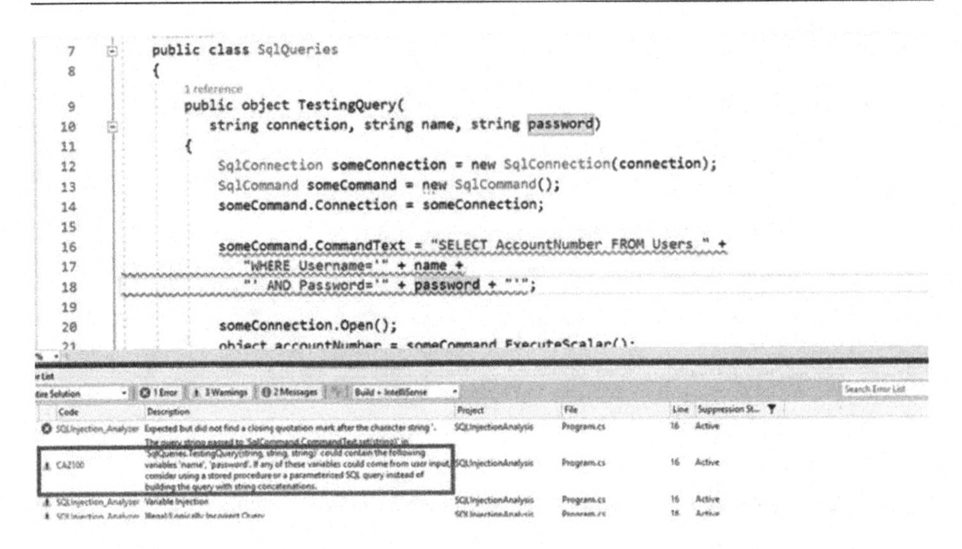

FIGURE 14.9 The code analysis results of the code analysis feature on Visual Studio Enterprise indicates the risk of SQL injection caused by flaws in the source code.

FIGURE 14.10 Recommendations for code refactoring to support secure coding practices.

Figure 14.9. However, the code analysis feature, a.k.a., CA2100: Review SQL queries for security vulnerabilities, does not address the type of SQL injection in the error list but recommends parameterized SQL query instead of building the query with string concatenations.

Figures 14.10 and 14.11 show that the proposed framework can recommend code refactoring relative to the SQL injection prevention practice in ASP.NET (C#). As a result, a software developer has some support to work toward secure coding practices resulting in SQL injection prevention during the implementation phase of the software development process.

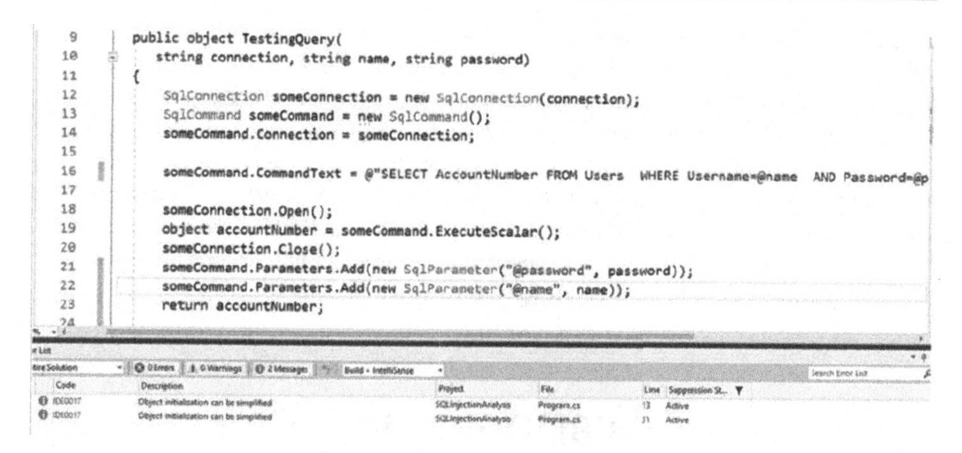

FIGURE 14.11 Security-improved version of the source code after refactoring by following the recommendations from the framework.

TABLE 14.6 Number of Vulnerable SQL Statements Detected by the Framework

TYPES OF SQL INJECTION VULNERABILITIES	*NUMBER OF DETECTED VULNERABLE SQL STATEMENTS*
Illegal/logically incorrect query	294
Union query	96
Piggybacked query	95
Others	5
Total	490 (98% of 500)

Sample data of 500 vulnerable SQL queries were extracted using the same method discussed in Section 14.4 to test the framework. Each query in the sample data was tested by putting the SQL statement behind the *someCommand.CommandText* command, which depicts line number 16 of the C# code in Figure 14.11 as well as the simulated parameters in C# code. The parameters were concatenated to the vulnerable SQL query as shown in Figure 14.9. The results show that the framework can detect 98.00% of the vulnerable SQL queries in the dataset. Table 14.6 presents the validation results. However, some of the vulnerable SQL statements cannot be detected since the statements are from the different types of SQL injection vulnerabilities that are not included in this research. These are left for future work.

14.7 CONCLUSIONS AND FUTURE WORK

Injection vulnerabilities are the consequences of the flaws in poorly designed software, which are sometimes inadvertently included in a source code. Both static and dynamic code analyses are important tools for a developer to prevent various types of injection vulnerabilities. This research developed a framework containing both compiler platform and machine learning models that can be integrated into Microsoft Visual Studio IDE to support SQL injection prevention by offering SQL injection vulnerability detection, SQL syntax validation, and recommendations for SQL statement refactoring at the early stage of software development. The framework can detect 98% of vulnerable SQL statements and can be expanded to identify more types of SQL injection and other injection vulnerabilities in the future. It is possible to integrate dynamic code analysis to the complier platform as well. This is left for future development.

BIBLIOGRAPHY

Chawla, N.V., K. Bowyer, L.O. Hall, and W.P. Kegelmeyer. 2002. "SMOTE: Synthetic minority over-sampling technique." *Journal of Artificial Intelligence Research* 16 (1): 321–357.

Clarke, J. 2012. *SQL Injection Attacks and Defense.* 2nd ed. Waltham: Syngress.

Deepa, G., and P.S. Thilagam. 2016. "Securing web applications from injection and logic vulnerabilities: Approaches and challenges." *Information and Software Technology* 74: 160–180.

Gertz, M. 2006. .NET compiler platform ("Roslyn"). Accessed November 1, 2015. https://archive.codeplex.com/?p=roslyn.

Hartley, D. 2012. "Reviewing code for." In *SQL Injection Attacks and Defense*, by J. Clarke, 89–138. Waltham: Elsevier, Inc.

Lee, I., S. Jeong, S. Yeo, and J. Moon. 2012. "A novel method for SQL injection attack detection based on removing SQL query attribute values." *Mathematical and Computer Modelling* (Elsevier) 55 (1–2): 58–68. Doi: 10.1016/j.mcm.2011.01.050.

Microsoft. n.d. Azure machine learning: Enterprise-grade machine learning service to build and deploy models faster. Microsoft Corporation. Accessed November 1, 2015. https://azure.microsoft.com/en-us/services/machine-learning/.

Microsoft. 2016. CA2100: Review SQL queries for security vulnerabilities. Microsoft Corporation. November 11. Accessed May 25, 2020. https://docs.microsoft.com/en-us/visualstudio/code-quality/ca2100-review-sql-queries-for-security-vulnerabilities?view=vs-2015&redirectedfrom=MSDN.

Microsoft. 2017. How to consume an Azure Machine Learning Studio (classic) web service. Microsoft Corporation. February 6. Accessed December 15, 2016. https://docs.microsoft.com/en-us/azure/machine-learning/studio/consume-web-services.

Microsoft. 2019. Two-class decision jungle. Microsoft Corporation. May 6. Accessed May 24, 2020. https://docs.microsoft.com/en-us/azure/machine-learning/studio-module-reference/-two-class-decision-jungle.

Microsoft. 2019. Two-class locally deep support vector machine. Microsoft Corporation. May 6. Accessed May 24, 2020. https://docs.microsoft.com/en-us/azure/machine-learning/studio-module-reference/two-class-locally-deep-support-vector-machine.

Microsoft. 2020. Microsoft azure machine learning algorithm cheat sheet. Machine Learning Algorithm Cheat Sheet for Azure Machine Learning designer. Microsoft Corporation. May 3. Accessed May 24, 2020. https://download.microsoft.com/download/3/5/b/35bb997f-a8c7-485d-8c56-19444dafd757/azure-machine-learning-algorithm-cheat-sheet-nov2019.pdf?WT.mc_id=docs-article-lazzeri.

Microsoft. 2020. Two-class boosted decision tree module. Microsoft Corporation. April 4. Accessed May 24, 2020. https://docs.microsoft.com/en-us/azure/machine-learning/-algorithm-module-reference/two-class-boosted-decision-tree.

Microsoft. 2020. Two-class decision forest module. Microsoft Corporation. April 22. Accessed May 24, 2020. https://docs.microsoft.com/en-us/azure/machine-learning/algorithm-module-reference/two-class-decision-forest.

Microsoft. 2020. Two-class neural network module. Microsoft corporation. April 22. Accessed May 24, 2020. https://docs.microsoft.com/en-us/azure/machine-learning/algorithm-module-reference/two-class-neural-network.

Microsoft. 2020. Two-class support vector machine module. Microsoft Corporation. April 4. Accessed May 24, 2020. https://docs.microsoft.com/en-us/azure/machine-learning/-algorithm-module-reference/two-class-support-vector-machine.

Robichaud, J. 2019. .NET compiler platform ("Roslyn") overview. January 8. Accessed May 27, 2019. https://github.com/dotnet/roslyn/wiki/Roslyn%20Overview.

Shar, L.K., and H.B.K. Tan. 2013. "Predicting SQL injection and cross site scripting vulnerabilities through mining input sanitization patterns." *Information and Software Technology* (Elsevier) 55 (10): 1767–1780. Doi: 10.1016/j.infsof.2013.04.002.

Zhu, J., J. Xie, H.R. Lipford, and B. Chu. 2014. "Supporting secure programming in web applications through interactive static analysis." *Journal of Advanced Research* (Elsevier) 5 (4): 449–462. Doi: 10.1016/j.jare.2013.11.006.

15 Multimedia Applications in Forensics, Security and Intelligence

M. Mubina Begum and C. Priya
Vels Institute of Science, Technology and Advanced Studies (VISTAS)

Contents

15.1 Introduction 270
15.2 Multimedia Application and Its Need 270
15.3 Forensics 271
15.4 Multimedia Applications in Security and Intelligence 272
15.5 Multimedia Encryption 272
15.6 Biometrics in Digital Rights Management 272
15.7 Digital Millennium Copyright Act 273
15.8 Secure Media Streaming and Secure Transcoding 273
15.9 Approaches to Multimedia Authentication 274
15.9.1 Active Image Authentication 274
15.9.2 Passive Image Authentication 274
15.10 Security Intelligence 275
15.11 A Glimpse at the Future 276
15.12 Conclusion 276
Bibliography 276

DOI: 10.1201/9781003051022-15

Abstract

A multimedia explosion of digital image, video, and audio material has occurred in recent years. End users can now easily create large volumes of multimedia content, store it, and distribute it through the Internet thanks to the rapid spread of mobile devices, which is followed by high-bandwidth Internet and cheaper storage. Meanwhile, our public spaces are being monitored by an increasing number of surveillance cameras. As a result, interactive multimedia has become embedded in our daily routines. With the proliferation of multimedia data, it has become important to protect it from unauthorized usage, identify and reconstruct illicit activities from it, and use it as a source of intelligence. However, the sheer volume of data poses serious challenges – tasks that were traditionally subjected to manual analysis are now well beyond the capabilities of forensic experts. There would bc a need for tools. Over the last few years, the multimedia research community has established a number of innovative solutions for dealing with video, photographs, audio, and other types of multimedia content, including information extraction, automated categorization, and indexing. Despite the fact that this work lays a solid framework for preserving and evaluating multimedia content, difficulties remain due to the material's scope, a lack of structure and metadata, and other application-specific constraints. It's finally time to customize, adapt, and broaden multimedia analysis for forensics, defense, and intelligence.

15.1 INTRODUCTION

A multimedia application is one that makes use of a variety of media channels, such as text, graphics, photographs, sound/audio, animation, and video. Hypermedia is one of the styles of multimedia applications. The terms multi and media are merged to create the word multimedia. The word "multi" means "multiple." All hardware or software used for communicating is referred to as media. Multimedia is an area that deals with the computer-controlled incorporation of text, diagrams, drawings, still and moving images (video), animations, and audio, among other things, to represent, store, transfer, and process all kinds of data.

15.2 MULTIMEDIA APPLICATION AND ITS NEED

Multimedia application is defined as the computer programs that incorporate various content types (e.g., text, pictures, graphics, sounds, and video) and allow users to communicate with them in a non-linear and interactive manner. Multimedia plays a vital part in today's culture, and everyone has to catch up with the times in today's society. Multimedia is an effective means of communicating since it makes it simple to talk and hear what others are saying. Then there's multimedia that incorporates animation, sound, film, and more.

15.3 FORENSICS

Forensic science is difficult to define precisely. Broadly speaking, it is an application of scientific techniques and principles to provide evidence to legal or relative investigations and determinations. For example, the detection of fingerprints at a crime scene is usually conducted by police officers who do not have science degrees, and recent court cases have questioned whether there is a true scientific principle underlying the discipline (Tilstone).

Students can enhance their study, critical thought, organization, persuasion, and oral communication skills by engaging in forensics. Preparation for careers in law, education, politics, journalism, religion, public relations, industry, and other areas that include critical thought and communication.

Cyber Forensics is perhaps one of the most exciting areas of computing today. Today, we'll talk about Multimedia Forensics, which is one of the most fascinating fields of Cyber Forensics.

The process of gathering, reviewing, and reporting on digital evidence in order to make it admissible in court is known as computer forensics. To establish a plausible case, forensic investigators must retrieve the probative evidence from the machines involved. The second step is Multimedia Forensics, in which analytical techniques are used to analyze the contents.

Multimedia content incorporates audio, video, pictures, and text. Because of the widespread usage of mobile devices, lower storage costs, and high bandwidth, Internet users are producing massive amounts of data. As a result of this expansion, interactive multimedia has risen to prominence. The volume of data is so high that it has outstripped the forensic experts' ability to efficiently examine and process it.

Multimedia Forensics is a set of techniques for analyzing multimedia signals such as audio, video, and photographs. Its goals are to

- Discover the origins of digital content.
- Identifying the data acquisition system that produced it.
- Validating the content's credibility.
- Knowledge extraction from multimedia signals.

Multimedia Forensics is based on the idea that inherent traces like digital fingerprints are left behind in digital media during both the creation phase and any other successively process. The pursuit of cybercrime detection has two main objectives. One to prevent the occurrence cybercrime in vulnerable institutions requiring security from loss, pilferage, and mishandling by accidental or intentional manipulations and the other detection and documenting cybercrimes through a disciplined methodology. With advancement of technology, computer is both the instrument of crime, as well as the location of evidence.

15.4 MULTIMEDIA APPLICATIONS IN SECURITY AND INTELLIGENCE

Owing to the increase in the variety and amount of data, multimedia security has improved dramatically over the years. Intelligent solutions for multimedia protection are in high demand in today's security setting. Biometrics, e-commerce, medical imaging, forensics, aerospace, and defense are only a few of the applications that include high-end computer protection systems. When it comes to high-resolution 2D/3D images and high-definition film, traditional cryptography, watermarking, and steganography fall short. Designing new protection algorithms for 3D graphics, simulations, and HD videos is in high demand. Traditional encryption methods are insufficient for today's needs because their security is limited when decoded.

15.5 MULTIMEDIA ENCRYPTION

Multimedia encryption is the key enabling technology for ensuring content confidentiality and preventing unauthorized entry. Traditional cryptographic algorithms cannot be used on multimedia data due to real-time limitations, massive amounts, and special characteristics. Recent years have witnessed an astounding development in the direction of format compliant, perceptual, and scalable encryption techniques that support advanced functionalities (Kulkarni et al.).

15.6 BIOMETRICS IN DIGITAL RIGHTS MANAGEMENT

Biometric recognition is a well-known research area that aims to provide more efficient solutions to the ever-growing growing human need for security. Biometrics refers to methods that can be used for uniquely recognizing humans based upon one or more intrinsic physical or behavioral characteristics. In information technology, in particular, biometrics is used as a tool for efficient and reliable identity management and access control. Biometrics is also used in surveillance applications in order to identify individuals. In the last decade, a large number of novel biometric modalities, such as palm, gait, and veins, as well as new methods for well-known biometric modalities, such as face, voice, and finger, have been proposed in the literature (Goudelis et al.).

15.7 DIGITAL MILLENNIUM COPYRIGHT ACT

The Digital Millennium Copyright Act (DMCA) is a copyright statute in the United States that incorporates two World Intellectual Property Organization treaties from 1996 (WIPO). The DMCA, which took effect on October 12, 1998, revised Title 17 of the United States Code to broaden the use of copyright, thereby restricting the responsibility of Internet content providers for copyright violations by their customers. It is expected that all users of any part of the Digital Media Law Project (DMLP) site will comply with applicable copyright laws. However, if the DMLP receives proper notification of claimed copyright infringement, our response to these notices will include removing or disabling access to material claimed to be the subject of infringing activity and/or terminating subscribers, regardless of whether we may be liable for such infringement under United States law or the laws of another jurisdiction (https://www.dmlp.org/digital-millennium-copyright-act-policy).

15.8 SECURE MEDIA STREAMING AND SECURE TRANSCODING

Two important desired capabilities for media streaming are media transcoding or adaptation and end-to-end security, and an important challenge lies in simultaneously enabling both capabilities. For example, it is beneficial to be able to efficiently stream and adapt encrypted media content at potentially untrusted nodes without breaking the end-to-end security. This may be desirable at a potentially untrusted (or vulnerable) streaming server (Apostolopoulos, 2004). To ensure maximum protection, data can be secured by the content producer and decrypted by the content user, with the media remaining encrypted in transit. This is referred to as end-to-end security.

Non-scalable H.264 video is packetized into secure scalable packets with unencrypted packet headers that provide R-D information about each packet's importance. The downstream transcoders will use this knowledge as a guide. According to preliminary research, one byte of knowledge in the unencrypted packet header might be adequate; however, this depends on the particular situation. Mid-network transcoders read each packet's unencrypted headers and choose whether to keep or discard it depending on its value and network constraints. The mid-network transcoder can perform R-D optimized adaptation across multiple packets of a single stream or across packets of multiple separate streams, which is a key feature of this approach (Liang et al., 2003).

15.9 APPROACHES TO MULTIMEDIA AUTHENTICATION

Since Internet content isn't only limited to text, but often comes in a wide range of formats, the forensic methods used to examine it must also vary in scope. The aim is to interpret photographs, text, audio, and video in order to create a logical piece of forensic evidence.

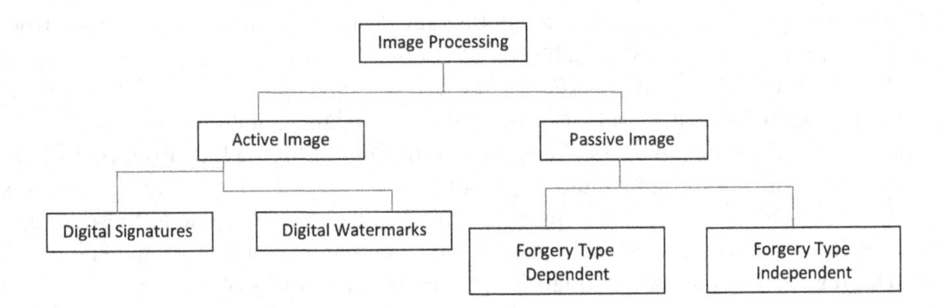

Multimedia Forensics divides its efforts between two main approaches:

- Active Image Authentication
- Passive Image Authentication.

15.9.1 Active Image Authentication

Active Image Authentication is a method of verifying the integrity of an image by embedding a known authentication code in the image at the time of creation or sending it with the image. Verifying this code verifies the image's authenticity. Digital Watermarking and Digital Signatures are two subcategories of Successful Authentication.

Digital watermarking – This technique involves embedding a digital watermark into an image at either the retrieval or processing stages.

Digital signatures – At the acquisition end of the file, digital signatures embed any secondary information that is normally derived from the image.

15.9.2 Passive Image Authentication

For accessing the integrity of an image, passive authentication, also known as image forensics, uses the only image with no prior knowledge. Passive authentication is based on the idea that while interfering with the picture cannot leave a visible trace, it is likely to change the underlying statistics. This means that digital forgeries can affect the image's underlying properties, such as quality, even if no physical evidence has been left behind.

Forgery-type dependent and independent techniques are two types of passive techniques.

Forgery-type dependent – These are only designed to detect particular types of forgeries, such as copy-move and image-splicing, and are based on the type of forgery performed on the image. There are two types of detection: copy-move detection and image-splicing detection.

Copy-move detection – Because of how simple it is to do, copy-move is the most common photo-tampering technique. It requires copying parts of an image and moving them to a different part of the image. The dynamic range and color of the copied area are consistent with the rest of the picture since it is part of the same image.

Image-splicing detection – The image-splicing process includes combining two or more images and greatly altering the original image to produce a forged image. Please keep in mind that when combining photos of different backgrounds, it can be difficult to hide the boundary and boundaries.

Forgery-type independent – These methods detect forgeries based on artifact traces left during the re-sampling process and lighting anomalies, regardless of the form of forgery. It is further divided into two groups: Retouching Detection and Lighting Conditions.

Retouching detection – This method is most commonly used for commercial and esthetic applications. Retouching is usually used to improve or minimize image features, or to create a compelling composition of two images that allows one of the images to be rotated, resized, or extended.

Lighting conditions – During tampering, images are taken in various lighting conditions and then combined. Combining images makes matching the lighting conditions incredibly difficult. This inconsistency in lighting in the composite image can be used to detect image tampering.

15.10 SECURITY INTELLIGENCE

The term Security Intelligence describes the practice of collecting, standardizing, and analyzing data that is generated by networks, applications, and other IT infrastructure in real time, and the use of that information to assess and improve an organization's security posture (https://www.sumologic.com/glossary/security-intelligence/#:~:text=The%20term%20Security%20Intelligence%20describes,improve%20an%20organization's%20security%20posture). The discipline of Security Intelligence includes the deployment of software assets and personnel with the objective of discovering actionable and useful insights that drive threat mitigation and risk reduction for the organization.

15.11 A GLIMPSE AT THE FUTURE

Above is the clear overview of the various issues that arise in the fields of forensics, defense, and intelligence when it comes to multimedia. They also serve as a foundation for future research in this fascinating area. A variety of study issues remain unresolved for the time being.

Criminals are widening their repertoire and using more sophisticated tactics, despite the fact that techniques for detecting and protecting illicit content are increasingly improving. To get around state-of-the-art detection and protection techniques, attackers only need to learn one tool, but research should consider all potential attacks and possibilities. Hence, advancing techniques along the established lines is of great importance.

Finally, there is still a lot of work to be done in multimedia analytics. Until recently, the majority of inquiries were conducted manually. The studies using automated methods are supported by the state-of-the-art techniques outlined in this issue and elsewhere.

These two areas must collaborate in order to conduct genuinely successful investigations. Developing advanced browsing, visualization, and data mining techniques that help every step of the investigative process is the best way to do this.

15.12 CONCLUSION

Further advancements in forensic science in combination with the introduction of new technology and methods that create an added value (innovation) for the end-user will definitely be able to cause a paradigm shift within the criminal justice system. However, according to Downes' law of disruption, technology changes exponentially while social, economic, and legal systems change incrementally (Downes, 2009). This makes it difficult for new technology to be introduced and implemented. Although Downes' law points at the fast developments in the digital world, it also holds for many other areas and especially for the criminal justice system as new legislation and extensive quality control measures are required to allow for new forensic methodology to be regularly applied.

BIBLIOGRAPHY

Available at https://www.sumologic.com/glossary/security-intelligence/#:~:text=The%20term%20Security%20Intelligence%20describes,improve%20an%20organization's%20security%20posture.

Available at https://www.dmlp.org/digital-millennium-copyright-act-policy.

J.G. Apostolopoulos. 2004. Secure media streaming & secure adaptation for non-scalable video. *2004 ICIP 04. 2004 International Conference on Image Processing*.

L. Downes. 2009. *The Laws of Disruption*. New York: Basic Books.

G. Goudelis, A. Tefas, I. Pitas. Intelligent multimedia analysis for emerging biometrics. Available at https://link.springer.com/chapter/10.1007/978-3-642-11756-5_5.

S. Kasinathan, C. Priya. 2020. Enabling the efficiency of blockchain technology in tele-healthcare with enhanced EMR. *2020 International Conference on Computer Science, Engineering and Applications* IEEE Explore.

N.S. Kulkarni, B. Raman, I. Gupta. Multimedia encryption: A brief overview. Available at https://link.springer.com/chapter/10.1007/978-3-642-02900-4_16.

Y. Liang, J. Apostolopoulos, B. Girod. 2003. Analysis of packet loss forcompressed video: Does burst-length matter? *2003 IEEE International Conference on Acoustics, Speech, and Signal Processing, 2003. Proceedings*, April, 2003.

A.A. Mapes, A.D. Kloosterman, C.J. de Poot. In press. DNA in the criminal justice system: the DNA success story in perspective. *J. Forensic Sci.* Doi: 10.1111/1556-4029.12779 [PubMed] [CrossRef].

W.J. Tilstone, W. Tilstone, K.A. Savage, L.A. Clark. Forensic science: An enclyclopedia of history, methods and techniques. Available at https://books.google.com.sg/books?id=zIRQOssWbaoC&printsec=frontcover&dq=forensics+science&hl=en&sa=X&ved=2ahUKEwi8iuz1hdzvAhXH8HMBHRCHC0gQ6AEwAnoECAEQAg#v=onepage&q=forensics%20science&f=false.

K.M. Tolle, D.S.W. Tansley, A.J.G. Hey (eds). 2009. *The Fourth Paradigm: Data-Intensive Scientific Discovery*. Microsoft Corporation. Available at http://research.microsoft.com/en-us/collaboration/fourthparadigm/4th_paradigm_book_complete_lr.pdf.

16 Advancements and Innovation in Digital Marketing and SEO

Anubha Jain and Chhavi Jain
IIS (deemed to be University)

Rahul G. Kargal and Salini Suresh
Dayananda Sagar College of Engineering

Contents

16.1 Introduction 280
16.2 Marketing 281
    16.2.1 Shift from Traditional Marketing to Digital Marketing 282
16.3 Digital Marketing 282
    16.3.1 Digital Marketing: Then and Now 283
    16.3.2 AI in Digital Marketing 285
16.4 Omni-Channel Marketing 287
    16.4.1 Augmented Reality (AR) and Immersive Technologies 287
    16.4.2 Augmented and Predictive Analytics 289
16.5 Marketing Automation 290
16.6 Social Media Marketing 291
    16.6.1 Social Media Stories 292
16.7 Mobile Marketing 293
    16.7.1 Mobile Website 293

DOI: 10.1201/9781003051022-16

16.7.2 Mobile Applications 293
16.8 Geo-Fencing Marketing 294
16.9 Influencer Marketing 296
16.10 Digital Advertising 297
16.10.1 Display Advertising 297
16.10.2 Audience Targeting 298
16.10.3 Programmatic Advertising 298
16.10.4 Search Advertising 298
16.10.4.1 Visual Search 299
16.10.4.2 Voice Search, Voice Assistants, and Smart Speakers 300
16.10.5 Banner and Video Advertising 301
16.10.6 Video Advertising 301
16.10.7 Social Media Advertising 302
16.10.7.1 Precise Targeting 302
16.10.7.2 Ad Placement 302
16.10.7.3 Ad Bidding 302
16.10.8 Mobile Advertising 304
16.11 Search Engine Optimization 305
16.11.1 SEO: Then and Now 305
16.11.1.1 Google Panda: The Game Changer Algorithm for Content 306
16.11.1.2 Google Penguin 308
16.11.1.3 Google Hummingbird 308
16.11.1.4 On-Site SEO 309
16.11.1.5 Off-Site SEO 309
16.11.1.6 SEO Best Practices 309
16.11.1.7 Title Tag 309
16.11.1.8 Meta Descriptions 309
16.11.1.9 URL 310
16.11.1.10 Content of Page 310
16.11.1.11 Image ALT Text 311
16.11.1.12 Page Ranking Factors 311
16.12 Benefits 313
Bibliography 314

16.1 INTRODUCTION

The amount of time that a digital consumer spends online is constantly on the rise. The digital population across the globe has augmented tremendously. With the digital landscape featuring a user base this vast, digital marketing as a phenomenon has been evolving and making rapid strides. Digital marketing is the promotion of products or services through the internet or any other electronic media. It involves managing online presence of business on websites, social media, smartphones, mobile apps, etc. It is used to grow the customer base. Companies and consumers both are adapting digital marketing. Digital marketing also enables customers in co-creation of many products. People who are actively using social media platforms account to 3.14 billion. Social media platforms such as Facebook, Instagram, and WhatsApp enable marketers to spread the word and effectively promote their brands. It can help any company by more customer engagement, increased customer retention, and generating more market share.

SEO or Search Engine Optimization (SEO) improves the visibility of a website. SEO allows marketers to understand how website visitors think and act while searching online. It is more focused on receiving the search results in a traffic-free or organic manner.

Search engines that are used mainly include Google, Yahoo, and Bing. Other less frequently used search engines include Baidu, Yandex, Naver, etc. Google dominates the search-engine market with an 86.86% market share as of July 2020. During an average search, on the top and side of the search page, ads can be seen. Organic results lie in the middle of the page. Search-engine marketers are more interested in ads. Bids are put on these ads on keywords. Every digital marketer wants its website to show on the first page of search results.

This chapter discusses some of the major advancements in the field of digital marketing and SEO.

16.2 MARKETING

Marketing is about understanding the needs of customers and providing solutions to those needs, which can satisfy the customers. Marketing is the activities, set of institutions, and processes that are involved in the creation, communication, delivery, and exchange of offerings that have value for customers, clients, partners, and society at large. Value is basically the net sum of all benefits and costs that a customer has to incur in order to have that offering. In addition to product benefits, any product also delivers service benefits and image benefits. Value creation is the benefits that the customer gets from buying the product. It is important for any marketer to communicate the product value to the customer. This created value in the format of a product or a service or an offering is communicated so that customers are aware of it. Once the customer becomes aware of the value that is being offered by the marketer, the customer might be interested to come and acquire the value from the marketer. So the marketer has to ensure that the delivery of all of this value is also planned properly. The place where the customer can come and buy this value is the place where the product/service is delivered. The product/service is delivered to the customer after an exchange of value.

Marketing is about the set of activities, institutions, and processes that are involved in the creation, communication, delivery, and exchange of offerings, which have value for customers, clients, partners, and the society. The marketing mix elements – the product, the price, the place, and the promotion – create a point of differentiation amongst different brands. The product is basically the offering that is given to a customer. The cost which is associated with the product exchange value is known as the price. The customer compares the product and its features with its price. Place is where the customers can come and purchase the product. It can be some physical store, online shopping portal, etc. The last 'P' of the marketing mix talks about the Promotion. Promotion is about making the customer aware of the product through advertising, sales promotion, etc.

16.2.1 Shift from Traditional Marketing to Digital Marketing

There are 4.57 billion internet users in the world. Thus there is a need for marketers to use the internet for digital marketing purposes. As of January 2020, India has 687.6 million internet users and across the globe, the worldwide digital population in April 2020 stood at 4.57 billion. The technological revolution has brought striking transformations in the world of marketing leading to the development of a phenomenon called digital marketing. In digital marketing also, marketers create, communicate, deliver, and exchange value for customers in digital space similar to traditional physical marketing. The three Ps of marketing are almost the same for both traditional and digital marketing. The major impact has been on the fourth 'P'- Promotion. Digital and social media have transformed the way companies promote their product/ services.

16.3 DIGITAL MARKETING

Digital marketing is the marketing that is done over the digital platform or digital media. Digital media can be used for the creation of products and services. Many companies gather ideas from customers over social media platforms and also through co-creation. Digital marketers use digital platforms for communication and promotion of products. The amount of time that a digital consumer spends online is constantly on the rise. As per a study conducted by Mindshare (a media agency) and Vidooly (an analytics platform), the average daily online media consumption in India prior to the lockdown stood at 1.5 hours. Lockdown during the COVID pandemic resulted in raising this figure to greater than 4 hours. Also, mobile data consumption skyrocketed during the lockdown with telecom operators witnessing a 30% spike in national consumption.

Digital marketing links consumers with sellers electronically using interactive technologies like emails, online forums and news groups, interactive television, mobile platforms, social media, etc. The emergence of digital platforms has made marketing more interactive in nature.

Digital marketing is the promotion of products or services through the internet or any other electronic media. It involves managing the online presence of business on website, social media, smartphones, mobile apps, etc. It is used to grow customer base. Companies and consumers both are adapting digital marketing. Digital marketing also enables customers in co-creation of many products. People who are actively using social media platforms account to 3.14 billion. Social media platforms such as Facebook, Instagram, and WhatsApp enable marketers to spread the word and effectively promote their brands. It can help any company by more customer engagement, increased customer retention and generating more market share. SEO improves the visibility of a website. SEO allows marketers to understand how website visitors think and act while searching online. It is more focused on receiving the search results in a traffic-free or organic manner. Other terminologies which are related to SEO are SEM and content writing. SEM or search-engine marketing is bidding for paid advertisements and search engines. Social Media Marketing encompasses free and paid social advertising and engagement practices both. Content marketing is focused on writing content for blogs, etc. Hat SEO is the best practice laid out by search engines like Google, etc., whereas black hat SEO tends to adapt manipulative practices which may also attract some penalties.

Search engines that are used mainly include Google, Yahoo, and Bing. Other less frequently used search engines include Baidu, Yandex, Naver, etc. Google dominates the search engine market with an 86.86% market share as of July 2020. During an average search, on the top and side of the search page, ads can be seen. Organic results lie in the middle of the page. Search engine marketers are more interested in ads. Bids are put on these ads on keywords. Every digital marketer wants its website to show on the first page of search result. Relevant keywords are used for this purpose.

16.3.1 Digital Marketing: Then and Now

The term 'Digital Marketing' was first used in the 1990s. According to some academicians Gugliemo Marconi, the inventor of radio, can be considered the world's first digital marketer, while others consider Ray Tomlinson, a computer engineer, as the Father of Digital Marketing because he sent an email for the first time in 1971.

The digital era started with the advent of internet and the development of the Web 1.0 platform. It was at the beginning of the 1990s. It is the first generation of the web. In this generation, static websites were built. Static websites are built of text and web pages, which are interconnected. They worked only as information portals. No direct communication or exchange was possible during this phase, between the user/reader and information provider. As Web 1.0 platform did not facilitate sharing of information by users, it was not useful for marketers for promoting their products. The year 1993 began the transition in digital marketing era (Figure 16.1).

The first successful mass-market browser – Netscape – was launched in the year 1994. An increase in the number of users gave rise to search engines like Yahoo in 1994 and Google in 1998. These search engines were helpful for users in finding the desired information regarding the products and services from their home or office. Increase in users also ignited the evolution of e-commerce sites – Amazon in 1994 and eBay in 1995. Thereafter, emails were used as a promotional tool for outbound marketing in addition to traditional modes of promotion such as TV, radio, telephone, and print media. Search engine traffic saw a remarkable growth of 6.4 billion in a single month during 2006. Thus, companies started using SEO techniques to generate better rankings. Now Web 1.0 is outdated and thus replaced by Web 2.0.

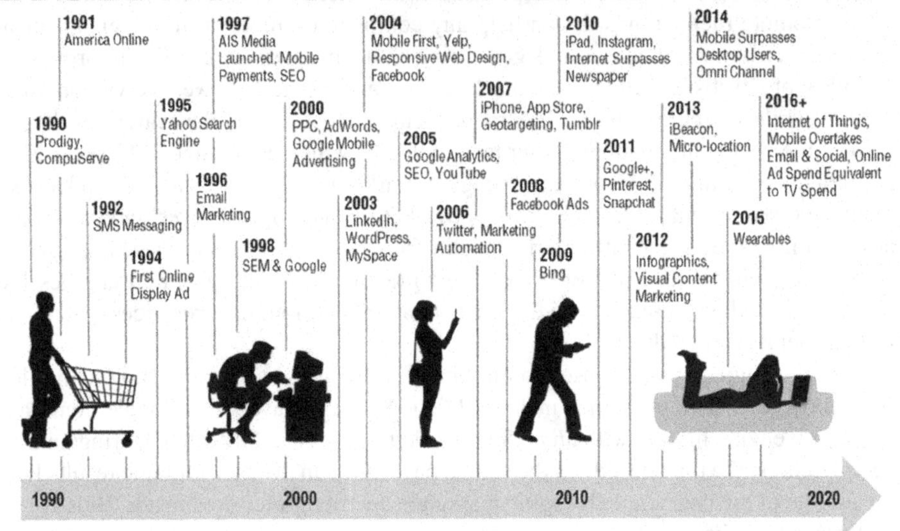

FIGURE 16.1 Evolution of digital marketing.

Web 2.0 is also referred to as social web. The main focus of this phase was not only on sharing information but also to engage people. This made Web 2.0 more dynamic and interactive. Now users can also generate the content. This way Web 2.0 made people active participants instead of passive users. It allowed interaction amongst users. Marketing through the internet generated 2.9 billion US $ in 2004. Thereafter, social networking sites like MySpace and Facebook evolved. LinkedIn, Twitter, Zynga, Google+, Pinterest, Quora, etc. were also the foundations for web's social layer. The most ingenious invention during this phase was the Like button of Facebook. It expressed user/reader's sentiment or expression of admiration to all connected people in the social circle. Soon companies identified the change in the trend and started pouring funds into social networking platforms for product promotion. Cookies were used to track browsing habits of frequent users. Data collected from cookies was further used to customise products and their promotion.

Web 3.0 depicts the third stage of website development. Its other names are Semantic Web, Intelligent Web, and Decentralized Web. This phase connects users to machines; it is about the deployment of Artificial Intelligence (AI) in web services. Web 3.0 tracks users' browsing habits, collecting and disseminating information in a smarter way which goes hand in hand with the rapidly changing buying trends of consumers. It also includes elements such as linked data, smart search capabilities, AI, and cloud computing to make technologies more engaging and useful for users. It stores real-time detailed information of users, which gives better meaning, relevancy, and value to users and marketers. Smartphones with incredible speeds and cloud computing have also made web experience better than before. Static videos are outpaced by live streaming videos. Crunch of time has boosted microblogging. Popular sites for microblogging

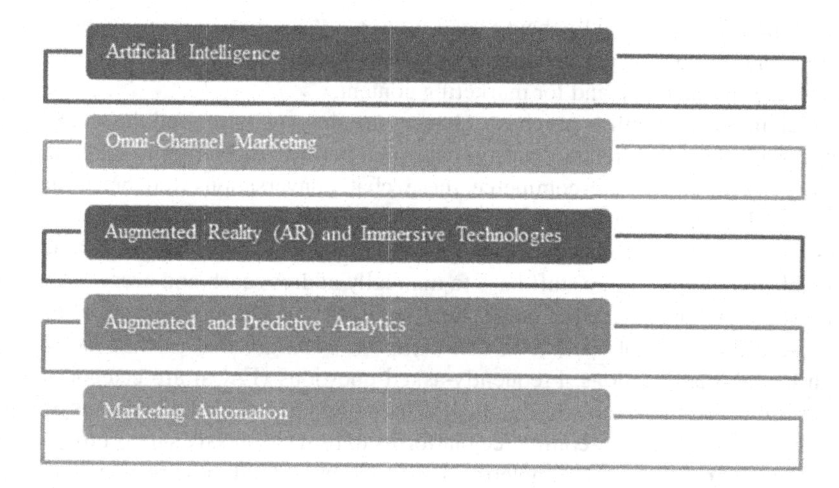

FIGURE 16.2 Advancements in digital marketing.

include Twitter, Jaiku, Plurk, etc. Trade shows are replaced by virtual reality world in a 3-D setting by Second Life and Funsites.

Web 3.0 will be most beneficial to marketers as they can use efficient and effective strategies to integrate intelligent data and use it for understanding consumers and customizing customer experiences. Companies can improve the relevance and effectiveness of digital marketing by using intelligent data. This data can be received from websites, mobile apps, email marketing, social media, etc. Digital marketing helps consumers by making products available 24×7 for them. The top three social sites used by digital marketers are LinkedIn, Twitter, and Facebook. Facebook has over 2.7 billion active users in the second quarter of 2020. Google analyses the content they receive through AdSense and targets advertisements based on users' interests. Web 3.0 is vital for the competitiveness of the organizations.

'Digital Darwinism' is a phenomenon that every corporation is wary of, where the early adopter of digital advancements emerges unscathed in the battle for digital survival and supremacy. While several advancements in digital marketing have been projected as 'game-changers' and have seen rapid adoption rates across global corporations, the following are prominent ones – Artificial Intelligence (AI), Omni-channel Marketing, Augmented Reality (AR) and Immersive Technologies, Augmented and Predictive Analytics, and Marketing Automation (Figure 16.2).

16.3.2 AI in Digital Marketing

As per Professor B.J. Copeland, the Director of the Turing Archive for the History of Computing, University of Canterbury, Christchurch, New Zealand and the author of *Artificial Intelligence* defines AI as "the ability of a digital computer or computer-controlled robot to perform tasks commonly associated with intelligent beings". AI is rapidly disrupting the marketplace by finding acceptance by corporations seeking to use customer data and build an automated ecosystem that enhances customer experience.

In digital marketing, AI is being used to create 'Chat bots' for customer communications and product recommendations, for sending personalized emails, to facilitate e-commerce transactions, and for marketing content.

When making an online purchase, chances are that you have used the 'chat' option to get a product or offer-related question clarified before proceeding to make the payment. During the advent of e-commerce, these 'chat' conversations were being managed by a human being, i.e., an actual person would chat with the consumer in real time and address the said query. Today, however, given the large consumer base, it is practically impossible to employ a large staff base to manually address each and every query. And this is where AI-enabled 'Chat bots' come to the rescue. 'Bots' are a short-form for robots. And this form of 'bot' is in fact computer software that runs at the backend of websites and applications. Frequently-asked-questions (FAQs) are categorized and answers are pre-authored to address them. Thereafter, AI-enabled algorithms are integrated into 'Chat bots' on e-commerce platforms to provide speedy support to customer when they are on the cusp of completing a transaction. Should these responses be inadequate to respond to a customer's query, an actual person can then take over to solve the matter.

So the next time you are on an e-commerce platform and trigger a chat conversation, chances are that the first set of responses that you receive are actually those being handed to you by an AI-enabled 'chat bot'. *Mastercard* worked with *Facebook* to integrate an AI-powered 'chat bot' that can analyse a customer's chat conversations and provide precise responses. In doing so, *Mastercard* is able to provide speedy resolutions to pay related issues and ensure a seamless transaction (Figure 16.3).

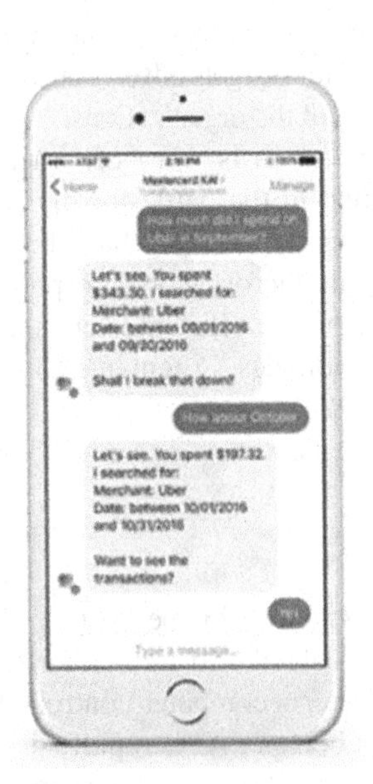

FIGURE 16.3 Artificial Intelligence (AI) in digital marketing.

16.4 OMNI-CHANNEL MARKETING

A digital marketing campaign features the use of several channels – Social Media Advertising, Search Ads, Display Ads, SEO, Shopping Ads, Display Advertising, YouTube Advertising, and Mobile Marketing. Given the ease and expanse of present-day digital accessibility, consumers (both retail and B2B) are now present in virtually every channel.

The general public, when searching for a place to dine, invariably ends up searching the term 'restaurants near me'. This translates to the usage of the Google Search console. Likewise, when browsing through one's feed on Facebook or Instagram, a user watches a 'carousel ad' from a retail brand. This is interpreted as the consumption of Social Media Ads or Shopping Ads. Likewise, when a person is reading a breaking news article on a media portal, he or she witnesses an ad from an e-commerce portal displaying a set of products. This is display advertising. Similarly, when watching a video on YouTube, at the beginning or midway through the video, a product ad plays out with the option for the user to 'skip' the ad. This is an example of YouTube Advertising. And when a mobile phone user receives an SMS with a discount code for an upcoming sale, Mobile Marketing is at play.

These examples only go to show the ubiquitous nature of digital marketing. And with brands increasingly realising that an average consumer spends a great deal of time online while browsing through social media content, reading and watching news and videos via mobile devices, it has become inevitable that they be present in all of these channels. Therefore, the phenomenon of relaying consistent marketing messages pertaining to a product or service across all digital channels has given birth to the term 'Omi-channel Marketing'.

Several retail and fitness brands such as *Reliance Trends* such as *Cult Fit* are prominent adopters of Omni-channel Marketing. One may visit their social media handles and watch their video ads on YouTube to observe the consistency in messaging.

When executing Omni-channel marketing campaigns, brands ensure that their message across all channels is singular and unified. Also, consistency and personalization techniques are used to ensure that a consumer watching a brand's message across all channels gets a unified message. The biggest benefit (See Figure 16.2) of Omi-channel marketing for brands is being able to connect and establish a rapport with their consumer base regardless of the channel that the consumer uses. Also, a study documented in the *2019 Omnisend Annual Report* states that the consumer engagement rate for Omni-channel campaigns was 18.96% when compared to 5.4% for single-channel campaigns. Similarly, the purchase frequency and consumer retention rates were 250% and 90% higher for Omni-channel campaigns when compared to single-channel campaigns (Figure 16.4).

16.4.1 Augmented Reality (AR) and Immersive Technologies

When *Pokémon Go* exploded on the digital gaming scene in 2016, users found themselves addicted to the experience of using their mobile devices to collect virtual

FIGURE 16.4 Omi-channel marketing.

Pokémon's while roaming the streets of their city. Little did gamers know at the time that the technology used by the developers of the game was called Augmented Reality (AR).

According to *Gartner's IT Glossary*, "Augmented reality (AR) is the real-time use of information in the form of text, graphics, audio, and other virtual enhancements integrated with real-world objects."

The success of *Pokémon Go* showed the joy that consumers experience when immersed in an ecosystem where the virtual world meets the real world. And digital marketers are quick to adopt this technology and integrate it into their marketing campaigns to heighten consumer experiences and thereby increase the probability of product sales.

The cosmetics brand *L'Oreal* has teamed up with app developers and has made available apps such as *Nail Genius* and *Makeup Genius* to consumers. Prospective buyers can now snap up their pictures, tap the shades of *L'Oreal* products on the app, and watch their skin tone change on the phone accordingly to the colour selected (Figure 16.5). Similarly, *Facebook* has now partnered with several cosmetic brands to provide an AR experience to consumers through the *Facebook* camera feature. This way, consumers can experience, from the comfort of their homes, products from brands such as *L'Oreal Paris, Maybelline, Yves Saint Laurent, and Giorio Armani.*

In India, a similar AR feature is offered by *Lenskart.com* where consumers can use a picture of theirs to virtually try spectacle frames before making a purchase. Such is the projected growth of AR that by 2022, *Gartner* estimates that 70% of all brands will be employing immersive technologies.

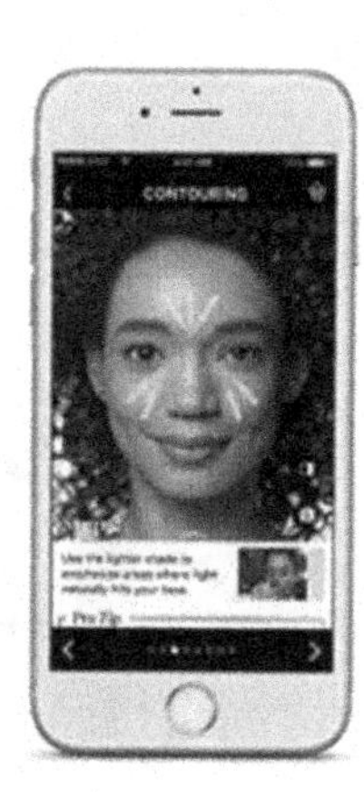

FIGURE 16.5 Augmented Reality (AR) & Immersive Technologies.

16.4.2 Augmented and Predictive Analytics

When e-commerce experienced a boom in the early 2000s, the ability to make product recommendations to consumers based on their preferences was seen as a catalyst to sales. *Amazon* moved quickly in that area and their early '*Recommendations*' feature provide to be a big hit amongst buyers. The idea was simple – gather enough data and information about the consumers' browsing behaviour when on the portal and then make recommendations based on those specific user patterns.

Advancements in this field have now elevated to such an extent that data mining and predictive modelling in combination with machine learning (ML) make it possible to study and identify buyer patterns, and in doing so make predictions with a great deal of precision.

Amazon has now elevated this technology to a whole new level by not just increasing the precision levels but also providing consumers with several associated product options. The '*Amazon Assistant*' is one such result of augmented and predictive analytics.

It is a browser extension, compatible with different browsers such as Chrome, Microsoft Edge, Mozilla Firefox, Internet Explorer, and Opera. When downloaded from the Google Play store, the extension resides on a consumer's browser. Thereafter, should a consumer spot an item on a portal or e-commerce site, the *Amazon Assistant* will at the outset check to see if the same item is available on Amazon and if possible for a lesser price. It also provides consumers with a 30-day price tracker while displaying the manner in which the price of that product has been trending over the past month. In effect, this feature enables *Amazon* consumers to save money while keeping them rooted to their e-commerce platform.

And this is not all, the extension also has a feature called *Amazon List*, which enables *Amazon* customers to add a product to their personalized *Amazon List* regardless of where they are on the worldwide web. *Amazon* is able to do all of this purely due to its advanced capabilities in being able to harness consumer data via analytics and in

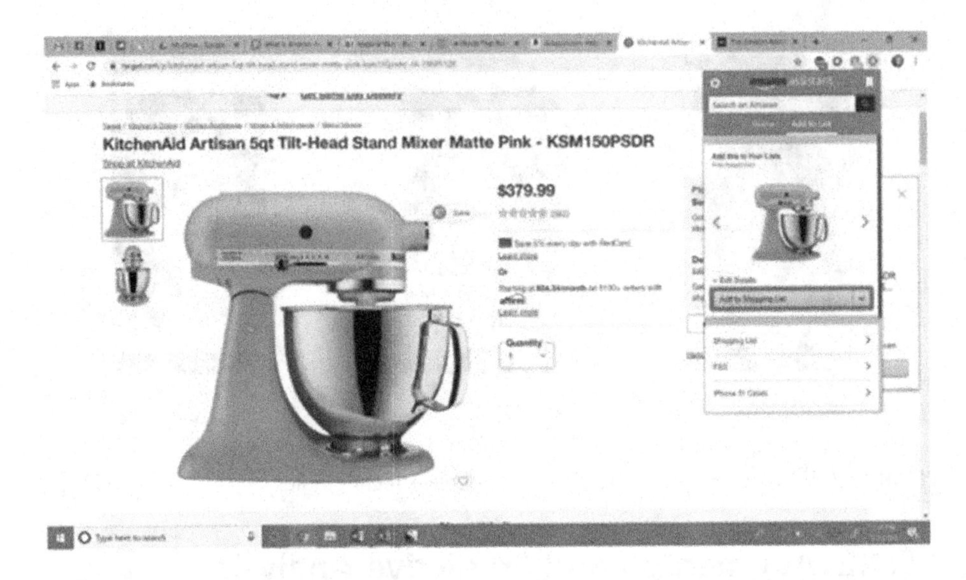

FIGURE 16.6 Predictive analytics.

turn use that to offer compelling purchase experiences. Therein lies the power of augmented and predictive analytics (Figure 16.6).

16.5 MARKETING AUTOMATION

A critical element of digital marketing is being able to understand the customer journey – from the time a prospective buyer visits a website, the products are browsed, to the eventual purchase. While this might appear easy at the outset, as businesses increasingly migrate to a digital environment, managing the sheer volume of leads and consumers will inevitably become a daunting task. Marketing automation software simplifies these action items for corporations operating in a digital ecosystem. From generating leads via new email and social media campaigns and engaging with them via remarketing efforts to driving sales and thereafter, engaging with them on an on-going basis, Marketing Automation Software is able to automate a bulk of all digital marketing functions (Figure 16.7).

Here's how it works. When a consumer visits a website and register on a web form, the automation software sends an email requesting the consumer to register. Most users comply. Some of these users may eventually go on to make a purchase using the shopping cart on the website. Occasionally, some users add products to their shopping cart but don't complete a purchase, a phenomenon called 'shopping card abandonment'. The Marketing Automation Software is now able to segregate all of these users into various categories and execute 'remarketing campaigns' by sending unique emails to each of these user-sets.

FIGURE 16.7 Marketing automation.

This advancement is so significant that several technology companies have developed sophisticated automation software and the Marketing Automation Software service has become an industry by itself and is expected to hit the $6.4 Billion mark by 2024. Further, as per the findings of the *2019 State of Marketing Automation Survey Report*, 75% of all businesses surveyed stated that they have employed Marketing Automation Software to manage their marketing efforts. Meanwhile, 14.5% of the companies studied indicated that the software was in fact increasing their sales productivity. Marketing Automation, therefore, is here to stay.

16.6 SOCIAL MEDIA MARKETING

According to data published by *Statista.com*, as of April 2020, there are 3.81 billion social media users across the globe. Meanwhile, as of January 2020, India had 400 million active social media users. With so many users constantly exchanging content on the multitude of social media platforms on offer, companies see an opportunity to build their brands, engage with their customer base, generate leads, and drive sales. Therefore, from *Facebook* and *Instagram*, to *WhatsApp, YouTube, Twitter, Snap Chat, LinkedIn*, and *Pinterest*, social media companies are constantly finding new ways to empower brands to use their platforms to advertise, create e-commerce eco-systems, and drive sales. This has given birth to the hugely popular practice of Social Media Marketing.

So, when do you state that Social Media Marketing is at play?

When browsing through one's *Facebook* or *Instagram* feed, chances are that you might view a post with a picture-set of Virat Kohli running while wearing a swanking red *PUMA* pair of shoes. Impressed, you invariably click the post and the embedded URL navigates you to the *PUMA* website. Once on the website, you view closely the shoes worn by the Indian captain, study their price, and are enticed to make a purchase by adding the product to the integrated shopping cart. This chain-reaction is the manner in which Social Media Marketing works its magic.

A game-changing advancement in Social Media Marketing was the advent of 'Social Media Stories'.

16.6.1 Social Media Stories

These are full-screen, vertical images and videos that appear in the 'stories' section of a social media mobile app. They appear for a duration of 24 hours before disappearing. The fact that they are limited for a day creates a sense of the present and sometimes even a semblance of scarcity. The 'stories' feature, first launched by *Snapchat* in 2013, has now been lapped up by *Instagram, Facebook, WhatsApp,* and more recently by *YouTube.*

The 'stories' feature offered by these platforms is highly immersive and mostly contain impromptu videos and pictures that provide a real-time update about an event or a day in the life of a brand ambassador. A study by *Statista.com* revealed that 76% of all *Facebook* users leverage the 'stories' feature via the *Facebook* app. For *Instagram,* this figure is even higher at 83%. *Instagram* has over 1 Billion active monthly users and the app generates a mind-numbing 500 million 'stories' each day.

Brands, therefore, use 'stories' to put out status updates about promotional offers, branding messages, polls, and flash sales alerts. Consumers of today such as personalized experiences and social media stories are proving to be a great way for brands to provide a casual flavour with an unfiltered view of what a brand really stands for (Figure 16.8).

FIGURE 16.8 Social media stories.

16.7 MOBILE MARKETING

The first mobile phone call was made in 1973 and it weighed a massive 1.1 kg while taking about 10 hours to charge. Finland released the first-ever 2G phones in 1991, and in 1993, the first text message was sent. Since those humble beginnings, mobile telephony has made incredible strides. In the late nineties and early 2000s, wallpapers and ring tones started appearing on mobile phones and with the foray of 3G, traditional feature phones made way for smartphones. Thereafter, since the launch of the iPhone, smartphones equipped with multimedia and high-speed data transfer capabilities have transformed the mobile landscape.

Accordingly to *Statista.com*, in 2020, there are 3.5 Billion smartphone users in the world. Meanwhile, India is expected to hit the 760 Million mark by 2021. Marketers leverage several formats to reach out to customers via the mobile channel. Traditionally, these have included voice, text, video, and display formats. Now, however, corporations are using mobile websites and mobile applications for the purpose of mobile marketing. Let us study and review each of these individually.

16.7.1 Mobile Website

Accordingly to *Statisca.com*, 51.53% of the global website traffic was generated via mobile devices. This data point alone is strong enough for brands to ensure that they have a mobile website. Additionally, brands now confirm that their mobile website is in fact a 'response site', i.e. one that has a single URL that automatically adapts to the dimension of the consumer's mobile device.

16.7.2 Mobile Applications

In early 2020, Forbes reported that across the world, there were about 8.93 Million mobile apps. Google Play had about 2.7 Million mobile apps listed on its app store in the second quarter of 2020. The Apple App Store, meanwhile, has 1.82 Million apps listed. From gaming and lifestyle to healthcare and entertainment, mobile apps offer a multitude of functionalities and features. Marketers, for their part, find this large ecosystem an irresistible canvass to conduct their marketing campaigns using a combination of video, audio, text, and several other aforementioned formats.

Voice

- Caller ring back tones (CRBT) – Somewhere in the mid-2000s, caller tunes became an obsession across India. Brands churned catchy ad jingles to go with their product or service. Thereafter, at an affordable rate of about INR15 per song, via telecom services provided, they handed consumers the option to replace the mundane mobile ring tone of theirs with their very own popular ad jingle. When a caller dialled a specific number and heard an ad jingle for a ring tone, the phenomenon of 'brand recall' worked its magic. Brands used this channel as a clever way to spread the reach of their products across the masses

- Outbound dialler (OBD) – While rural India continues to grapple with illiteracy, consumerism has made inroad to the rural hinterlands of the country. Brands, therefore, see OBD as a viable option when reaching out to the vast rural marketplace. Pre-recorded audio messages are sent out to mobile numbers for greater reach. Alternate mobile marketing options in this genre include employing interactive voice response (IVR) systems and providing 'missed call' numbers to elicit consumer engagement.

Text

- Short message service (SMS) Push – Mobile automation software now makes it possible for brands to send out bulk-SMSs to a large consumer base. Additionally, these messages are embedded with URLs to landing pages or call-back numbers – all strategically placed with the motive of driving sales.
- Tag footers – Brands found a way to buy small space, i.e., about 20–60 characters, that appear at the end of a non-advertorial message received by consumers. These ads, more often than not, contain URLs to landing pages that entice consumers to make a purchase via an e-commerce site.

Video

- Reels – With the advent of 3G and then 4G technologies, video content has been all the rage. Brands curate viral content that lasts for a brief 15-second duration on Instagram Reels or upwards of 30-seconds on YouTube Reels and broadcast it on social media handles.
- Pre-app video pre-rolls – Sometimes when a user opens a mobile app, a short video plays out an advertorial message. This is the Pre-App Video Pre-roll format.

Display

- In-app ads – If you are a smartphone user and have used an app for news, gaming, or shopping, chances are that you have noticed a banner displaying a brand's product or service. This banner is either static or interactive in nature and is aimed at enticing a consumer to navigate away from the app that they were using to the landing mobile-site of the advertising brand – again, with the motive of making a sale.

16.8 GEO-FENCING MARKETING

The mobile device has empowered the consumer in unimaginable ways, primary amongst them is the manner that it equips a consumer to be able to make purchases on the go, regardless of location. Marketers are now leveraging this capability and executing campaigns while using a technology called Geo-fencing Marketing.

Here's how it works. A smartphone is GPS equipped and therefore, the location of a user can be spotted and identified by software systems meant for GPS tracking. And when a brand recognises, via a mobile user's GPS co-ordinates, that a potential consumer of theirs is in close proximity to the brick-and-mortal store, a text message or a push notification containing a promotional message is sent out to that consumer inviting him or her to the store. Further, if that prospective consumer has previously made a purchase at the store, the message is further customized with his or her name. And in doing so, the personalization element and a good promotional offer is bound to get the consumer interested in visiting the store and making a purchase.

Bain & Company collaborated with *Harvard Business School* to conduct a study that revealed that a 5% customer retention rate could hike profits from 25% to upwards of 95%. Similarly, accordingly to *Forrester Research*, 90% of all brands felt that creating a good experience was vital for brand sales.

Marketers thus realize that through Geo-fencing Marketing, customer engagement, and retention practices can be employed to propel brand recall and sales in equal measure.

The pizza chain *Pappa John's* executed a marketing campaign while leveraging Geo-fencing technology during the highly popular Super Bowl game (American Football) in the United States. Their campaign offered free pizza for customers of their loyalty programme while focusing on four locations across an average distance of 2.5 miles. The successful campaign generated sixty-eight thousand impressions while mobilising 469 purchase actions (Figure 16.9).

FIGURE 16.9 Geo-fencing marketing.

16.9 INFLUENCER MARKETING

With the explosion of social media usage, the ability of brand advocates to relay messages and influence the choice of consumers, to increases the visibility of brand exponentially. Moreover, for that reason alone, Influencer Marketing as a concept and practice has become extremely popular, especially across social media platforms.

An 'influencer' is an individual that specialises in a niche genre and maintains dedicated social media channels, mobile sites and websites, while advocating related content to a select audience group of followers. Brands, therefore, flock to influencers to spread the word about their products and services to a select consumer base. Mandira Bedi can be termed as an influencer for health and fitness in keeping with her dedication to fitness routines and diet. Likewise, Mallaika Arora is sought after by fashion labels in keeping with her ability to advocate the latest fashion trends. Meanwhile, Virat Kohli and Rafael Nadal attract sports brands such as PUMA and NIKE, respectively, owing to their popularity in sports. The key for brands, however, is to ensure that they identify an 'influencer' that resonates with both their inherent brand messaging and the target audience.

MediaKix.com's 2019 'Influencer Marketing Survey' indicated that 80% of all marketers found Influencer Marketing to be an effective strategy. Also, 89% of the brands indicated that the return-on-investment (RoI) from Influencer Marketing was similar to, if not even better, when compared to other marketing channels. *AdWeek* reported that Influencer Marketing in 2020 is expected to be a USD10 Billion industry. Influencer Marketing, therefore, is highly effective just as exciting as it is and thus here to stay for the foreseeable future (Figure 16.10).

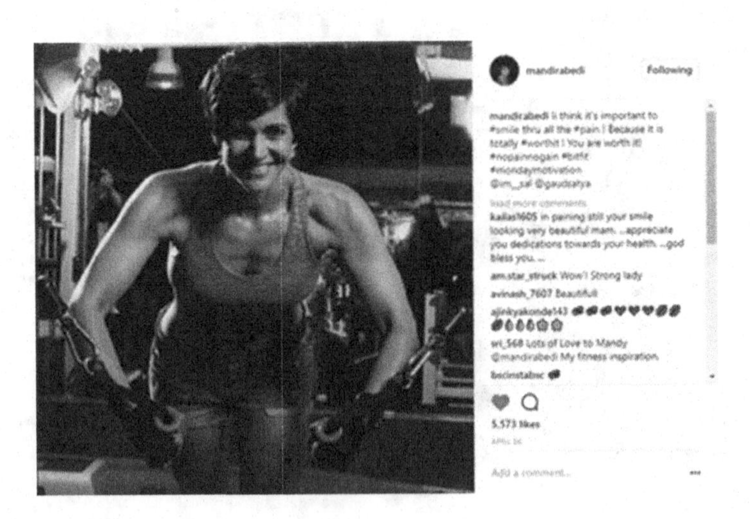

FIGURE 16.10 Influencer marketing.

16.10 DIGITAL ADVERTISING

Online marketing using platforms like websites, streaming media, and other means is referred to as digital advertising.

16.10.1 Display Advertising

It was Henrik Ibsen, a Norwegian playwright, who first said, "a thousand words leave not the same deep impression as does a single deed." Ibsen passed away in 1906 and thereafter, his quote was paraphrased into the now popular "A picture is a worth a thousand words".

Display Advertising draws inspiration from Ibsen's words and it now forms an integral weapon in a digital marketer's arsenal. Display Advertising is in fact the action of relaying advertorial messages through visually rich media. When browsing through a news portal and you witness 'Banner ads' from Induslnd Bank or Amazon, you're watching a form of Display Ad.

Display Ad is high on impact factor and helps brands generate large volumes of leads by directing traffic to their websites and triggering sales. When Display Ads reside on a website that belongs to a given brand, it is called 'Native Ad'. Likewise, when a Display Ad is placed on another website, it is termed a 'Non-native ad' (Figure 16.11).

Display Ads are categorized as Image Ads, Rich Media Ads and Video Ads. While Image Ads contain static images of products and services, Rich Media Ads comprise of interactive elements such as GIFs and animation. Video Ads are very similar to TV

FIGURE 16.11 Display advertising.

commercials, only here, Video Ads are embedded in YouTube videos and inserted in popular apps to garner eyeballs.

The latest advancements in Display Advertising are Audience Targeting and Programmatic Advertising.

16.10.2 Audience Targeting

Historically Digital Advertising meant identifying which websites were visited frequently by consumers seeking a certain product/service and then placing related ads there. This strategy is termed 'Contextual Advertising'. The thought process though has now changed. For example, a buyer seeking NIKE running shoes visits a NIKE website or a retail website such as AJIO that lists branded shoes. Traditional 'Contextual Advertising' saw shoe brands targeting such websites dedicated to shoes and placing ads on them. Now, however, advertisers are able to target consumers whose interests lie in shoes (e.g., consumers who have browsed for shoes twice over the past week) and not necessarily target shoe websites. This way, regardless of where a prospective shoe buyer is on the World Wide Web, an advertiser is able to find that buyer and place ads strategically where this buyer can see and click on. Such a strategy is termed 'Audience Targeting'.

16.10.3 Programmatic Advertising

Much such as buying ad-space in print media (newspapers and magazines) and on roadside hoardings, purchasing ad-space on websites comes at a price. And given the magnitude of websites on the internet, an auction process is used by ad exchanges and ad publishing platforms to determine the real-time cost of placing an ad on a given website and facilitating a seamless process.

Given the sheer volume of websites around the internet and the number of users, advertisers rely on Marketing Automation Software to guide them in this real-time auction process and for finding the best Audience Targeting strategy, i.e., the right target customer and the right websites. This real-time process of buying ads via an auction and having them strategically placed across websites while leveraging AI is called Programmatic Advertising.

According to *eMarketer*, 86.2% of all digital ads in 2020 are being bought via the automated process of Programmatic Advertising.

16.10.4 Search Advertising

Google is the world's largest search engine, and its holds 92.18% of the search market share. And there is a good reason for this as over 3.4 Billion searches are performed by global users each day on Google. Brands realize the marketing potential of 'search' and thus Search Advertising is integral to a Digital Marketing strategy.

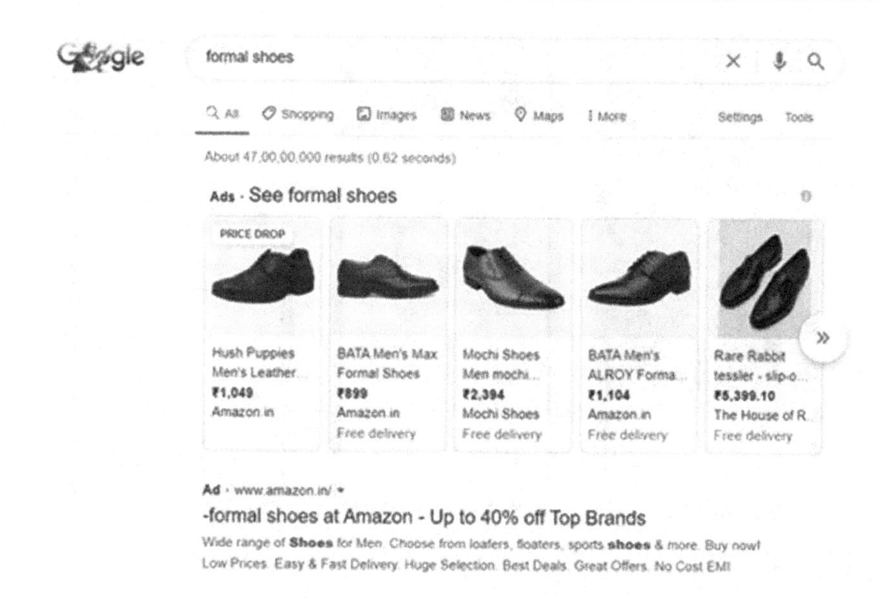

FIGURE 16.12 Search advertising.

Search Advertising is the technique of getting an ad displayed every time a search is performed on a search engine. These ads are displayed every time they match a search query undertaken on the said search engine, which uses complex algorithms to ensure that only an ad that is related to a search is displayed (Figure 16.12).

Search Advertising is also known as search engine marketing, pay-per-click marketing, or cost-per-click marketing. The reason for these terminologies stems from the fact that an advertiser will not be charged a fee until such time that a user clicks on an ad displayed via a search result.

The interesting advancements in the area of Search Advertising are Visual Search, Voice Search, and Smart Speakers. Here we are having the brief description of these technologies.

16.10.4.1 Visual Search

While text search enables users to search via text, Visual Search makes it possible for users to perform searches through real-world images, i.e., photographs. For example, users can take a picture of a shoe on their camera phones, and an application's embedded AI capability will provide the user with the associated product information.

Several companies have integrated this advancement into their overarching product capabilities. This includes Google Lens, Pinterest Lens, Bing Visual Search, Snapchat Camera Search, and Amazon StyleSnap. The usage of this technology in the e-commerce industry is extremely potent.

While all of this might sound like science fiction, this is actually made possible by a specific field of study called 'Sensory Search' which leverages advancements in the area of Computer Vision and ML. While Computer Vision enables mobile devices to

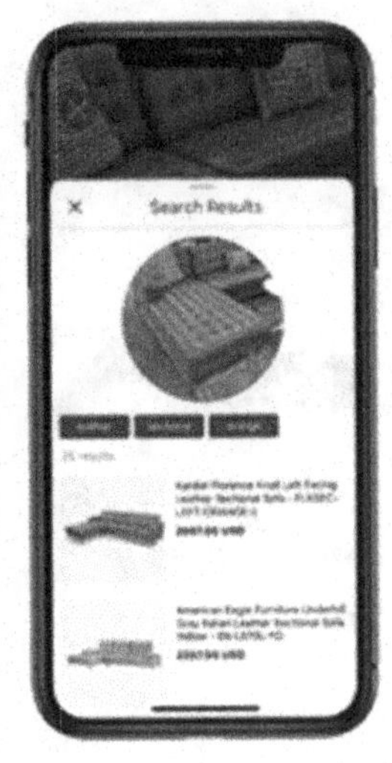

FIGURE 16.13 Visual search.

see, ML provides the requisite information to the mobile device to help understand what exactly it is seeing.

Accordingly to a 2018 study by *ViSenze*, an AI company, Millennials and Gen Z users amounting to upwards of 62% sought to use visual search when compared to other forms of search (Figure 16.13).

16.10.4.2 Voice Search, Voice Assistants, and Smart Speakers

Voice Search enables a user to search for information from the internet using a simple voice command. Natural Language Processing (NLP) Technology makes this possible and applications like Google Voice Search have revolutionized the function of 'search'.

A 2019 study of 1,700 adults in the United States by *Perficient Digital* showed that after the mobile browser, Voice Search is the next most sought-after search option. In India, the Voice Search volumes grew by a staggering 280% in 2018. As per a 2019 report from Microsoft, the top-two voice searches were 'searching for a quick fact' (68%) and 'asking for directions' (65%).

Voice Assistants are an extension of Voice Search and in fact aid the search process. It is a digital assistant that uses speech synthesis, voice recognition, and NLP to provide answers to questions and queries posed by users. When the iPhone 4S came out in October 2011, it had the voice assistant *Siri* and this came to be the first digital virtual assistant on a smartphone. Thereafter, the *Google Assistant* became extremely popular on the Android platform and Cortana on the Windows 10 computer.

Slowly but surely, Smart Speakers emerged on the landscape. Amazon brought out the *Echo* and the *Dot*, both of which featured the Voice Assistant *Alexa*. Similarly, Google's Smart Speaker product is *Google Home*, featuring *Google Assistant*.

As per *Comscore*, by the end of 2020, 50% of all searches will be voice searches. In 2019, the global Smart Speaker market was worth USD8.4 Billion and by 2025, this market is expected to reach USD 15.6 Billion.

16.10.5 Banner and Video Advertising

Banner Advertising, also called Display Advertising, comprises static or animated images or rich media content placed in a prominent section of popular websites. The motive of a brand in the wilful display of these ads is the capture of a consumer's attention in a manner such that he or she is directed to the e-commerce portal of the brand with the eventual intent of making a purchase.

The brand that places an ad pays the website that hosts the ad through various methods – Cost per Click (CPC), Cost per Milli (CPM), Cost per Lead (CPL), or Cost per Acquisition (CPA).

- Cost per click (CPC) – In this model, the advertiser pays the host website only when a visitor clicks on their ad. The motive here is to drive traffic via the ad to the advertiser's website.
- Cost per milli (CPM) – Here the advertiser pays for every one thousand impressions served on a host website. When a user browsing a website witnesses an ad on the mobile screen, an impression is said to have been served. An advertiser seeking to reach a large audience usually prefers CPM as the motive is to garner eyeball and not necessarily drive traffic back into a website.
- Cost per lead (CPL) – A lead is said to have been generated with a user filling up a form, downloading a brochure, signing up for an email newsletter, etc. And when an advertiser generates a lead via a banner ad running on such a model, a CPL campaign is said to be in play.
- CPA – When an advertiser runs a CPA model, money changes hands only when a customer is acquired.

16.10.6 Video Advertising

Since the advent of 3G and 4G, video advertising has skyrocketed. A study stated that 70% of consumer had shared a brand's video at least once and 72% of companies said that their conversion rate has improved thanks to video advertising.

YouTube remains the second most used 'search engine' and its hosting of a ton of video content presents avenues for video marketers to exploit. Also, social media platforms provide several formats for video advertising as well and these range from 'live' video formats on Facebook, Instagram, and LinkedIn to catchy short-form videos like Instagram Reels.

A 2019 study revealed that video advertising budgets are up almost by 25% and the spending on brands is increasing by upwards of 75%. Also, brands are using a combination of real footage and animation to ensure that the content is more appealing and captivating for consumers (Figure 16.14).

FIGURE 16.14 Video advertising.

16.10.7 Social Media Advertising

Earlier in this chapter, we have already talked about the widespread usage of several social media platform and the manner in which brands leverage them for their marketing campaigns. What makes advertising on social media an exciting proposition for brands is the fact that instead of the conventional 'spray and pray' tactic where brands put out an ad and simply hope that users watch it, Social Media Ads enable an ad to precisely target their consumer base.

16.10.7.1 Precise Targeting

Platforms like Facebook, Instagram, and YouTube are able to track consumer behaviour and content choices made by their consumers in keeping with their app activity and engagement behaviour. Thereafter, through ad features like 'Customer Audiences', advertisers are able to send out ads precisely to the audience demographics and geographies of their choice (Figure 16.15).

16.10.7.2 Ad Placement

A TV ad being aired at the right time on the right channel makes a huge difference. Similarly, the placement of a Social Media Ad is a critical aspect. Social Media platforms have various sections on their app platforms that brands can leverage to have their ads placed. From the 'news feed' to the 'stories' sections, brands can select the positioning of their choice. Also, they can decide the time and duration when an ad is displayed to the selected target segment (Figure 16.16).

16.10.7.3 Ad Bidding

Facebook ads alone are hugely voluminous, with over 3 million advertisers putting up ads on this platform alone each month. Ad Bidding, therefore, ensures that a seamless

FIGURE 16.15 Social media advertising - precise targeting.

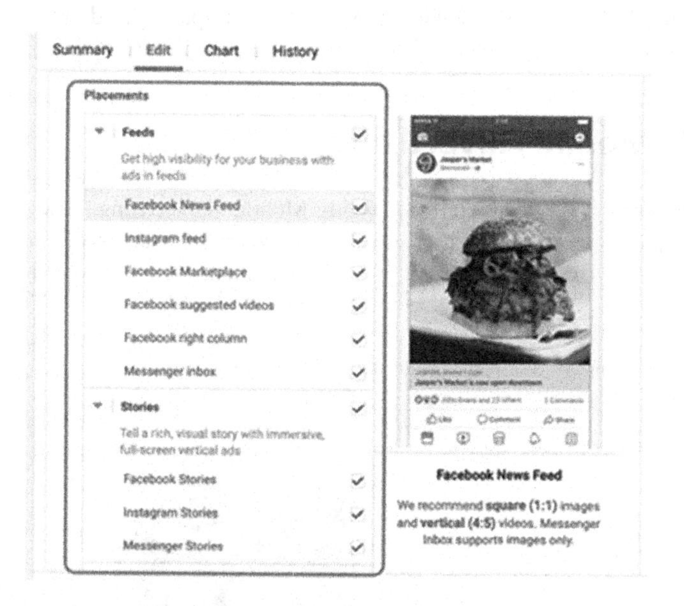

FIGURE 16.16 Social media advertising - ad placement.

process is in place for all interested advertisers to follow when advertising on social media platforms.

While it is tempting for social media platforms to simply provide greater eyeballs to the brand willing to spend the most money, they adopt a slightly different approach – one that involves a 'Value Score' – to ensure that there is a level playing field for advertisers to compete on for prime digital space.

Social Media platforms determine a Value Score, which is arrived at by computing three distinct elements – the Relevance Score, the Estimated Action Rate, and the Advertiser Bid Rate.

- Relevance score – In keeping with the number of people that have seen a certain ad, the Relevance Score determines the success rate of an ad. Most social media platforms maintain a base limit for a number of impressions (e.g. 500 impressions) after which, the performance of the ad in terms of impressions and engagement determines its Relevance Score. The greater a Relevance Score, the greater the success of the ad
- The estimated action rate (EAR) – The motive of social media ads is to get users to take action, i.e. click on it and arrive at a website and make a purchase action. Therefore, social media platforms compute an EAR to evaluate the bid rate of an ad
- The advertiser bid rate – this refers to the amount and budget that a brand is willing to spend and allocate on a given ad campaign. Budgets can be allocated daily or spread across the lifetime of an ad (Figure 16.17).

16.10.8 Mobile Advertising

Mobile Marketing is often confused with Mobile Advertising. The former is an all-encompassing terminology that envelops the various aspects of marketing via a mobile device and includes the use of various content formats, websites, and social

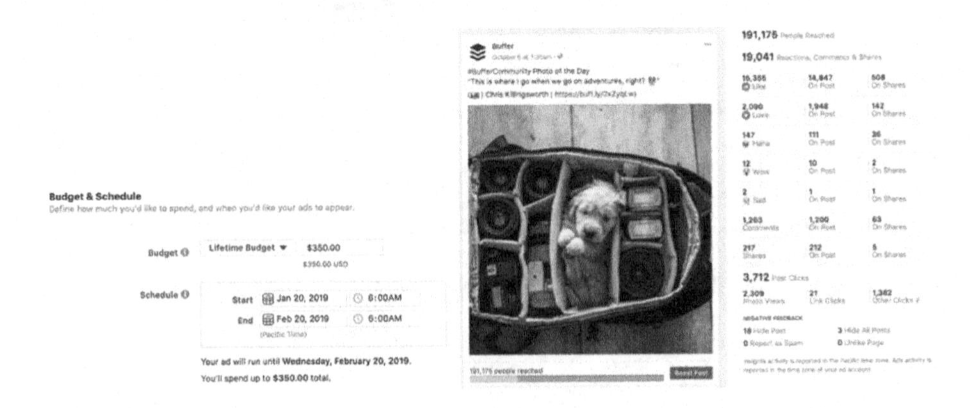

FIGURE 16.17 Social media advertising - ad bidding.

media platforms. Mobile Advertising, however, specifically addresses the idea of sending out advertorial messages to consumers via mobile devices.

Mobile Advertising, therefore, provides various options for brands to consider. This includes Text and Banner Ads, Push Notifications, click-to-download ads, click-to-message ads, and click-to-call ads.

- Text and banner ads – Plain text SMSs with website URLs and rich media ads on mobile devices entice users to click on them. In doing so, a user is lured onto a brand's website where their products and services are displayed in great detail.
- Push notifications – Brands send out pop-up notifications to users on their mobile devices. A user may not necessarily be using an app at that point of time but this unique feature ensures that a user is reminded about the brand's app and the benefits they can avail when using the said app.
- Click-to-download ads – Brands send out such ads when a user clicks and directs them to an app store like Google Play or the Apple App Store. The motive here is not to make an outright sale but to get users to download the app of a brand and open up their entire mobile ecosystem to the consumer.
- Click-to-message ads and click-to-call ads – these are mostly text messages equipped with a call-back number or a number where consumers can send a message to connect directly with a brand. These ads are mostly used by brands to drive consumer engagement via contents.

16.11 SEARCH ENGINE OPTIMIZATION

SEO is the process of increasing or improving the *quantity* and *quality* of traffic to your website through *organic search engine results*. It involves improving one's site to increase its visibility for relevant searches. The better the visibility of the pages in search results, the more likely is to harvest attention and attract prospective and existing customers to the business.

16.11.1 SEO: Then and Now

Gerard Salton is often considered the father of modern search technology. He created the idea of a search engine and developed the information retrieval system called SMART, which stood for Salton's Magic Automatic Retriever of Text.

The term hypertext was coined by Ted Nelson. ARPANET was created by Advanced Research Projects Agency Network, which allowed information transmission over long distances. Alan Emtage created the first search engine, Archie in 1990. A template indexing method was also created by Alan to index documents, images, sounds, and services on the network. Archie was not capable enough to use keywords to find related documents like modern search engines do. With the growing popularity of Archie, two more search engines were created – Veronica and Jughead. First bot or spider, World

Wide Web Wanderer, was created in 1993. Due to server lag, websites crashed and it became a problem and caused distrust for bots. Later ALIWEB was developed which allowed owners for submitting their site for inclusion in search index.

By late 1993 three new search engines, Jumpstation, WWW Worm, Repository Based Software Engineering spider or RBSE were created. But neither performed well enough.

Jerry Yang and David Filo created the search engine Yahoo. They created a directory of websites by manual compilation. This directory was named David and Jerry's Guide to the World Wide Web, which was later known as Yahoo. Google creators – Larry and Sergey – realised that ranking of websites should be made as per the number of links it has. It was known as Page Rank. Bill Gross realized that whenever a user enters a search query, they were telling the search engine their interest area and products they were interested in. This information was very important for digital marketers. He identified that search engines can sell this information and earn money by telling the keywords to companies, which users were looking for. This helped companies in adding specific keywords to their websites for generating more traffic. He launched a site – overture.com in 1998, based on sponsored links for specific keywords. In the year 2000, Google released Adwords. They further separated paid advertisements from organic search results.

Web has become integral to companies in their mission and lines of products, search engine rankings have become particularly crucial for strategic sales and marketing. SEO is the act of increasing the extent and nature of traffic to your site using organic search engine results. Organic traffic is any traffic that you don't need to pay for. The quality and quantity of traffic on the search engine results pages (SERPs), is decided by the number of visitors who are genuinely interested in the products or services the site offers. Advertisements make up a critical section of many SERPs. The entire process of searching starts through crawling and indexing the web pages. The search engines have built up their own robots called spiders to crawl the web. These robots move between different pages and examine their substance and decide if it is a pertinent site or not. These crawling and indexing spiders can reach the many billions of documents that are interlinked. Once these pages are found by the engines, they interpret the code and store it, to be remembered later. When a search is initiated, search engines scan the billions of documents that have been stored and perform two important tasks. Initially it returns the results most relevant to your query, and then, ranks those results in order of perceived significance (Figure 16.18).

The "Google Zoo" consists of Panda, Penguin, and Hummingbird, which are Google's algorithms that help determine where sites rank on the search results page. Each of these animals has a different function that can affect your website rankings in different ways. For example, Panda's job is to remove low-quality content from search results, Penguin works to remove sites that are spammed, and Hummingbird assesses if your content is what readers want to read.

16.11.1.1 Google Panda: The Game Changer Algorithm for Content

Google Panda algorithm was initially released in 2011 with a definite purpose of rewarding high-quality websites and reducing the incidence of low-quality websites in Google's organic search engine results. Originally it was also called "Farmer."

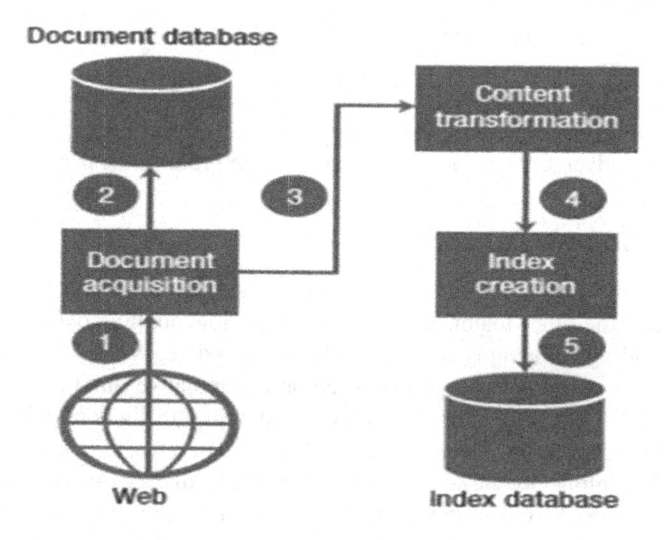

FIGURE 16.18 Search Engine Optimization (SEO).

'Content farms' were one of Google's main targets with Panda. Many sites produced content of low quality that tended to rank simply because of the sheer amount of copying. This was a huge concern for Google, which was always endeavouring to deliver high-quality results for an ideal consumer experience. Google ogled content spammers through the Panda algorithm and commendably removed the content farms. The Panda algorithm ranks higher for websites with good and unique content and original content whereas the low-quality pages are ranked lower in search engine results. Search results pages (SERPs) are stripped of poorly conceived, spammy content, enabling websites of higher quality to remain at the top. The algorithm compares several ranking signals in contrast to the human quality rankings. Google Panda updates turned out once a month for the first 2 years, but Google stated in March 2013 that future updates would be incorporated into the algorithm and thus would be less obvious and incessant. Google released a "slow rollout" of Panda 4.2 in 2015.

Some of the problem areas in Google SERPs, handled by Panda algorithm updates are discussed below:

a. Low-quality content – Pages that are worthless and with flat information are considered as low-quality pages.
b. Thin content – The page holds information that does not add value so as to improve the quality of SERP. Pages only with a few phrases, with a mass of words that are incoherent without proper grammar and spelling, are also considered to have thin content.
c. Content farming – Low-quality pages, gathered from other websites.
d. Duplicate content – Reproduced content that shows up in more than one place on the Internet. If your website has multiple pages containing the same text with no or little difference, duplicate content problems also can occur on your own website.

e. Lack of credibility – Content generated by sites that are not deemed credible or authenticated
f. Automatically generated content – The spun content is automatically created to seal webpage gap with keywords with inferior information.
g. On-site adverts: Pages flooded by adverts compromise the user experience.

16.11.1.2 Google Penguin

Google Penguin is a Google's search algorithm update introduced in 2012. Besides Google Panda, the other major initiative to combat low-quality content and black hat SEO link-building campaigns is Google's Penguin update. While Panda dealt mainly with the quality of content, Penguin focused on manipulative link-building techniques that are used to artificially boost the amount and nature of links pointing back to the website of an organization. Since its introduction, Penguin has been revised quite a few times, thus becoming Google's key algorithm in 2016. An excessively optimized site is considered by Google as one in which the SEO strategies surpass the real essence of the site. Penguin ensures that the rankings of your website would never be adversely affected by doubtful links from mysterious sources.

Penguin drops the page ranging with bizarre usage of keywords or over-aggressive link-building methods which are employed in order to manipulate the search engine rankings. The Penguin code basically looks for aggressive link-building practises and penalize your website or perhaps even totally delete itself from its search index. Even though links are important for website SEO, maintaining high-quality content in the sites is the best way to improve search rankings.

The Google Penguin is fully automatic and makes sure only natural and sustainable links are connected on your aspects that provoke a Google's Penguin are unethical practice of connecting the site with links of inferior quality, links that emerge from spam sites and stuffing the site with keywords.

16.11.1.3 Google Hummingbird

While Panda and Penguin were modifications to a part of the old algorithm, Hummingbird is the new algorithm with the biggest modifications implemented considering the mobile market surge in recent times. Conversational search is introduced to the Hummingbird algorithm, which is intended to pay attention to the meaning of a phrase, instead of specific keywords, this is because while searching on mobile, many users are now using voice rather than typing. The core of Google Hummingbird lies in the concept of semantics or meaning.

Semantic search is the idea of enhancing search engine results by having to focus on user intent. In a broader sense, the subject of a search relates to other data or its context. Effectively, semantic search aims to determine what a user really means and then serve relevant results, rather than just a string of keywords,

The SEO white-hat has two significant classes – the first one being on-page optimization, which deals with the structure and content of a website. The second SEO class of white hats is off-page optimization, which tackles best practises for the integration of both external inbound and outbound connections.

16.11.1.4 On-Site SEO

On-site SEO otherwise known as on-page SEO is an approach to optimize elements on a website. On-site SEO corresponds to optimising a page's source code for both content and HTML. In addition to helping search engines to interpret page content, accurate on-site SEO also aids the users to recognise quickly and evidently the page and whether their search query is answered. In principle, good on-site SEO makes sure that a visitor to the page is served with accurate content according to the search query. On-site SEO's definitive goal can be understood as trying to make it easy for both search engines and users to comprehend what a page is about, recognise that page as relevant to a search query, and find that page appropriate on a SERP and worthy of ranking it good.

16.11.1.5 Off-Site SEO

Off-Site Optimization (Link Building) is defined as a process that enables the website to enhance its ranking on the results page of the search engine. These factors operate outside the bounds of web pages. Such ranking factors are not controlled entirely by the publisher. Off-page optimization deals extensively with SEO link building. All internet search engines contemplate *Back Linking* (BL) in their ranking algorithms. The back link is a hyperlink to the target site from an external site. The effectiveness of the text used in the back link or the anchor text is considered by all popular search engines. Off-page SEO is a long-term and time-consuming process.

16.11.1.6 SEO Best Practices

SEO Best Practices are designed and implemented to help enhance the search engine rankings of the website. Some of the important terms related to SEO best practices are

- Title tag
- Meta descriptions
- URL
- Content of page
- Image alt text

16.11.1.7 Title Tag

The title tag proclaims to the search engines the gist of the page and the relevance of the page with respect to the keywords or keyword phrases. **Title tags ought to be unique for every single page in the format,** *Primary Keyword – Secondary Keyword | Brand Name*. Also, it is recommended to avoid the duplication of the title tags

16.11.1.8 Meta Descriptions

Meta descriptions are extremely important in getting users to click through from the SERP to your website, while not as important in search engine rankings. Meta descriptions should use keywords judiciously, but more pertinently, a credible description should be provided that a user would prefer to click on. The SERPs will showcase keywords that the user has searched for, such as title tags, increasing the probability of the user clicking on the website. The best practise is to write compelling *Meta descriptions* with recommended length of 150 to 160 characters, avoiding any duplicate descriptions and quotes.

16.11.1.9 URL

The URL should be obvious, appropriate, captivating, and authentic as possible to be easily understood by both your users and search engines. A definitive yet concise URL gives a fair idea about the page content. Good readability can be brought in by using proper word separation through hyphens but avoiding the usage of underscores or spaces. The issue with page duplication can be minimized by using lowercase letters. The value of inbound links to the homepage is spread out through each of these URLs. This implies if there are multiple links to these different URLs on the homepage, the major search engines give credit to them independently, not in a joint manner.

Canonical tag informs the search engines that a particular URL denotes the master copy of a page. By means of the canonical tag complications initiated by identical or duplicate content that appears on multiple URLs can be prevented. Essentially, the canonical tag states which version of a URL will pop up in search results (Figure 16.19).

16.11.1.10 Content of Page

Content of website pages is the most obvious reason for high traffic to the site. Create content that is unique and highly relevant to that of keyword phrase. For each medium

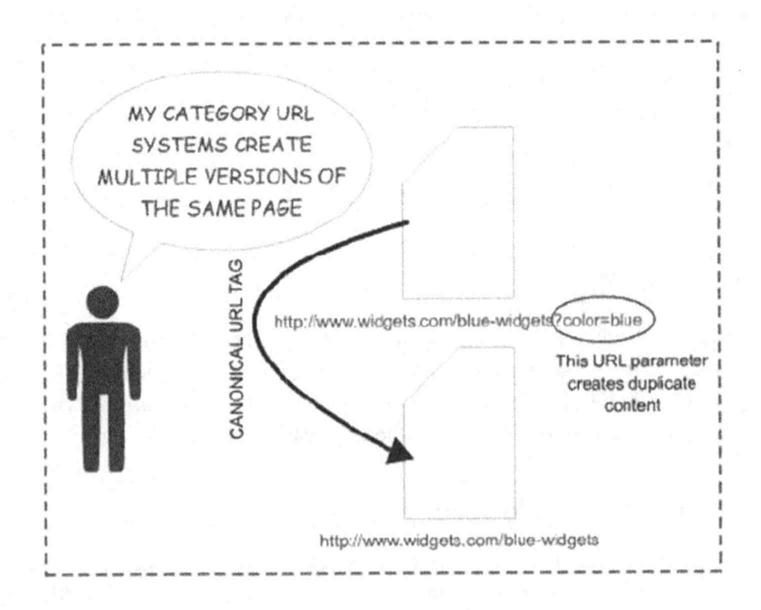

FIGURE 16.19 Canonical URL tag.

like news releases, blog posts, guest social media posts, etc., create unique content. Appropriate content pages must be a very specific subject, generally a product or object, and highly relevant. A unique page content supplies demand and should essentially be linkable. One of the most important factors in SEO is building quality 'backlinks'. The amount of backlinks is an indication of the viewership or relevance of the website. Apply internal linking where links from other pages refer back to the page inside your website. Internal linking is imperative since it reinforces the anchor text and lets users and search engines navigate the website.

16.11.1.11 Image ALT Text

An Alt text or alternative text refers to an image's name. Suitable alt tags should be used in all images. It subsists in the HTML code and is not generally evident on the page itself. Alt text increases accessibility. Images lacking alt text cause difficulties for screen readers (used mainly by visually impaired users) since there's no way to converse. When selecting alt text, emphasize on building useful, information-rich content that appropriately uses keywords and is in the context of the content of the page. Avoid filling keywords with alt attributes since it leads to a negative user experience and may induce spam n your site.

16.11.1.12 Page Ranking Factors

PageRank algorithms rank the websites on the search results; the greater the PageRank, search engines consider it more important. A combination of on-page methods and off-page techniques are employed for the improvement of page ranking. Even though on-page methods do website optimization, off-page methods cannot be undervalued as it is very significant to create traffic and therefore enhance ranking.

Some of the page ranking factors deliberated by search engines are as follows:

a. Site security: Site security is a very important ranking factor that denotes the use of HTTPS encryption with SSL certificates for a secure connection between a website and its users. This adds an additional layer of security that shields information exchanged among users and the website. Search engines want users to be driven to the most trusted sites, and HTTPS encryption demonstrates that a site is secure.
b. Crawlability

   Crawlability defines the ability of search engines to access and crawl content on a page.

   Crawlability lets search engines scan a website and evaluate its content, so as to determine the gist of the page and how must be ranked. The crawling software is used by search engines to read the page and create entries in an index that can search. According to the search query, the search engine scans the index and ranks the pages appropriately. The site with less crawlability has the least chances of getting entry in an index and hence will reduce the probability of visibility in search results.

c. Mobile-friendliness

Mobile-friendliness corresponds to how a web page will look like and performs on a mobile device when someone views it. Mobile-friendly sites provide a good user experience with a responsive design that adjusts the content and be visually appealing on every screen size. Mobile devices contribute to 52.2% of internet traffic and the number is increasing. Since more searches are carried out on mobile devices than on desktops, both search engines and users need to be mobile-friendly.

d. Page load speed.

Page load speed is an important SEO ranking factor that relates to user experience. Slow loading sites deliver a bad user experience. Search engines understand that people want to find answers as quickly as possible, so they prefer to show users websites that load rapidly. For mobile sites also this holds true, as Google said that its Speed Update will certainly consider load speed a ranking factor for mobile searches. A site speed checker can be used to make sure that your site loads quickly. If your site is slow, incorporate website caching, compressing files, limiting the number of redirects, and bringing in other measures to speed up load time.

e. User engagement

The user's reaction to the results of the search is used to determine the page ranking. Google employs an AI tool called Rankbrain to do this. The user engagement can be measured through the factors like *Click-Through Rate* (CTR) (the ratio of people clicking on the result of a search engine once it is given to them), *Time on Site* (the time spend on a page by anyone having found it through search), and *Bounce Rate* (the number of people who exit the page after viewing only one page they find through search). User engagement can be heightened by using high-end website design. An intuitive and simple site architecture that loads rapidly to optimize the user experience.

f. Exact target keywords

Strategic content creation by identifying accurate keywords can more traffic to the site. With Voice Search being popular, search engines need to map for long phrases and questions in addition to shorter generic keywords. The search engines increase the ranking of the pages with content that matches the search intent. The exact keywords that reflect the domain of the business can drive useful traffic and thereby leads. Assigning more than one page on your site under the same keyword can mislead search engines and lead to keyword cannibalization.

The choice of keywords can be based on the following aspects:

- The relevance of the keyword to the content of the website.
- Probability of users finding the site when they search using the specified keywords.
- Possibility of converting the traffic to the site to useful leads and thereby meeting the company's objective.

g. Structured Data

Using structured data is yet another way to tell the search engines what a web page really is, so as to rank it appropriately. Microdata added to the backend of a web page that states to the search engines how to categorize and analyse the content (e.g., recognising a mailing address, the title of the book, recipe, or other relevant information) is called structured data or schema mark-up.

16.12 BENEFITS

In this chapter, thus far, we have outlined the various Digital Marketing and SEO advancements that have taken place, its features, and the manner in which they have transformed the digital landscape. While their benefits have been mentioned as well, we will summarize them in a nutshell for you here.

Digital Marketing

- Global reach – The virtual nature of the digital world and the sheer magnitude of consumers that engage in the digital landscape makes it possible for even a local brand to read a global audience. Through search engines, website, and social media platforms, a small-town brand in rural India can aspire to have consumers in New York.
- Cost control – From pay per click to cost per acquisition, the various models of publishing digital ads enable even a small firm with a limited budget to compete in a crowded and competitive landscape. Also, the ability to control daily budgets and the ad duration provide firms with the unique opportunity to test campaigns and determine what works best for them.
- Measurable results – Ad analytics are so advanced that brands are able to determine the nature of action that a consumer performs with a specific ad. This aids a brand to measure the effectiveness of a campaign and make modifications going forward.
- Targeting & personalization – One of the biggest advances of digital marketing is the level of understanding of a consumer that a brand is able to garner. From demography and buyer behaviour to geographic spread, a brand knows enough and more about the consumer, and in doing so, it is able to single out select consumer categories and create campaigns with unique messages for each of them with the intent of impressing them with their values.
- Changes to campaigns on the go – Brands love digital marketing not just for the scale and variety they offer but also for the fact that brands are able to make modifications to their 'live' campaigns at any given point of time. Also, should a campaign not achieve the desired results, a brand can pause it until such time that they come up with an alternative campaign to test.
- Improved conversion rate – Digital marketing automation software offers features such as conversion rate optimization (CRO) to enable advertisers to select the right model and thereafter optimize it for greater results.

Search Engine Optimization (SEO)

- SEO enhances a site's user experience – SEO strategy involves the usage of content keywords that are very specific to a given brand. Therefore, when a brand user is on a site, the brand-specific content engages the user in a manner that ensures that the user spends a great amount of time on the site, which in turn enhances brand loyalty. For example, the website for Royal Enfield has several pages with stories of rider excursions and their experiences. A prospective buyer, therefore, gets a first-hand account from the Royal Enfield community and is hooked to the rider experience that the motorcycle offers.
- SEO helps in attracting quality traffic and increases brand visibility – Most traditional marketing techniques involve spraying marketing messages and hoping that an interested consumer grasps it and walks into a store. SEO, on the other hand, ensures that the content is very specific to a given brand in a manner such that the search engines direct only those users expressing interest in related content and keywords to the said brand's website. In doing so, SEO attracts only those consumers that are specifically interested in a given brand's content and the said products and services. For example, a running shoe website is likely to contain keywords like 'running', 'shoes', 'fitness', 'health', etc. Thus, consumers searching for such categories are more likely to arrive at the said website and are more likely to make a purchase given that the website offers products related to their area of interest.
- No paid ads needed in SEO – A vital point to recall here is that the process of SEO is unlike that of publishing an ad. Putting out an ad involves the spending of money, while the process of SEO does not involve the publishing of ads. Thus, SEO limits the spending of a marketing budget to a large extent.
- SEO aids public relations activities – Given that SEO draws related consumers to a website in large numbers, it is a great tool for public relations. When a large number of consumers who are interested in a brand's products and services congregate on a website, the opportunities for word-of-mouth advocacy increase exponentially.

BIBLIOGRAPHY

Adamic, L. A., and B. A. Huberman. 2006. "The nature of markets in the World Wide Web." Working paper, Xerox Palo Alto Research Center. http://www.parc.xerox.com/istl/groups/iea/www/webmarkets.html (accessed March 12, 2014).

Bhattacharya, J. "What is marketing automation: Definition, benefits & uses." Singlegrain.com. https://www.singlegrain.com/marketing-automation/what-is-marketing-automation-definition-benefits-uses/ (accessed February 4, 2020).

Britannica. "BJ Copeland." Britannica.com. https://www.britannica.com/contributor/BJ-Copeland/4511 (accessed September 10, 2020).

Bush, V. "The Atlantic". Theatlantic.com http://www.theatlantic.com/magazine/archive/1945/07/-as-we-may-think/303881/ (accessed February 2, 2020).

Collins, K. "Omnichannel marketing automation statistics for 2019." Clickz.com. https://www.clickz.com/onmichannel-marketing-automation-statistics-for-2019/231381-2/231381/ (accessed April 5, 2020).

Content Pilot. 2007. *Foundational Search Engine Optimization--Best Practices*. Content Pilot.

Craig, J. "What is at the core of an omni-channel marketing solution?" Martechadvisor.com. https://www.martechadvisor.com/articles/customer-experience-2/what-is-at-the-core-of-an-omni-channel-marketing-solution/ (accessed March 19, 2020).

Gartner. "Information technology Gartner glossary." Gartner.com. https://www.gartner.com/en/-information-technology/glossary (accessed April 20, 2020).

Gudivada, V. N., Rao, D. and Paris, J. 2015. "Understanding search-engine optimization." *Computing Practices* 48, no. 10: 43–52.

Omnisend. "Marketing automation statistics 2019." Omnisend.com. https://www.omnisend.com/resources/reports/omnichannel-marketing-automation-statistics-2019/ (accessed March 25, 2020).

Panetta, K. "Gartner top 10 strategic technology trends for 2020." Gartner.com. https://www.gartner.com/smarterwithgartner/gartner-top-10-strategic-technology-trends-for-2019/ (accessed March 31, 2020).

PTI. "Average daily online media intake in India soars to over 4 hours from 90 minutes: Study." timesofindia.indiatimes.com. https://timesofindia.indiatimes.com/gadgets-news/average-daily-online-media-intake-in-india-soars-to-over-4-hours-from-90-minutes-study/articleshow/75277623.cms (accessed April 30, 2020).

Shenoy, A. 2016. *Introducing SEO: Your Quick-Start Guide to Effective SEO Practices*. Apress.

Statista. "Number of internet users in India from 2015 to 2020 with a forecast until 2025." Statista.com. https://www.statista.com/statistics/255146/number-of-internet-users-in-india/ (accessed October 30, 2020).

U.S. Census Bureau. 2013. "Health insurance coverage status and type of coverage by sex, race, and Hispanic origin." Health Insurance Historical Table 1. http://www.census.gov/hhes/hlthins/historic/hihistt1.html (accessed October 31, 2020).

Advanced Wireless Solutions (Case Studies on Application Scenarios)

17

Jyoti Prabha
University College of Engineering & Technology

Contents

17.1 Foreword and Preamble to Wireless Technologies 318
17.2 Applications of Wireless Networks 320
17.3 Internet of Things and Advanced Scenarios 321
17.4 Key Cases and Applications of IoT 322
17.4.1 Smart Homes 322
17.4.2 Healthcare System 322
17.4.3 Traffic Management 323
17.4.4 Smart Farming 323
17.4.5 Business Automation 323
17.4.6 Defense Application 323
17.4.7 Woman Security Bands 323
17.4.8 Connected Cars 323
17.5 Key Technologies and Standards with Wireless Environment 324
17.6 Key Features of Wireless Environment 325
17.7 Advanced Cases and Technologies with Internet of Things (IoT) 326

DOI: 10.1201/9781003051022-17

17.8 Cloud Platforms for MQTT 327
17.8.1 CloudMQTT 328
17.8.2 DIoTY 328
17.8.3 Cloud Integration with Node-RED 329
17.8.4 Dynamic Key-Based Communication in IoT Scenario 330
17.9 Conclusion 330
Bibliography 332

Abstract

Internet of Things (IoT) is the advance wireless network that we have used to enforce higher degree of security. It can be used if Mobile Ad Hoc Network is further integrated with IoT or advance wireless. Security and integrity is the main issue in IoT-based network environment in which interception-free secured communication is required. To enforce and integrate the higher degree of security, there is a need to implement IPv6 for (IoT) scenarios with secured hash-based cryptography in the key's generation and authentication. The IPv6-based approach can be enabled with fully secured algorithms and non-vulnerable toward the interceptions. With the increasing implementations of IoT in diversified domains, it becomes necessary to work out the security aspects of IoT with the secured routing of packets, so that the intrusion cannot take place and all the transmission can be fully secured. RPL is the IPv6-based protocol for IoT. It is primarily integrated for IPv6 over Low-power Wireless Personal Area Networks. It works with the dynamic creation of Destination-Oriented Directed Acyclic Graph having unidirectional as well as bi-directional communication. It is having multiple instances with the localized behavior for higher optimization. RPL enables each node in the framework to pick if packets are to be sent upward to their root or downward to their child nodes.

17.1 FOREWORD AND PREAMBLE TO WIRELESS TECHNOLOGIES

A wireless network is a technology that is used to establish connections between different network-integrated devices without the use of wires. In a wireless network, a connection is established based on the different types of signals that are transmitted over the air by the sender and receiver (Lin et al., 2016). The wireless network uses different types of waves like radio waves, electromagnetic waves, and infrared waves to transfer the data over the network. Wireless networks convert digital data into signals and transfer over the internet with the help of the physical layer of the OSI model. It connects remote locations like different countries, cities, offices, and homes, and reduces the cost that occurs due to the installation of a wired network (Yetgin et al., 2017) and assorted scenarios presented in Figure 17.1.

In 1969, a university located in Hawaii provided ALOHANET as the first wireless network. Later on Ethernet 802.3, 2G cell phone, Wi-Fi protocol 802.11, and VoIP 803.11 are introduced. Wireless networks provide a fast, reliable network that overcomes the various problems of wired networks. Home users' and organizations' demands for

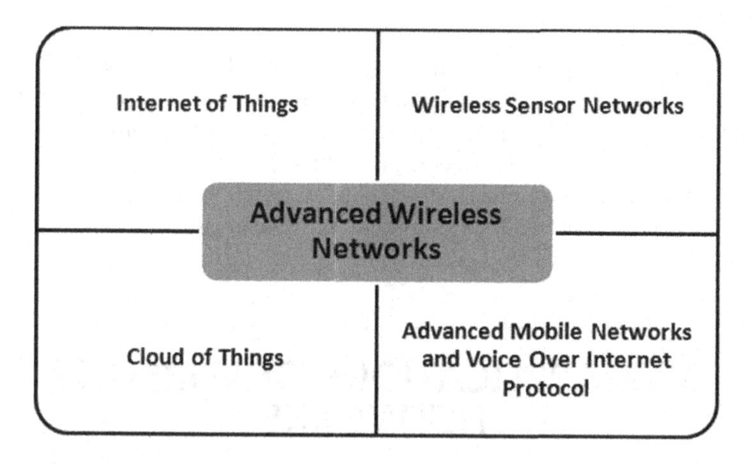

FIGURE 17.1 Advanced wireless networks and taxonomy.

wireless networks increase rapidly because they use different types of wireless technologies. These technologies include 2G, 3G, 4G, and 5G mobile communication that provide a fast and large amount of data transfer over the networks. Today's most mobile users use the LTE standard of 4G technologies, which provides faster communication as compare to 3G technologies. A wireless network provides benefits over the wired network, where the installation of a wired network is difficult because of the sea, rivers, and streets. Wireless networks can be established by installing sending and receiving terminal at both ends (Rashid & Rehmani, 2016).

Wireless networks can be categorized into different types depending on the area that is covered by the network. It can be a personal area network (PAN), local area network (LAN), ad hoc network, metropolitan area network (MAN), wide area network (WAN), mobile network, and space network. Wireless PAN network implementations are used for small-area communication. Bluetooth and infrared light waves are used for wireless PAN communication. TV remote and connecting wireless headphones with laptop and mobile are an example of a PAN. Wireless LAN (WLAN) network is used to connect two or more devices within the local area. It uses IEEE 802.11 standard for connectivity. It is used to connect two or more buildings within organizations. The router that connects the rooms of the home is an example of the LAN network. The wireless MAN (WMAN) network connects different LAN networks and follows the IEEE 802.16 standard (Noel et al., 2017). The wireless WAN network is used to cover large areas like providing connectivity between different cities and branches of a company located in different cities. Parabolic dishes, access points, base station gateways, and microwave signals are used. The mobile network also called a cellular network uses different radio frequencies for different land areas called cells. Each cell uses its own radio frequency to prevent interference with its neighbors. It provides large geographic area coverage when different frequency cells are joined. Space network provides connectivity among the spacecrafts and is used by NASA.

A wireless network also faces some limitations like interference, fading in multipath, reflection, and absorption, the problem of the hidden node, and the problem of resource sharing. In wireless networks, interference can be created by other devices near to communication terminal. Due to interference, data can be lost or corrupted. In multipath fading, because of reflection, signals can be canceled when following a different path. Electromagnetic waves can be absorbed by other materials, and the receiver cannot receive wireless signals. In wireless network resource sharing, problems can occur because one resource can be shared by multiple devices over the wireless network (Han et al., 2016; Shinghal & Srivastava, 2017).

17.2 APPLICATIONS OF WIRELESS NETWORKS

Wireless networks provide a variety of applications because of their easy-to-install behavior, less cost-effectiveness as compared to a wired network, and providing fast and reliable data communication with the help of different types of waves over different frequencies (Qiu et al., 2017).

These days wireless network provides support to every sector, either it is related to making calls, establishing a connection between devices, managing organizations' operations, banking sector, education sector, healthcare sector, security, location tracking, or home utilities. Some of the applications of wireless networks are discussed below (Erdelj et al., 2017):

1. Making calls – Wireless networks are widely used in the field of communication. Most people use cell phones to make calls. These cell phones use wireless networks for calling. The wireless network uses radio waves to transmit signals. It helps to provide wireless connectivity to a large geographical area. People can perform a lot of work like bill payment, shopping, email checking, and transfer data.
2. Connecting devices – Wireless networks are used to establish a connection between different devices like connecting wireless headphones and fitness devices with a mobile phone using low-energy radio waves called Bluetooth. Wireless networks are also used to operate home utility devices like TV, air conditioner, and stereo systems with the help of a control system that uses infrared waves.
3. Weather information – Wireless network helps in providing weather information from different areas with the help of satellite communications. It provides information about rain, thunderstorm, and earthquakes and helps various departments to take necessary steps according to these conditions.
4. Healthcare sector – In the healthcare sector, wireless communication helps to provide information about internal parameters of the human body using various devices like MRI and ECG machines. It also helps the doctors to manage patient records online.

5. Education sector – In the education sector, it provides accessibility to the internet everywhere with the help of a WLAN. Students and staff can access the internet on laptops without any wired connection. It helps in sending and receiving various study-related documents, surf the internet, and manage online student databases.
6. Security system – Wireless network helps in providing security with the help of installing wireless cameras at sensitive locations. Users can check the live footage of sensitive areas using mobile phones. A wireless network also helps in providing security information about borders to the army with the help of satellite communication.
7. Organization operations – Large organizations can provide connectivity between their different departments using a WLAN network. Each department can transfer data required by the other departments. For example, the stock department can send information to the sales department about the remaining stock of products.
8. Location tracking – Wireless network plays a major role in tracking locations. It helps in tracking the location of vehicles, ships, airplanes, and space shuttles using GPS technology.

17.3 INTERNET OF THINGS AND ADVANCED SCENARIOS

Internet of Things (IoT) means connecting various physical devices with the help of the internet and transfer data between these devices to perform various types of operations without human involvement. IoT commonly deals with sensors, wireless networks, cloud, and artificial intelligence. As the usage of the internet grows, so applications based on IoT grow rapidly. Most of the IoT-based applications remotely access devices using wireless networking and transfer real-time data between these devices with the help of cloud storage. IoT is very useful in areas where working condition is very difficult for human – like a chemical industry in which most of the work done with the help of IoT devices. As the increase in availability of the internet, the application based on IoT also increases day by day. In IoT, "Things" can be referred to as software, hardware, services, and data that perform smart work when they are connected.

In IoT, each device is assigned a unique IP address that is used to send and receive data over the network. IoT provides automation in organizations that reduce human intervention, reduce processing time, and provide error-free data. IoT is very useful in a real-time system where data is transferred in real time to take necessary decisions. Some communication devices like sensors, actuators, and RFID are used. Sensors convert the physical conditions into signals and send information to microprocessors like Arduino or Raspberry Pi for taking further actions. Actuators convert the signals into actual movements and act according to received signals. RFID is the unique key that is transfer over the internet for unique identification.

17.4 KEY CASES AND APPLICATIONS OF IoT

IoT provides a lot of applications in a different area like defense, education, medical field, healthcare, and traffic. It helps in automation of day-to-day activities. Some of the common applications of the IoT are shown in Figure 17.2 and given below.

17.4.1 Smart Homes

IoT provides automation at home. It helps in energy saving, water saving, and controlling various appliances at home, and also provides security to homes. IoT uses smart switches that automatically turn on and off and provides a smart lock that detects faces and fingerprints to open the door. It also provides a water monitoring device to control water wastage in homes.

17.4.2 Healthcare System

IoT provides various devices that monitor the health of patients remotely. These devices check blood pressure, heart rate, and temperature of patients and send data to a doctor

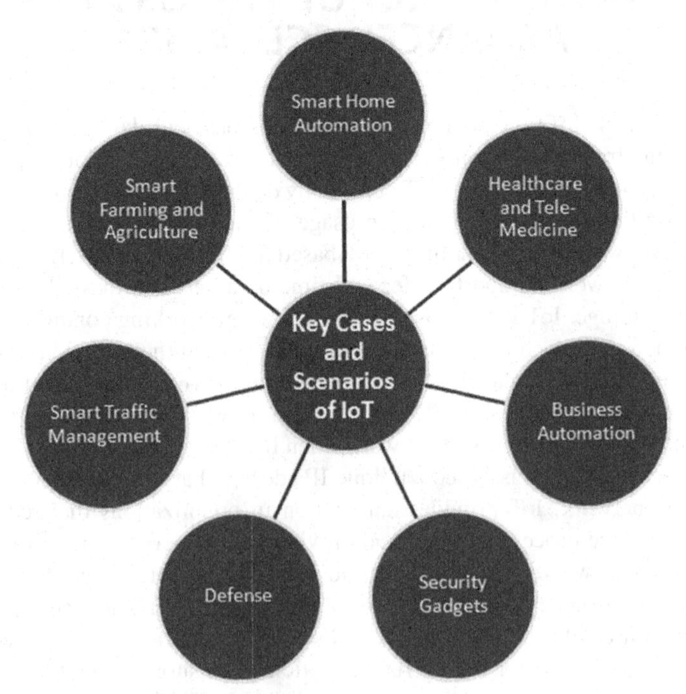

FIGURE 17.2 Internet of Things and associated cases.

for future analysis. Healthcare devices also include fitness bands that automatically check all the parameters of the human body and send data over the doctor's mobile phone. The healthcare system also includes smart ambulances. These ambulances automatically provide essential medicine to a patient when the patient is on the way to the hospital.

17.4.3 Traffic Management

IoT also provides help in traffic management. It provides smart parking area, smart toll plaza, smart road map, and smart vehicle control devices. It also provides a smart traffic tracking system with the help of GPS.

17.4.4 Smart Farming

IoT helps farmers to know about the weather condition, humidity in the soil, and soil moisture. This data can be used to make further decisions about the crops and helps the farmer to improve crop quality.

17.4.5 Business Automation

Nowadays most businesses automate their production process, sales process, and marketing process with the help of IoT. It not only provides internet connectivity between various devices, but it also provides accuracy, speed, and efficiency in work through different devices, so that organizations can increase their profit ratio.

17.4.6 Defense Application

In defense, IoT also takes a vital role. With the help of IoT, enemy ships and airplanes can be easily traced. With the help of drones, defense personnel can easily capture the images of enemy locations.

17.4.7 Woman Security Bands

IoT provides wearable devices that provide security to women. Different types of wearable bands are built that send signals to a nearby police station if the woman is in trouble.

17.4.8 Connected Cars

These are used with the advanced wireless monitoring as are the next generation automobiles industry revolutions for assorted applications in wireless environment. Most of the companies work on these concepts with advanced scenarios. Connected cars contain various types of sensors that automatically manage car operations and provide ease to the passengers.

17.5 KEY TECHNOLOGIES AND STANDARDS WITH WIRELESS ENVIRONMENT

- Radio communication system
- Radio channels with ITU RR
- Land mobile radio/professional mobile radio
  - dPMR
  - OpenSky
  - EDACS
  - DMR
  - TETRA
  - P25
- Cellular networks: 0G, 1G, 2G, 3G, beyond 3G (4G), future wireless
- Cordless telephony system: digital enhanced cordless telecommunications
- Short-range P2P
  - IrDA
  - Wireless microphones
  - Near-field communication
  - EnOcean
  - RFID
  - DSRC
  - Remote controls
  - TransferJet
  - Wireless USB
- Wireless sensor networks
  - PAN
  - TransferJet
  - ZigBee
  - Bluetooth
  - EnOcean
  - Ultra-wideband
- Wireless networks
  - HiperMAN
  - WLAN
  - WMAN
  - HiperLAN
  - LMDS
  - WiMAX
  - Wi-Fi

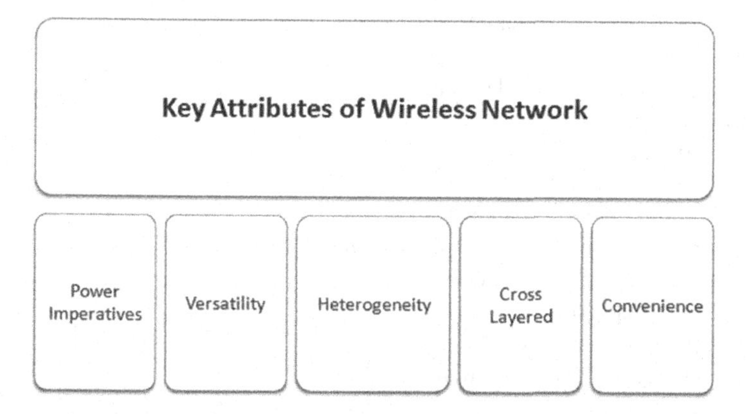

FIGURE 17.3 Key attributes of wireless sensor networks.

TABLE 17.1 Comparison Aspects of Wireless Networks Technology

	WIMAX	*WI-FI (A)*	*WI-FI (B)*	*WI-FI (G)*	*BLUETOOTH*
Limitations	Cost	Cost factor	Speed	Cost and range both	Range issues
Frequency (in GHz)	2–66	5	2.4	2.4	2.45
International standard	802.16	802.11a	802.11b	802.11g	802.15
Advantages	Speed, range	Speed	Low cost	Speed	Low cost
Speed (in Mbps)	80	54	11	54	0.72
Range parameter (meters)	50	50	100	100	10

The enormous key attributes of wireless networks are presented in Figure 17.3.

The assorted wireless technologies are analyzed with comparative evaluations and presented in Table 17.1.

17.6 KEY FEATURES OF WIRELESS ENVIRONMENT

- Dynamic and effective load balancing
- Scalability
- Autonomous
- Dimensional approaches
- Lightweight terminals

- Dynamic network topology
- Multi-hop routing
- Assorted channel-based communication
- Ease to deployment
- Network access control
- Network scalability
- Cavernous analytics patterns
- Decreased dependence on infrastructure
- Mobility and quality of service
- Speed of deployment
- Distributed operation
- Portability and transportation

Security is one of the key aspects of wireless implementations. This is shown in Figure 17.4 with assorted algorithms and underlined in Table 17.2.

17.7 ADVANCED CASES AND TECHNOLOGIES WITH INTERNET OF THINGS (IoT)

In the era of high-performance network transmission including 5G networks, there is a need to integrate the scalability and compatibility (Jeschke et al., 2017). As per the reports from the prominent data analytics portal Statista.com, the fifth-generation 5G wireless networks shall touch 200 million connections. The 5G network is to be adopted very soon in the global market in most of countries with the speed of network reaching 2 Gbps. To work on the high-performance network transmission in multiple devices

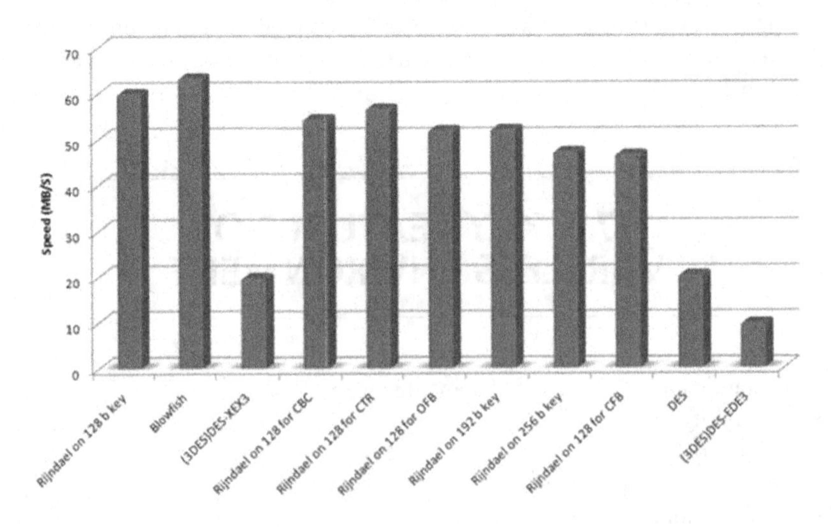

FIGURE 17.4 Security approaches and their performance in Mbps.

TABLE 17.2 Algorithms and Related Speed

ALGORITHM	*MB/SECOND*
Rijndael on 128-b key	60.010
Blowfish	63.386
(3DES)DES-XEX3	19.783
Rijndael on 128 for CBC	54.447
Rijndael on 128 for CTR	56.710
Rijndael on 128 for OFB	51.925
Rijndael on 192-b key	52.145
Rijndael on 256-b key	47.229
Rijndael on 128 for CFB	46.601
DES	20.340
(3DES)DES-EDE3	9.748

and gadgets, the speed and compatibility is the key concern. With a large number of devices and gadgets with different configurations and internal settings, there is a need to adopt uniformity, so that different devices and gadgets can have the connectivity with high performance and accuracy rate. The work is presenting the use case scenarios and implementation patterns in advanced IoT- and cloud-based networks using Message Queuing Telemetry Transport (MQTT) (Li et al., 2018; Sethi & Sarangi, 2017).

Cloud of Things refers to the integration of communication channels and protocols with advanced networks of IoT and cloud computing, and these are quite effectual for advanced implementations (Ning et al., 2018; Farahzadi et al., 2018). For this, the MQTT-based communication is required for secured and privacy-aware communications with lightweight data transmission. MQTT broker is the centralized platform and repository where the data signals from the publishers are collected and then transmitted to the subscribers. In traditional web-based applications, these platforms are known as web servers. In case of IoT-based applications, these are referred to as MQTT brokers (Naik, 2017; Soni & Makwana, 2017).

MQTT contains two key services for transmission and communication of data between IoT devices and gadgets. These services are subscribing and publishing. The sender of data signals is known as the publisher. Technically, sending data from one device to another is referred as publishing. The receiver device in IoT transmission is known as the subscriber (Kang et al., 2017; Gündoğan et al., 2018). For example, the data is to be transmitted from a temperature sensor to a mobile gadget of a farmer. In this case, the temperature sensor is publishing the data while the gadget of farmer is the subscriber of the data.

17.8 CLOUD PLATFORMS FOR MQTT

The MQTT broker platforms are available for installation on dedicated systems including Raspberry Pi, Arduino, laptops, and server. Besides the specialized installation, a number of cloud platforms are available that are used as the MQTT broker, so that the developers and research scholars can directly work on their research projects effectively.

17.8.1 CloudMQTT

CloudMQTT is one of the widely used MQTT brokers that can be used for interfacing of IoT gadgets and connection in the devices. The implementations of IoT scenarios including smart home or smart office automations can be done using CloudMQTT (Yi et al., 2016; Dizdarević etal., 2019).

It contains a free plan titled "Cute Cat" in which five users/connections can be established and the IoT devices can be attached without linking any credit card. With the existing Google Account, the instances can be created on the platform of CloudMQTT (Luoto, 2019; Meloni et al., 2016) as shown in Figure 17.5.

Once the authentication details are obtained on CloudMQTT, these credentials can be used by the IoT gadgets or smartphones. A number of Android apps are available on Google Play Store as "IoT MQTT Dashboard" as in Figure 17.6. After installation of IoT apps, these IoT MQTT apps directly communicate with the MQTT broker platforms, so that the IoT gadgets, smartphones, and cloud brokers can communicate with each other in real time.

17.8.2 DIoTY

DIoTY is another cloud-based MQTT broker platform that contains a free account that can be used for interfacing the IoT gadgets. DIoTY provides the scripts so that the connection of MQTT broker can be done with different programming environments and presented in Figure 17.7.

Application programming interfaces are available for multiple programming languages and platforms.

The following are the platforms that can be programmed for publishing as well as subscribing the data and signals:

- Python
- Java

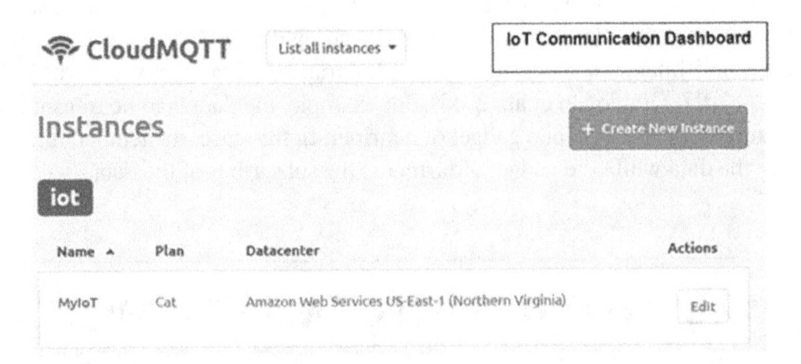

FIGURE 17.5 Instances in CloudMQTT.

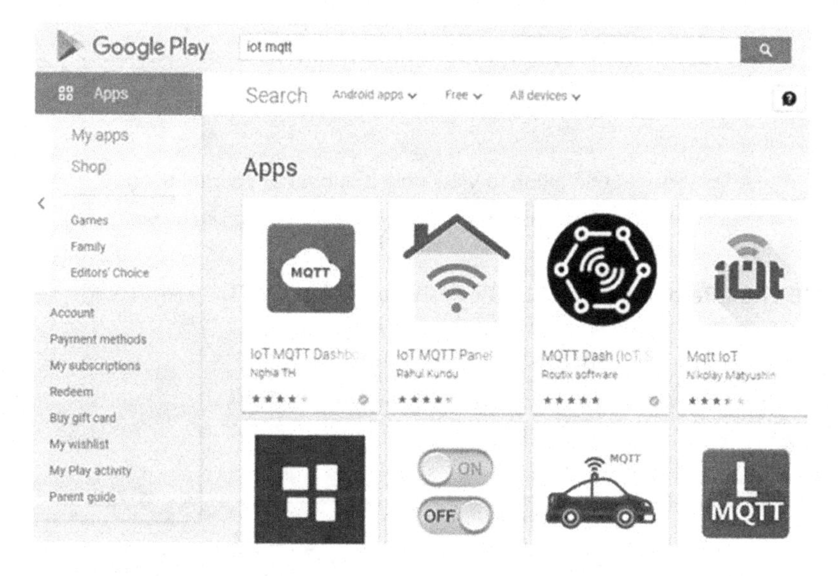

FIGURE 17.6 IoT MQTT apps on Google Play Store.

FIGURE 17.7 Dashboard of DIoTY.

- Arduino
- C#
- Node.JS

The best freely available python library is the Paho Python client. The scenario of fetching data is presented in Figure 17.8.

17.8.3 Cloud Integration with Node-RED

Node-RED is one of the widely used and high-performance platforms that contain the features for connectivity and programming with IoT-based devices with web-based interface as in Figure 17.9.

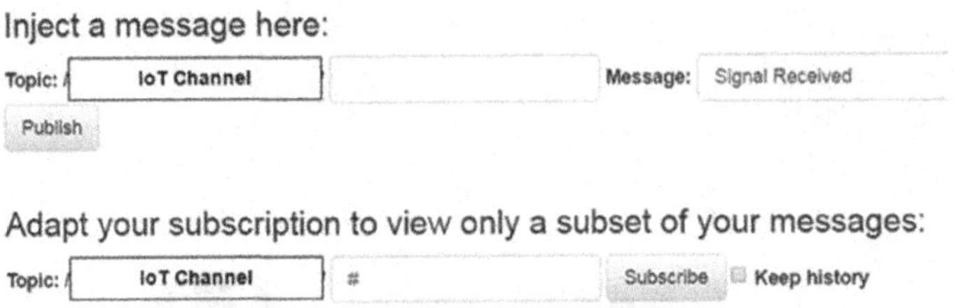

Inject a message here:
Topic:
IoT Channel
Message:
Signal Received
Publish
Adapt your subscription to view only a subset of your messages:
Topic:
IoT Channel
#
Subscribe
Keep history

FIGURE 17.8 Receiving data on DIoTY Dashboard using MQTT.

Sheet 1
input
inject
catch
status
link
mqtt
http
websocket
tcp
udp
fred
sensor1
soldier cloudmqtt server
connected
sensors/#
connected
msg.payload

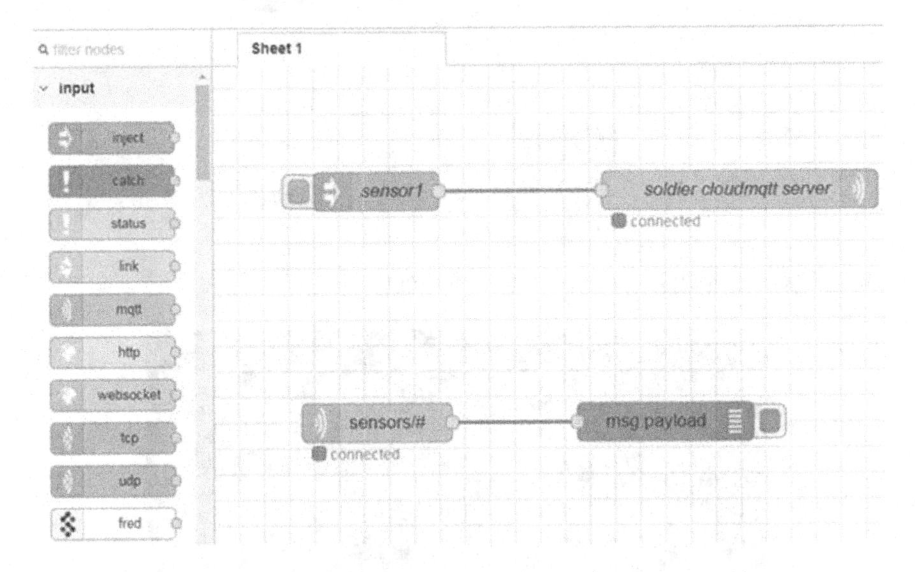

FIGURE 17.9 Node-RED-integrated cloud with MQTT broker.

17.8.4 Dynamic Key-Based Communication in IoT Scenario

The implementation of dynamic key-based communication in IoT is presented with simulation as in Figures 17.10 and 17.11.

To perform the advanced implementations of IoT and Cloud of Things-based scenarios, the simulation scenarios are created and then the performance evaluations are done using testbeds.

17.9 CONCLUSION

The research scholars, academicians, and the professionals can perform advanced implementations associated with microcontroller-based applications for real-time applications

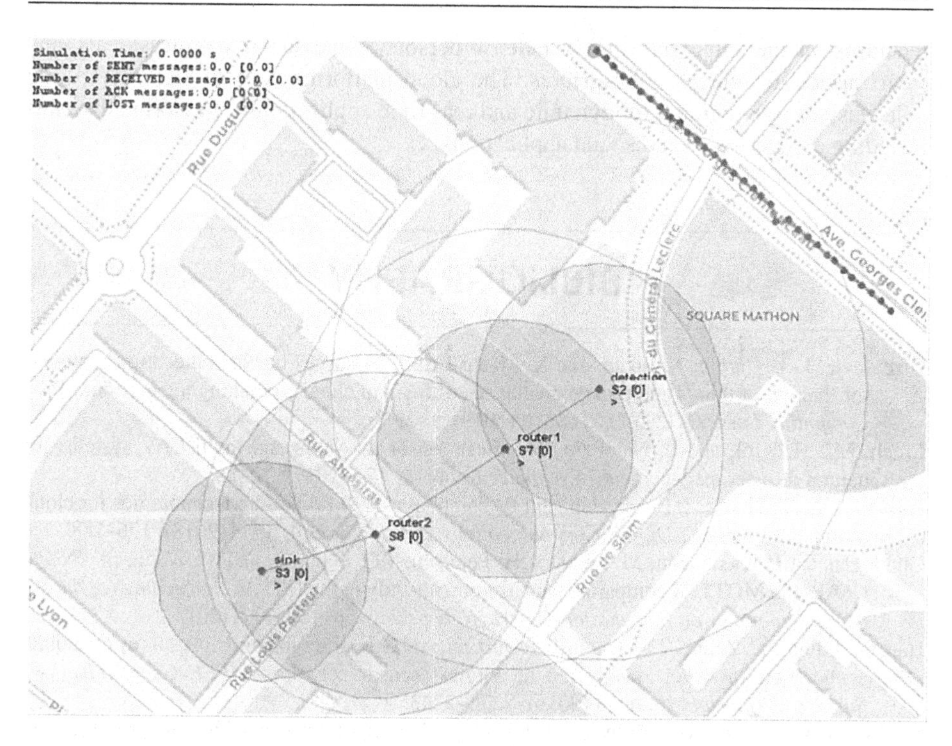

FIGURE 17.10 Advanced communication in IoT simulation scenario.

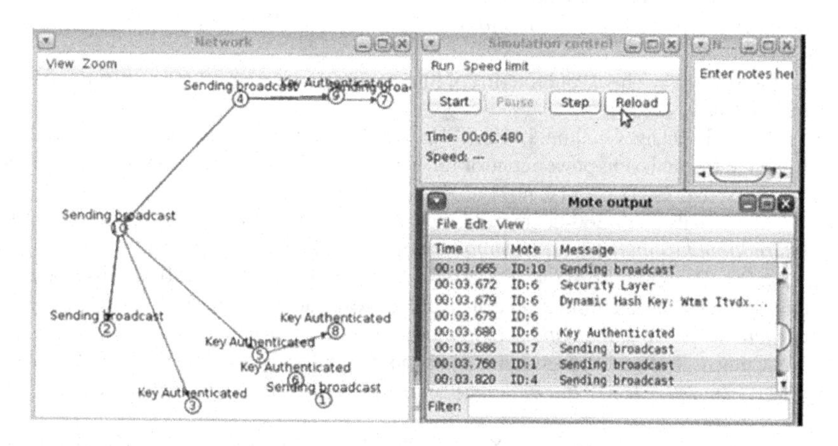

FIGURE 17.11 IoT-integrated advanced implementation pattern.

including telemedicine, gadgets for defense personnel and security professionals, and smart agriculture-integrated Agribots. The cloud platforms with MQTT service are quite easy to be integrated for dynamic and real-time applications for assorted domains including e-governance and social applications.

BIBLIOGRAPHY

Dizdarević, J., F. Carpio, A. Jukan, and X. Masip-Bruin. "A survey of communication protocols for internet of things and related challenges of fog and cloud computing integration." *ACM Computing Surveys (CSUR)* 51, no. 6 (2019): 1–29.

Erdelj, M., M. Król, and E. Natalizio. "Wireless sensor networks and multi-UAV systems for natural disaster management." *Computer Networks* 124 (2017): 72–86.

Farahzadi, A., P. Shams, J. Rezazadeh, and R. Farahbakhsh. "Middleware technologies for cloud of things: A survey." *Digital Communications and Networks* 4, no. 3 (2018): 176–188.

Gündoğan, C., P. Kietzmann, M. Lenders, H. Petersen, T.C. Schmidt, and M. Wählisch. "NDN, CoAP, and MQTT: A comparative measurement study in the IoT." In *Proceedings of the 5th ACM Conference on Information-Centric Networking*, pp. 159–171, (2018).

Han, G., J. Jiang, C. Zhang, T.Q. Duong, M. Guizani, and G.K. Karagiannidis. "A survey on mobile anchor node assisted localization in wireless sensor networks." *IEEE Communications Surveys & Tutorials* 18, no. 3 (2016): 2220–2243.

Jeschke, S., C. Brecher, T. Meisen, D. Özdemir, and T. Eschert. "Industrial internet of things and cyber manufacturing systems." In *Industrial Internet of Things*, pp. 3–19. Springer, Cham, (2017).

Kang, D.H., M.S. Park, H.S. Kim, D.Y. Kim, S.H. Kim, H.J. Son, and S.G. Lee. "Room temperature control and fire alarm/suppression IoT service using MQTT on AWS." In *2017 International Conference on Platform Technology and Service (PlatCon)*, pp. 1–5. IEEE, (2017).

Li, S., L. Da Xu, and S. Zhao. "5G Internet of things: A survey." *Journal of Industrial Information Integration* 10 (2018): 1–9.

Lin, S., F. Miao, J. Zhang, G. Zhou, L. Gu, T. He, J.A. Stankovic, S. Son, and G.J. Pappas. "ATPC: Adaptive transmission power control for wireless sensor networks." *ACM Transactions on Sensor Networks (TOSN)* 12, no. 1 (2016): 1–31.

Luoto, A. "Log analysis of 360-degree video users via MQTT." In *Proceedings of the 2019 2nd International Conference on Geoinformatics and Data Analysis*, pp. 130–137, (2019).

Meloni, A., P.A. Pegoraro, L. Atzori, P. Castello, and S. Sulis. "IoT cloud-based distribution system state estimation: Virtual objects and context-awareness." In *2016 IEEE International Conference on Communications (ICC)*, pp. 1–6. IEEE, (2016).

Naik, N. "Choice of effective messaging protocols for IoT systems: MQTT, CoAP, AMQP and HTTP." In *2017 IEEE International Systems Engineering Symposium (ISSE)*, pp. 1–7. IEEE, (2017).

Ning, Z., X. Kong, F. Xia, W. Hou, and X. Wang. "Green and sustainable cloud of things: Enabling collaborative edge computing." *IEEE Communications Magazine* 57, no. 1 (2018): 72–78.

Noel, A.B., A. Abdaoui, T. Elfouly, M.H. Ahmed, A. Badawy, and M.S. Shehata. "Structural health monitoring using wireless sensor networks: A comprehensive survey." *IEEE Communications Surveys & Tutorials* 19, no. 3 (2017): 1403–1423.

Qiu, T., A. Zhao, F. Xia, W. Si, and D.O. Wu. "ROSE: Robustness strategy for scale-free wireless sensor networks." *IEEE/ACM Transactions on Networking* 25, no. 5 (2017): 2944–2959.

Rashid, B., and M.H. Rehmani. "Applications of wireless sensor networks for urban areas: A survey." *Journal of Network and Computer Applications* 60 (2016): 192–219.

Sethi, P., and S.R. Sarangi. "Internet of things: Architectures, protocols, and applications." *Journal of Electrical and Computer Engineering* 2017 (2017).

Shinghal, D., and N. Srivastava. "Wireless sensor networks in agriculture: For potato farming." *Neelam, Wireless Sensor Networks in Agriculture: For Potato Farming* (September 22, 2017).

Soni, D., and A. Makwana. "A survey on MQTT: A protocol of internet of things (IoT)." In *International Conference on Telecommunication, Power Analysis and Computing Techniques (ICTPACT-2017)*, (2017, April).

Yetgin, H., K.T.K. Cheung, M. El-Hajjar, and L.H. Hanzo. "A survey of network lifetime maximization techniques in wireless sensor networks." *IEEE Communications Surveys & Tutorials* 19, no. 2 (2017): 828–854.

Yi, D., F. Binwen, K. Xiaoming, and M. Qianqian. "Design and implementation of mobile health monitoring system based on MQTT protocol." In *2016 IEEE Advanced Information Management, Communicates, Electronic and Automation Control Conference (IMCEC)*, pp. 1679–1682. IEEE, (2016).

Printed in the United States
by Baker & Taylor Publisher Services